METHODS IN
ENZYMOLOGY

Cryo-EM, Part A

Sample Preparation and Data Collection

METHODS IN ENZYMOLOGY

Editors-in-Chief

JOHN N. ABELSON AND MELVIN I. SIMON

Division of Biology
California Institute of Technology
Pasadena, California

Founding Editors

SIDNEY P. COLOWICK AND NATHAN O. KAPLAN

VOLUME FOUR HUNDRED AND EIGHTY-ONE

Methods in
ENZYMOLOGY

Cryo-EM, Part A

Sample Preparation and Data Collection

EDITED BY

GRANT J. JENSEN
*Division of Biology and Howard Hughes Medical Institute
California Institute of Technology
Pasadena, California, USA*

AMSTERDAM • BOSTON • HEIDELBERG • LONDON
NEW YORK • OXFORD • PARIS • SAN DIEGO
SAN FRANCISCO • SINGAPORE • SYDNEY • TOKYO
Academic Press is an imprint of Elsevier

ELSEVIER

Academic Press is an imprint of Elsevier
525 B Street, Suite 1900, San Diego, CA 92101-4495, USA
30 Corporate Drive, Suite 400, Burlington, MA 01803, USA
32 Jamestown Road, London NW1 7BY, UK

First edition 2010

For information on all Academic Press publications
visit our website at elsevierdirect.com

ISBN: 978-0-12-374906-2
ISSN: 0076-6879

Printed and bound in United States of America
10 11 12   10 9 8 7 6 5 4 3 2 1

Contents

Contributors

Priyanka D. Abeyrathne
C-CINA, Biozentrum, University of Basel, Basel, Switzerland

Christopher J. Ackerson
Colorado State University, Department of Chemistry, Fort Collins, Colorado, USA

D. A. Agard
The Howard Hughes Medical Institute, University of California, San Francisco, California, USA

Christopher P. Arthur
The Scripps Research Institute, La Jolla, California, USA

Lindsay A. Baker
Molecular Structure and Function Program, The Hospital for Sick Children, and Department of Biochemistry, The University of Toronto, Ontario, Canada

Ariane Briegel
Division of Biology, and Howard Hughes Medical Institute, California Institute of Technology, California Boulevard, Pasadena, California, USA

Mohamed Chami
C-CINA, Biozentrum, University of Basel, Basel, Switzerland

Joshua S. Chappie
Laboratory of Molecular Biology, National Institute of Diabetes and Digestive and Kidney Diseases, National Institutes of Health, Bethesda, Maryland, USA

Songye Chen
Division of Biology, California Institute of Technology, California Boulevard, Pasadena, California, USA

Yifan Cheng
The W.M. Keck Advanced Microscopy Laboratory, Department of Biochemistry and Biophysics, University of California San Francisco, San Francisco, California, USA

Radostin Danev
Okazaki Institute for Integrative Bioscience, National Institutes of Natural Sciences, Okazaki, Japan

Sacha De Carlo
Department of Chemistry, Institute for Macromolecular Assemblies, City University of New York, City College Campus, New York, USA

David DeRosier
Brandeis University, Waltham, Massachusetts, USA

Megan J. Dobro
Division of Biology, California Institute of Technology, Pasadena, California, USA

Danijela Dukovski
Department of Cell Biology, and Howard Hughes Medical Institute, Harvard Medical School, Boston, Massachusetts, USA

Kenneth N. Goldie
C-CINA, Biozentrum, University of Basel, Basel, Switzerland

James F. Hainfeld
Nanoprobes, Incorporated, Yaphank, New York, USA

Richard K. Hite
Department of Cell Biology, Harvard Medical School, Boston, Massachusetts, USA

Grant J. Jensen
Division of Biology, and Howard Hughes Medical Institute, California Institute of Technology, California Boulevard, Pasadena, California, USA

Deborah F. Kelly
Department of Cell Biology, Harvard Medical School, Boston, Massachusetts, USA

Abraham J. Koster
Department of Molecular Cell Biology, Faculty of Biology and Institute of Biomembranes, Utrecht, The Netherlands

Mark S. Ladinsky
Division of Biology, California Institute of Technology, Pasadena, California, USA

Huilin Li
Biology Department, Brookhaven National Laboratory, Upton, and Department of Biochemistry & Cell Biology, Stony Brook University, Stony Brook, New York, USA

Zongli Li
Department of Cell Biology, and Howard Hughes Medical Institute, Harvard Medical School, Boston, Massachusetts, USA

Alasdair W. McDowall
Division of Biology, California Institute of Technology, Pasadena, California, USA

Linda A. Melanson
Gatan, Inc., Pleasanton, California, USA

Kuniaki Nagayama
Okazaki Institute for Integrative Bioscience, National Institutes of Natural Sciences, Okazaki, Japan

Radosav S. Pantelic
C-CINA, Biozentrum, University of Basel, Basel, Switzerland

Jürgen M. Plitzko
Max-Planck-Institut für Biochemie, Abteilung Molekulare Strukturbiologie, Martinsried, Germany

Richard D. Powell
Nanoprobes, Incorporated, Yaphank, New York, USA

John L. Rubinstein
Molecular Structure and Function Program, The Hospital for Sick Children, and Department of Biochemistry; Department of Medical Biophysics, The University of Toronto, Ontario, Canada

Andreas D. Schenk
Department of Cell Biology, Harvard Medical School, Boston, Massachusetts, USA

Cindi L. Schwartz
Boulder Laboratory for 3-D Electron Microscopy of Cells, Department of MCD Biology, University of Colorado, Boulder, Colorado, USA

J. W. Sedat
The W. M. Keck Advanced Microscopy Laboratory, Department of Biophysics, University of California, San Francisco, California, USA

Fred J. Sigworth
Department of Cellular and Molecular Physiology, Yale University, New Haven, Connecticut, USA

Henning Stahlberg
C-CINA, Biozentrum, University of Basel, Basel, Switzerland

Holger Stark
MPI for Biophysical Chemistry, and Göttingen Center of Molecular Biology, University of Göttingen; Max-Planck-Institut für Biophysikalische Chemie, Göttingen, Germany

Jingchuan Sun
Biology Department, Brookhaven National Laboratory, Upton, New York, USA

Thomas Walz
Department of Cell Biology, and Howard Hughes Medical Institute, Harvard Medical School, Boston, Massachusetts, USA

Liguo Wang
Department of Cellular and Molecular Physiology, Yale University, New Haven, Connecticut, USA

Elizabeth M. Wilson-Kubalek
The Scripps Research Institute, La Jolla, California, USA

Shawn Q. Zheng
The Howard Hughes Medical Institute, University of California, San Francisco, California, USA

PREFACE

In this, the fifty-fourth year of *Methods in Enzymology*, we celebrate the discovery and initial characterization of thousands of individual enzymes, the sequencing of hundreds of whole genomes, and the structure determination of tens of thousands of proteins. In this context, the architectures of multienyzme/multiprotein complexes and their arrangement within cells have now come to the fore. A uniquely powerful method in this field is electron cryo-microscopy (cryo-EM), which in its broadest sense, is all those techniques that image cold samples in the electron microscope. Cryo-EM allows individual enzymes and proteins, macromolecular complexes, assemblies, cells, and even tissues to be observed in a "frozen-hydrated," near-native state free from the artifacts of fixation, dehydration, plastic-embedding, or staining typically used in traditional forms of EM (Chapter 3, Vol. 481). This series of volumes is therefore dedicated to a description of the instruments, samples, protocols, and analyses that belong to the growing field of cryo-EM.

The material could have been organized well by two schemes. The first is by the symmetry of the sample. Because the fundamental limitation in cryo-EM is radiation damage (Chapter 15, Vol. 481), a defining characteristic of each method is whether and how low-dose images of identical copies of the specimen can be averaged. In the most favorable case, large numbers of identical copies of the specimen of interest, like a single protein, can be purified and crystallized within thin "two-dimensional" crystals (Chapter 1, Vol. 481). In this case, truly *atomic* resolution reconstructions have been obtained through averaging very low dose images of millions of copies of the specimen (Chapter 11, Vol. 481; Chapter 4, Vol. 482; Chapters 5 and 6, Vol. 483). The next most favorable case is helical crystals, which present a range of views of the specimen within a single image (Chapter 2, Vol. 481 and Chapter 7, Vol. 483) and can also deliver atomically interpretable reconstructions, although through quite different data collection protocols and reconstruction mathematics (Chapters 5 and 6, Vol. 482). At an intermediate level of (60-fold) symmetry, icosahedral viruses have their own set of optimal imaging and reconstruction protocols, and are just now also reaching atomic interpretability (Chapters 7 and 14, Vol. 482). Less symmetric particles, such as many multienyzme/multiprotein complexes, invite yet another set of challenges and methods (Chapters 3, 5, and 6, Vol. 481; Chapters 8–10, Vol. 482). Many are conformationally heterogeneous, requiring that images of different particles be first classified and then averaged

(Chapters 10 and 12, Vol. 482; Chapters 8 and 9, Vol. 483). Heterogeneity and the precision to which these images can be aligned have limited most such reconstructions to "sub-nanometer" resolution, where the folds of proteins are clear but not much more (Chapter 1, Vol. 483). Finally, the most challenging samples are those which are simply unique (Chapter 8, Vol. 481), eliminating any chance of improving the clarity of reconstructions through averaging. For these, tomographic methods are required (Chapter 12, Vol. 481; Chapter 13, Vol. 482), and only nanometer resolutions can be obtained (Chapters 10–13, Vol. 483).

But instead of organizing topics according to symmetry, following a wonderful historical perspective by David DeRosier (Historical Perspective, Vol. 481), I chose to order the topics in experimental sequence: Sample preparation and data collection/microscopy (Vol. 481); 3-D reconstruction (Vol. 482); and analyses and interpretation, including case studies (Vol. 483). This organization emphasizes how the relatedness of the mathematics (Chapter 1, Vol. 482), instrumentation (Chapters 10 and 14, Vol. 482), and methods (Chapter 15, Vol. 482; Chapter 9, Vol. 481) underlying all cryo-EM approaches allows practitioners to easily move between them. It further highlights how in a growing number of recent cases, the methods are being mixed (Chapter 13, Vol. 481), for instance, through the application of "single particle-like" approaches to "unbend" and average 2-D and helical crystals (Chapter 6, Vol. 482), but also average subvolumes within tomograms. Moreover, different samples are always more-or-less well-behaved, so the actual resolution achieved may be less than theoretically possible for a particular symmetry, or to the opposite effect; extensively known constraints may allow a more specific interpretation than usual for a given resolution (Chapters 2–4 and 6, Vol. 483). Nevertheless, within each section, the articles are ordered as much as possible according to the symmetry of the sample as described above (i.e. methods for preparing samples proceed from 2-D and helical crystals to sectioning of high-pressure-frozen tissues; Chapter 8, Vol. 481). The cryo-EM beginner with a new sample must then first recognize its symmetry and then identify the relevant chapters within each volume.

As a final note, our field has not yet reached a consensus on the placement of the prefix "cryo" and other details of the names of cryo-EM techniques. Thus, "cryo-electron microscopy" (CEM), "electron cryo-microscopy" (ECM), and "cryo-EM" should all be considered synonyms here. Likewise, "single particle reconstruction" (SPR) and "single particle analysis" (SPA) refer to a single technique, as do "cryo-electron tomography" (CET), "electron cryo-tomography" (ECT), and cryo-electron microscope tomography (cEMT).

GRANT J. JENSEN

METHODS IN ENZYMOLOGY

VOLUME XXXV. Lipids (Part B)
Edited by JOHN M. LOWENSTEIN

VOLUME XXXVI. Hormone Action (Part A: Steroid Hormones)
Edited by BERT W. O'MALLEY AND JOEL G. HARDMAN

VOLUME XXXVII. Hormone Action (Part B: Peptide Hormones)
Edited by BERT W. O'MALLEY AND JOEL G. HARDMAN

VOLUME XXXVIII. Hormone Action (Part C: Cyclic Nucleotides)
Edited by JOEL G. HARDMAN AND BERT W. O'MALLEY

VOLUME XXXIX. Hormone Action (Part D: Isolated Cells, Tissues, and Organ Systems)
Edited by JOEL G. HARDMAN AND BERT W. O'MALLEY

VOLUME XL. Hormone Action (Part E: Nuclear Structure and Function)
Edited by BERT W. O'MALLEY AND JOEL G. HARDMAN

VOLUME XLI. Carbohydrate Metabolism (Part B)
Edited by W. A. WOOD

VOLUME XLII. Carbohydrate Metabolism (Part C)
Edited by W. A. WOOD

VOLUME XLIII. Antibiotics
Edited by JOHN H. HASH

VOLUME XLIV. Immobilized Enzymes
Edited by KLAUS MOSBACH

VOLUME XLV. Proteolytic Enzymes (Part B)
Edited by LASZLO LORAND

VOLUME XLVI. Affinity Labeling
Edited by WILLIAM B. JAKOBY AND MEIR WILCHEK

VOLUME XLVII. Enzyme Structure (Part E)
Edited by C. H. W. HIRS AND SERGE N. TIMASHEFF

VOLUME XLVIII. Enzyme Structure (Part F)
Edited by C. H. W. HIRS AND SERGE N. TIMASHEFF

VOLUME XLIX. Enzyme Structure (Part G)
Edited by C. H. W. HIRS AND SERGE N. TIMASHEFF

VOLUME L. Complex Carbohydrates (Part C)
Edited by VICTOR GINSBURG

VOLUME LI. Purine and Pyrimidine Nucleotide Metabolism
Edited by PATRICIA A. HOFFEE AND MARY ELLEN JONES

VOLUME LII. Biomembranes (Part C: Biological Oxidations)
Edited by SIDNEY FLEISCHER AND LESTER PACKER

VOLUME 345. G Protein Pathways (Part C: Effector Mechanisms)
Edited by RAVI IYENGAR AND JOHN D. HILDEBRANDT

VOLUME 346. Gene Therapy Methods
Edited by M. IAN PHILLIPS

VOLUME 347. Protein Sensors and Reactive Oxygen Species (Part A: Selenoproteins and Thioredoxin)
Edited by HELMUT SIES AND LESTER PACKER

VOLUME 348. Protein Sensors and Reactive Oxygen Species (Part B: Thiol Enzymes and Proteins)
Edited by HELMUT SIES AND LESTER PACKER

VOLUME 349. Superoxide Dismutase
Edited by LESTER PACKER

VOLUME 350. Guide to Yeast Genetics and Molecular and Cell Biology (Part B)
Edited by CHRISTINE GUTHRIE AND GERALD R. FINK

VOLUME 351. Guide to Yeast Genetics and Molecular and Cell Biology (Part C)
Edited by CHRISTINE GUTHRIE AND GERALD R. FINK

VOLUME 352. Redox Cell Biology and Genetics (Part A)
Edited by CHANDAN K. SEN AND LESTER PACKER

VOLUME 353. Redox Cell Biology and Genetics (Part B)
Edited by CHANDAN K. SEN AND LESTER PACKER

VOLUME 354. Enzyme Kinetics and Mechanisms (Part F: Detection and Characterization of Enzyme Reaction Intermediates)
Edited by DANIEL L. PURICH

VOLUME 355. Cumulative Subject Index Volumes 321–354

VOLUME 356. Laser Capture Microscopy and Microdissection
Edited by P. MICHAEL CONN

VOLUME 357. Cytochrome P450, Part C
Edited by ERIC F. JOHNSON AND MICHAEL R. WATERMAN

VOLUME 358. Bacterial Pathogenesis (Part C: Identification, Regulation, and Function of Virulence Factors)
Edited by VIRGINIA L. CLARK AND PATRIK M. BAVOIL

VOLUME 359. Nitric Oxide (Part D)
Edited by ENRIQUE CADENAS AND LESTER PACKER

VOLUME 360. Biophotonics (Part A)
Edited by GERARD MARRIOTT AND IAN PARKER

VOLUME 361. Biophotonics (Part B)
Edited by GERARD MARRIOTT AND IAN PARKER

3D Reconstruction from Electron Micrographs: A Personal Account of Its Development

David DeRosier

Contents

Abstract

Prior to the development of 3D reconstruction, images were interpreted in terms of models made from simple units like ping-pong balls. Generally, people eye-balled the images and with other knowledge about its structure, such as the number of subunits, proposed models to account for the images. How was one to know if the models were correct and to what degree they faithfully represented the true structure? The analysis of electron micrographs of negatively stained viral structures led to the answers and 3D reconstruction.

Thinking my manuscript would have to be a scholarly history of 3D reconstruction, I turned down Grant Jensen's initial request to write an introductory article. For sure I would get parts wrong, which would annoy various older colleagues and I would wind up in the dog house. Worse, younger readers would take my version for gospel. Quoting Sarah Palin, I told Grant thanks but no thanks. He countered, "However I was imagining something more personal rather than a comprehensive history of our field. I was imagining a description of the early days, for instance, as you experienced them, and reflections on how the field has developed and/or where it is likely headed. You know, it could be something like 'in the bad old days our computer filled the whole room and it took a week to calculate a

Brandeis University, Waltham, Massachusetts, USA

Methods in Enzymology, Volume 481
ISSN 0076-6879, DOI: 10.1016/S0076-6879(10)81016-X

512 × 512 Fourier transform'. And Aaron Klug kept bugging me about. . ."
I take this to mean that I am being given free rein (or is it reign?) and that Grant
will take the blame. My recollection of some events will not be perfect or
perhaps correct and it may be colored by my experience. What follows is not
to be taken as an accurate history but rather as a way of understanding how and
why the method of 3D reconstruction developed as it did in my view.

I want to begin with the way images were interpreted before the 1960s,
what methods were introduced along the way that led to 3D reconstruction,
and what hurdles lay in the path to realizing 3D reconstruction.

As a new Ph.D., I arrived at the Nobel-Laureate-laden Medical
Research Council's Laboratory of Molecular Biology in Cambridge,
England in the summer of 1965. I was drawn there to work with Aaron
Klug because of his article with Don Caspar on the design of icosahedral
viruses. Their quasi-equivalence theory explained how large viral capsids
could be built from identical protein subunits, which would interact almost
equivalently. The capsid designs were derived in a formal sense by folding a
hexagonally symmetric sheet of subunits. Somewhere deep within,
I believed that large complexes and structures would be important elements
in cells and that understanding viruses was a beginning point.

Prior to my arrival in Cambridge, I had little experience with electron
microscopy and knew nothing about image analysis. I had seen ping-pong
ball models of various macromolecular assemblies derived by eyeballing
electron micrographs, which seemed to be a plausible method to interpret
images, but I could not see how one could be sure that a particular model was
correct (Fig. 1). Being a naïve graduate student, I had assumed that the
investigators, given their experience with their structures, could correctly
interpret images. The acrid controversy over the structure of polyoma–
papilloma viruses set me right.

1. VIRUSES

Prior to Aaron and John's analysis, there were two camps regarding the
structure of the polyoma–papilloma family of viruses: one camp argued that
the viruses had 92 capsomers and the other argued that it had 42. Capsomers
are rings of five or six subunits. In images of negatively stained virus particles,
capsomers are seen as blobs that have either five or six nearest neighbors.
Since the subunits making up the capsomers were not resolved, the number
of neighbors was taken to indicate the number of subunits in each ring, which
turned out much later not to be the case for this family of viruses. The 42
camp counted capsomers in puddles, which they hypothesized arose as virus
capsids fell apart. They counted an average number around 40 (Fig. 2).

The other camp counted capsomeric "bumps" on intact particles at least
on those few images where counts could be made. They thought 42 was too

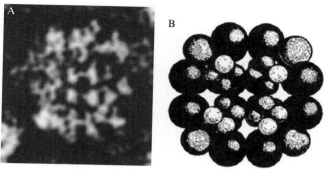

Figure 1 (A) Electron micrograph of a negatively stained preparation of the *E. coli* pyruvate dehydrogenase enzyme complex. The view is down the fourfold axis of the dihydrolipoyl acetyltransferase component of the complex, which lies at the heart of the complex. (B) Ping-pong ball model of the complex down the same axis. The two outer enzymes, the pyruvate dehydrogenase and the dihydrolipoyl dehydrogenase, are represented by the outermost balls. The images, which are taken from a review by Reed and Cox (1970), were reproduced with permission from Elsevier.

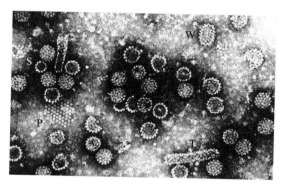

Figure 2 Electron micrograph of a negatively stained preparation of polyoma virus from the work of Finch and Klug (1965). The field shows a variety of polymorphic forms of the capsid protein: P for puddles, T for narrow tubes, W for wide tubes, and S for small isometric closed shells. They are also filled viral particles and empty viral shells. Images reproduced with permission from Elsevier.

small and argued for 92. The difficulty of counting capsomers on particles is that most intact particles did not present clear images of all the capsomers. Instead, the images looked disordered with confusion of detail; only an occasional particle seemed to reveal almost all the capsomers clearly. Even where one could make a reasonable count, one did not know whether to double the count in order to account for the other side of the particle or perhaps double the number of interior capsomers but not the number of peripheral capsomers. Given the uncertainty of dealing with puddles or with the matter of what part of a count to double, how was one to get the definitive answer? And how was one to explain why so many of the images seemed not to show capsomers clearly?

The Caspar–Klug theory of capsid design (Caspar and Klug, 1962) showed the way; the theory predicted that symmetrical shells would have $10T + 2$ capsomers, where $T = h^2 + hk + k^2$ and h and k are any integers. Plugging in integers, Caspar and Klug generated the possible values; $T = 1$, 3, 4, 7, 9, 13, etc. Thus, 42 ($T = 4$) and 92 ($T = 9$) were in accord with the theory. What is key to interpreting the images, however, is that the patterns of 5- and 6-coordinated capsomers are different for designs with different values of T (Fig. 3).

Thus, an alternative to counting subunits was to determine the pattern, which Aaron and John did (Klug and Finch, 1965). The key was to locate one-sided images, that is, images where the negative stain was thin enough to reveal the capsomeric detail on the side of the virus closest to the grid. Aaron and John found a small percentage of such one-sided images. They showed that every one-sided image had the same $T = 7$ pattern, and thus they argued that there are neither 42 nor 92 but rather 72 capsomers in the capsid.

Aaron and John came under immediate crossfire from both camps who thought their own method was as valid as Aaron and John's. It seemed clear to me, still a neophyte, that Aaron and John were right. But their case was not airtight. While John and Aaron's analysis was correct for the images they could analyze, these images only accounted for a small fraction of the particles. What was wrong with the rest of the particles? John and Aaron argued that the one-sided images were rare, that most images were two-sided, and that the confusion of detail in the two-sided images was due to the superposition of the near and far sides of the virus capsid. To prove this, they produced analog, two-sided images from a $T = 7$ model. They did so for many orientations showing that the confusing and varying patterns seen in micrographs could be accounted for in detail by their $T = 7$ model (Finch and Klug, 1965). That still did not satisfy everyone. Others produced two-sided images for their $T \neq 7$ models and produced patterns reminiscent of the two-sided virus images, but if one looked at the details, the other models failed. Even though John and Aaron could account for essentially all the two-sided images, the case made by John and Aaron was not totally airtight. Perhaps, there was more than one structure that could give rise to the same patterns seen in two-sided images. Just because they had a structure that explained everything did not mean that their answer was unique. Intuitively, Aaron and John argued that tilts would prove the uniqueness of the model (Klug and Finch, 1968). They tilted virus particles in the microscope, tilted the model by the same amount and in the same direction, and showed that the images derived from the model accounted for the images seen in the microscope. Intuitively, they must have a unique model (Fig. 4).

A

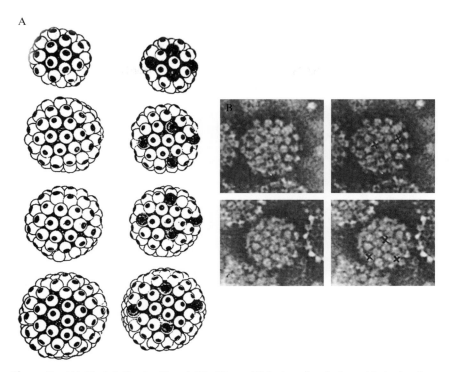

Figure 3 (A) Models for the $T = 4$, 7L, 7R, and 9 designs for viral capsids (going from top to bottom). The models on the right are repeats of those on the left, except that the positions of the five-coordinated capsomers are marked with a black circle. The unmarked capsomers are six coordinated. Note that the pattern of moving from one five-coordinated capsomer to the next nearest five-coordinated capsomer is different for each of the four structures. For $T = 4$, the pattern is a straight line move from 5 to 6 to 5. For $T = 7$, the move is like a knights move in chess, 5 to 6 to 6 in a line and then a move left (L) or right (R) to the next five-coordinated capsomer. For $T = 9$, it is again a straight line move from 5 to 6 to 6 to 5. (B) Electron micrographs of two one-sided images of negatively stained polyoma virus on the left. On the right, the clearly five-coordinated capsomers are marked with Xs. The pattern is clearly $T = 7$. The images are taken from the work of Klug and Finch (1965). Figures reproduced with the permission from Elsevier.

Here was the situation. Aaron and John had devised a method for solving the structures of negatively stained particles by electron microscopy: generate a model, use it to generate two–sided images, and show that upon tilting, the images of the model could account for the actual images seen when the structures were tilted in the same way. Don Caspar added to the method with his analysis of images of turnip yellow mosaic virus (Caspar, 1966). He took the most accurate physical model of the virus, which was made from corks to simulate protein subunits, coated it in plaster of Paris to simulate the

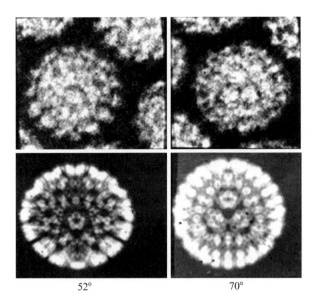

52°   70°

Figure 4 The top two, two-sided images are obtained from a single polyoma virus particle that has been tilted by 18°. The bottom two images are from a $T = 7$ model of the virus that has been tilted by the same amount and in the same direction. Note that the patterns of capsomers on the model and virus correspond. The images are taken from the work of Klug and Finch (1968) and are reproduced with permission from Elsevier.

negative stain, and took radiograms of the model with an X-ray machine to obtain two-sided images. The agreement was remarkable but not quite perfect. This then was the method for solving structures from micrographs of negatively stained specimens (Fig. 5).

Aaron and John put the method to use on electron micrographs of fraction 1 (aka rubisco). One of my first jobs as an apprentice structural biologist was to help with the model imaging (i.e., the grunt work). John had taken micrographs of negatively stained preparations, and Aaron and John recognized the octahedral symmetry of the complex. They had the workshop build a model of their best guess of the true structure. My job was to slather the model with plaster and make radiograms. The features in the radiograms did not agree well with the EM images. Aaron and John decided on alterations to the model. So, I knocked all the plaster off the model and had the shop alter the wooden doweling subunits. When the shop was done, I replastered the model and made new radiograms. The fourfold view was better, but the threefold or perhaps the twofold view was worse. After a few weeks of trial and error, Aaron and John could not get the details right for all views. The reason why this structure was harder to solve than that of a virus had to do with the structure itself. In the virus images, it is easy to see the side views of the capsomer by looking at the edge of a particle and to see

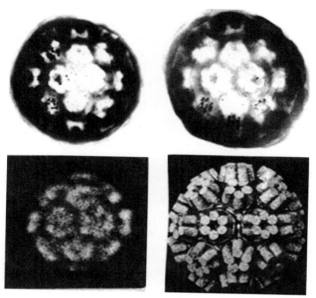

Figure 5   The top shows an X-ray radiogram of the plaster of Paris coated model of turnip yellow mosaic virus shown at the bottom. The view is down a twofold axis. The middle frame shows an actual electron micrograph of a negatively stained preparation. The view of the virus particle is again down the twofold axis. The bottom frame is an image of the model. The subunits are represented by corks. The image is from the work of Caspar (1966) and is reproduced with permission from Elsevier.

the top views by looking at the center of a particle. With fraction 1, the length of the subunit and the lower symmetry with its smaller radius made it impossible to get a clear side or top view. It was impossible to mentally extract the end-on view of a subunit in the middle of the image from the superposed features coming from sideways–oriented subunits along the periphery and vice versa. Images of the subunits simply overlapped one another. Moreover, the smaller size meant that there were no recognizable one-sided views. The project was eventually abandoned, and we turned to the analysis of helical structures.

2. HELICAL STRUCTURES

A couple of years before my arrival at Cambridge, Roy Markham and his colleagues had generated improved images from electron micrographs by photographically averaging repeating features within the images (Markham *et al.*, 1963, 1964). The effect of averaging was startling (Fig. 6).

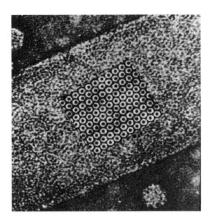

Figure 6 Tubular (helical) polymorph (often called a polyhead) of the capsid protein found in a preparation of turnip yellow mosaic virus. The inset shows the result of photographically averaging the repeated capsomers in the tube. In the original formulation of the method, the correctness of the shifts was judged by the appearance of the resulting average. In this example carried out later, the shifts used for photographic superposition/averaging were deduced from the optical diffraction patterns of the image, as introduced by Klug and Berger. The image is taken from the work of Hitchborn and Hills (1967) and reproduced with the permission of Science.

They began with an image in which features were not clearly seen and produced an image that revealed the regular features in beautiful detail. In order to correctly enhance detail, they had to guess at which directions and by which amounts to shift the image between exposures. They judged the correctness of the shifts subjectively by the appearance of hidden detail. Such trial and error and the subjectivity of picking in the process moved Aaron and Jack Berger to introduce optical diffraction, which in essence tried all possible directions and amounts of shift; if the structure was periodic, one would get a series of sharp reflections whose positions could be used to deduce the direction and amount of image shifting to enhance structural detail (Klug and Berger, 1964). The amplitudes of the reflections above the noise provided an objective way to determine that the periodicities did not arise by chance from the noise in the images (Fig. 7).

Optical diffraction thus removed the subjective nature of determining the direction and length of shifts needed. But Aaron had the idea to select and then recombine the diffracted rays to generate a filtered image directly rather than by the more laborious process of shifting and exposing to generate photographically an averaged image. I was set to the task of getting this method, called optical filtering, to work.

With two 150 cm focal length lenses, the optical diffractometer with its vertical I-beam construction was too tall to fit on one floor. It had instead been put into an old dumbwaiter shaft. The high-pressure mercury arc lamp

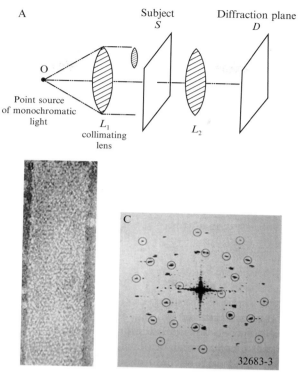

Figure 7 (A) Schematic of an optical diffractometer. A point source of light on the left is converted into a plane-parallel beam by the lens L_1. The plane parallel beam is diffracted by the subject, and the diffracted rays are brought to a focus on the diffraction plane by lens L_2. The modified image is taken from the work of Klug and De Rosier (1966) and is reproduced with permission from Nature Publishing Group. (B) An electron micrograph of a negatively stained T4 polyhead, a helical polymorph of the main capsid protein. The capsomers are not clearly seen, although regular periodicities can be seen by viewing the image at a glancing angle. (C) Optical diffraction pattern of the image in (B). The appearance of two sets of sharp reflections related by a vertical mirror line means that the image is a two-sided, flattened tube. The circles denote the set of reflections coming from one side of the tube. The diffraction pattern provides clear evidence that the confusion in the image arises from the superposition of details in the two sides. The two images in (B) and (C) were taken from the work of Yanagida et al. (1972) and are reproduced with permission from Elsevier.

used for illumination sat about level with my feet on the ground floor and the sample and camera were about chest high in the basement. The system was folded by a single mirror. We wanted to add a lens just after the diffraction plane. This lens would recombine the diffracted rays into an image. To generate a filtered image, we would replace the film in the diffraction camera with a filter mask that would pass only selected rays into the lens and hence into the filtered image. We were unsure what kind of lens to use. We wanted a lens that would correctly pass the rays without altering the all-important phase relationships needed to get the correct details in the filtered image. We consulted Professor Edward Linfoot, the head of the Cambridge Observatory. We stopped in at Linfoot's house just after tea time. In one room in his house, he had set up a workshop where he could grind and test his own optical components. He showed us around in his "museum" of optical devices. After our tour, we settled down to the business of what kind of lens we needed. We told him that we needed to select rays from our approximately 3 cm-sized diffraction patterns and recombine them into a filtered image. We wanted the shortest focal length lens possible for space reasons. He said that we needed a 30 cm focal length achromat because lenses are good to about f10, where the f number is the focal length divided by the usable diameter, which as we had said, was 3 cm. He said that with monochromatic light, we did not need an achromat but that commercial achromats were generally better optically than commercially available singlet lenses. He said that camera lenses having smaller f numbers and shorter focal lengths were generally not so good because the purpose of the smaller f numbers was not to produce greater resolution but rather to have a large aperture to reduce exposure time without degrading the image sharpness. We bought two commercial achromats, which Linfoot tested for optical quality. We used the better of the two for our optical filtering apparatus. We applied the filtering method (Klug and De Rosier, 1966) to several helical structures to images of T4 polyheads taken by John Finch (De Rosier and Klug, 1972; Yanagida et al., 1972). The filtered images were pretty spectacular. For the helical T4 phage tail, we generated filtered images that showed only the front or only the back half of the structure, but the method was only a partial answer. The images were still two-dimensional, and although we could separate the front from the back, the features at different radii in the image of one half were still superposed. We needed something that could filter different radii—at least that is how we thought of it at that time (Fig. 8).

3. 3D RECONSTRUCTION OF HELICAL STRUCTURES

Ken Holmes had embarked on solving the structure of tobacco mosaic virus, a helical virus. Ken was collecting X-ray generated diffraction amplitudes from oriented fibers of the virus. He was using the method of isomorphous replacement to get phases (Holmes and Leberman, 1963). It might seem at first blush that this could not work because although the rod-shaped virus particles had their axes aligned, they could be rotated by different amounts around those axes, be staggered at different axial distances,

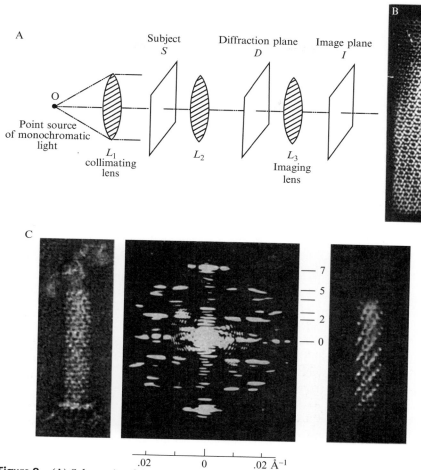

Figure 8 (A) Schematic of the optical filtering apparatus taken from the work of Klug and De Rosier (1966) and is reproduced with permission from Nature Publishing Group. The instrument was adapted from the optical diffractometer by adding an imaging lens L_3, which combines the diffracted rays into an image of the subject. A filter mask, which was placed in the diffraction plane, permitted one set of the mirrored pair of sets of diffracted rays to pass into the imaging lens, L_3, and produce a filtered image of one side of the tube. (B) The filtered, one-sided image of the polyhead is shown in Fig. 7B. The circles in Fig. 7C show the positions and sizes of the holes in the filter mask. The image is taken from the work of Yanagida *et al.* (1972) and is reproduced with permission from Elsevier. (C) An isolated T4 phage tail, its optical transform, and a one-sided filtered image. Although one side is removed, there is still superposition of features from different radii. What we said to ourselves was that some way was needed to separate radial features. The figure is taken from the work of DeRosier and Klug (1968) and is reproduced with permission from Nature Publishing Group.

and at random be turned right-side up or upside down. The reason why it is possible to solve the structure from oriented fibers is that the transforms of helical structures have special properties. Their diffraction patterns are invariant with respect to rotation about the axis, to axial translation, and to up versus down orientation. The reason is that helically symmetric structures diffract planes of beams whose amplitudes are cylindrically symmetric and whose phases linearly vary with the rotation angle about the helical axis. Thus, when Ken measured amplitudes, it was as if he were measuring amplitudes from a single TMV particle. (Actually, the situation is more complicated due to layer plane overlap, which depends on the detailed symmetry operations, but I will not cover that here.) Ken could thus use isomorphous derivatives to determine phases much as one does for a p1 crystal.

Ken's work brought a possibility to our attention: if we could measure the amplitudes and phases of the optical diffraction pattern for any one T4 phage tail image, we could generate its 3D Fourier transform and then calculate the inverse transform to see the particle in 3D.

We measured the amplitudes from the optical diffraction patterns, and we decided to obtain the phases optically by making holograms. In essence, we added a plane wave to the diffraction pattern. The resulting interference generated a fine set of fringes across every layer line reflection. The positions of the maxima encoded the phases (Fig. 9).

Measuring the absolute positions of the fine fringes along layer lines, however, was a nightmare that seemed to take forever. After too many hours of mind-bewildering measurements, we had the phases and, using Ken Holmes' Fourier–Bessel inversion program, we computed our first 3D map. The subunit looked rather like a radially oriented dog's leg. We were excited, but we needed to do more particles.

4. DIGITAL IMAGE PROCESSING

Needless to say, I was not enthusiastic about measuring phases optically. It was not just that it was slow tedious work; it was that I thought I would lose my mind gathering the one set of phases. The only thing that kept me going on the first set of phases was the thought of seeing the first 3D reconstruction from an electron micrograph. Aaron was sympathetic and we decided to look into the determination of phases by calculating the transforms of digitized images. Some of the crystallographers, who calculated Fourier transforms regularly, were not encouraging saying that a transform of the size of an image was impractical. We were unfazed because the alternative of measuring phases optically was too awful to consider adopting. Luckily, Uli Arndt of the lab had recently constructed a flying

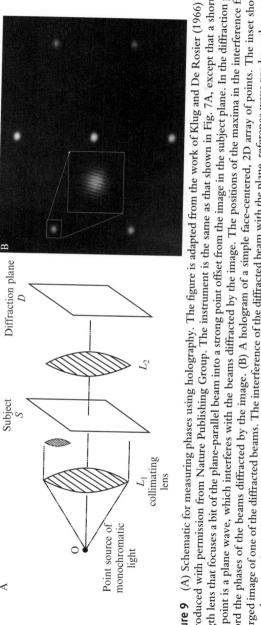

Figure 9 (A) Schematic for measuring phases using holography. The figure is adapted from the work of Klug and De Rosier (1966) and is reproduced with permission from Nature Publishing Group. The instrument is the same as that shown in Fig. 7A, except that a short focal length lens that focuses a bit of the plane-parallel beam into a strong point offset from the image in the subject plane. In the diffraction plane, this point is a plane wave, which interferes with the beams diffracted by the image. The positions of the maxima in the interference fringes record the phases of the beams diffracted by the image. (B) A hologram of a simple face-centered, 2D array of points. The inset shows an enlarged image of one of the diffracted beams. The interference of the diffracted beam with the plane, reference wave produces the set of fine fringes that cross the diffracted beam. Measuring the absolute position of the maxima in one beam relative to those in the other beams enables one to determine the phases. The figure is adapted from the work of Matthews (1972).

spot densitometer, which Tony Crowther programmed to scan X-ray diffraction patterns. He obligingly wrote a program to scan our images. We needed additional programs to process the digitized images. I had never written a computer program before and taught myself both machine and FORTRAN programming to carry out the work. The densitometer was under the control of the lab's Argus computer, a room-sized device programmed in octal using punched paper tape. Octal was not a language; it was the bit pattern fed into the instruction register encoded in octal rather than binary. Argus had real-time capabilities or perhaps it was time-sliced; I have forgotten the details, but it did more than one thing at a time. It ran several X-ray diffractometers, a magnetic tape deck, the punched paper tape reader, a teletype, and the flying spot densitometer. In those days, there was always the danger that your program could do bad things and overwrite someone else's program or data or simply hang the machine. Programming was more nerve wracking especially if one accidentally ruined the X-ray diffraction data being collected, for example, by Max Perutz, who still personally mounted hemoglobin crystals and collected data. To ease the situation, time was set aside for programmers to test and debug their programs. Max, if he wished, could still run his diffractometer but with the understanding that bad things could happen. I needed to do a bit of programming on Argus for our work. Even after "fully" debugging my programs and therefore running them outside the testing time, I was apprehensive because of the programmer's rule: there is always one more bug.

We put the optical densities from our images onto large reel magnetic tapes driven by Argus. Our Fourier transforms were calculated at the Imperial College in London on an IBM 7090, which had 32 k of memory and could do 33,000 divisions per second. Programs were loaded onto punch cards, and indeed even our image data was dumped from magnetic tape onto punch cards. There were two computer runs a day; one of the computer staff would take the cards and tapes into London on the train in the morning and would return with line printer output, cards, tapes, etc., in the afternoon. We would pour over the output and scramble to get ready for the afternoon trip to London. The output would return later that evening. When we began our processing of digitized images, no FORTRAN programs existed to box images from the rough scanned data, to calculate Fourier transforms from images, to correct the phase origins, to pick data off of layer lines, etc. We had to write these ourselves, and luckily we had a lot of help from good programmers, especially Tony Crowther.

I got a copy of a book on FORTRAN. It was not a book on how to write FORTRAN programs but rather a manual that told me what each FORTRAN statement did. It did not tell me why I might want such a statement or when to use it. Some statements were obvious like *if* statements or *do* loops. Others like *equivalence* statements seemed idiotic; why would you want two different and differently named chunks of data to be in the same memory locations where they would overwrite each other? Programming proved amusing and frustrating, and with time, I found uses for all the FORTRAN statements, even the idiotic ones. With two runs a day, debugging my Fourier transform program was slow.

The first draft of my Fourier transform program timed out; I had to set a time limit for the execution of my job, which prevented any one program from using all the time allotted for the whole lab. My program timed out producing no output other than the timed out error message. I had no idea where in the program the end had come or what to fix in the program. Naively I inserted a print statement at the start of the outermost loop in the transform calculation so that I might see how far the computation had gone before it timed out. The program again timed out, but with the print statement, I thought I could figure out how far it had gotten. The output was an unhelpful one. It had timed out somewhere just after entering the first of about 100 loops. Clearly, whatever bugs might be present in the code, I needed to make things faster. Eventually, I sped up the program so that it could do a 100 by 40 transform in a reasonable time, probably measured in minutes. I had produced a couple of other crude programs, enough to get amplitudes and phases from the transforms, but all the corrections for phase origin and so on were done by hand. The first transform was exciting; it looked like a digital version of the optical transform (Fig. 10).

I extracted the phases and amplitudes along the layer lines and calculated our second 3D map of a single phage tail. The first map of the particle, which we had done with optically measured phases, was somewhat different from the second, all digital maps presumably due to differences in the phases. The bigger features were similar but the finer details were changed. The computer-generated phases were more rigorously obtained and we believed them, but we needed to look at 3D maps generated from images of other T4 tail particles. We computed about half a dozen maps, which varied quite a bit because of differences in stain penetration, shrinkage, and flattening. Yet, the underlying structure was quite evident. It was early 1967, and it was an exciting time.

Here is how our early digital processing went. We selected for densitometry the best particles or segments of particles based on their optical transforms. Then, we had our densitometered images dragged into London. After lunch, we rushed to our output, an analog display of the image. There were no graphic monitors; we simulated a graphical display using the overprinting capability of the line printer. There was a FORTRAN format statement that prevented the line printer from advancing the paper and allowed us to print one line on top of another. By choosing the right characters to print and print on top of one another, we generated a poor-man's gray scale for each pixel and thus could print a gray-scale image of what we had scanned. The line printer image was immense, about 2 by 3 ft

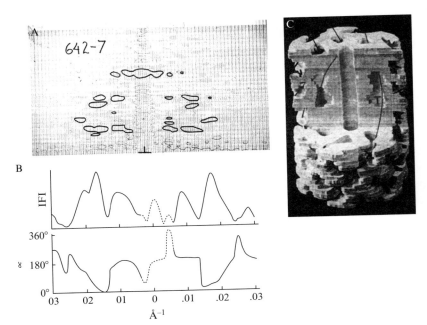

Figure 10   (A) Line printer output of the amplitudes from the Fourier transform of a digitized T4 phage tail (middle image in Fig. 8). The amplitudes are scaled from 0 to 9 so that there is one character for each reciprocal pixel. Only the top half of the transform is shown; the bottom half is identical. Taken from the work of DeRosier and Moore (1970) and reproduced with permission from Elsevier. (B) A trace through the first layer line above the equator in Fig. 10A. The top trace shows the relative amplitude. The bottom trace shows the phases. Both the amplitude trace and phase trace have approximate mirror symmetry about the point 0 as expected for an even order layer line. (C) Balsa wood model of a three-dimensional reconstruction of the T4 phage tail. (B) and (C) are taken from the work of DeRosier and Klug (1968) and are reproduced with permission from Nature Publishing Group.

and required some distance to comprehend the levels of gray. We set up a cork board at one end of the hall outside our office, pinned up the image, and backed down the hall until our eyes saw the levels of gray rather than the overprinted characters. To mask off the area we wanted to transform, we picked one corner of the particle in the image. We walked back up the hall keeping our eye fixed on the desired point. We marked it with a large red china marker. We went back down the hall to work on the next corner. After four trips up and down the hall, we had the coordinates of the four corners of our mask. The image on punched cards was then carted into London to be windowed and transformed. Since transforms were expensive, we saved the transform on punch cards. Our output for viewing the Fourier transforms was again done with the line printer. The amplitudes were scaled from 0 to 9 so that only one character was needed per reciprocal pixel.

The amplitudes were put out as one map (see Fig. 10A to see such a map). The phases were scaled in 10° intervals, and a separate map was generated with one character per phase using the characters A–Z and 0–9 for the 36 possible phase intervals of 10°. We happily contoured the amplitudes by hand onto clear plastic and then put the plastic contours over the phase map. We then handpicked phases along the layer lines. We computed 3D maps for six different particles, which were similar but had differences due to variable stain penetration and shrinkage. We averaged them crudely by overlaying the density contours and drawing a conservative subunit.

5. 3D Reconstruction of Asymmetric Structures

In early 1967 while working on the phage tail reconstructions, Aaron and I had no idea how to use micrographs to generate 3D images of nonhelical particles. The use of Fourier–Bessel transforms somehow hid the essential connection between the transform of an image and the transform of the 3D structure until, as often happens and for no particular reason, the answer came in a real "aha" moment. The Fourier transform of an image, which is a projection of the 3D structure, is a central section of the 3D Fourier transform of the structure. By combining data of different views, we could build up the 3D transform and hence generate a 3D image of the structure. The helix is a special case because one view of a helical object can present many views of that helical object, and thus, one view can be enough to generate a 3D structure.

The section/projection theorem was known to crystallographers and to us. It is hard to say why its application was not totally obvious from the start, but one factor may lie in the terminology used to talk about images of negatively stained virus particles. Images were described as one-sided or two-sided. The difficulty in interpreting the two-sided images lay in the confusion of detail arising from a superposition of features on the near and far sides of the particle. Realizing how to do 3D reconstruction certainly involved recognition that the terms "two-sided" and "superposition" meant that the images were projections with the obvious implications for 3D reconstruction (Fig. 11).

Figuring out the minimum number of views one needed was straightforward by analogy with the collection of data in X-ray crystallography. In Fourier space, we needed to collect data spaced by no more than $1/D$, where D is the dimension of the structure. The higher the resolution, $1/d$, the smaller the angle needed between views. If we could tilt through 180°, the total number of tilts would be simply $n \approx \pi D/d$. So, for an asymmetric structure like the ribosome, about 30 views would be needed for a resolution of about 3 nm.

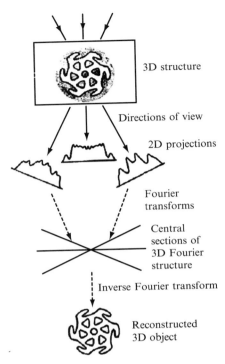

3D structure

Directions of view

2D projections

Fourier
transforms

Central
sections of
3D Fourier
structure

Inverse Fourier transform

Reconstructed
3D object

Figure 11 A schematic representation of the principle of 3D reconstruction from electron micrographs. One collects views of the unknown structure over a wide angular range either by tilting a single particle or by using different particles, which are viewed in different directions in the electron micrograph. The Fourier transforms of these views, which are projections, are central sections of the 3D transform of the 3D structure. By combining all the data into a properly sampled 3D transform, one can calculate an inverse transform to obtain the 3D structure. The figure is taken from the work of DeRosier and Klug (1968) and is reproduced with permission from Nature Publishing Group.

Thus at one fell swoop, we had answered all those questions posed by the work on the polyoma–papilloma viruses. We knew how many tilts it took to prove a structure, which depended on the dimension of the structure and the resolution of the model. Better still, we had a way to directly generate the 3D structure from the images rather than having to guess at a model, make it, plaster it, take radiograms, and then guess why the agreement was not perfect.

Even though we could not tilt a single particle through 180°, a field of particles usually offered a variety of views, and we could use those to fill up Fourier space. That was the good news. The bad news was that we needed Fourier coefficients on a regular grid of positions and not on a set of irregularly spaced points, which undersampled some regions of Fourier space and oversampled other regions. The algorithms to correct uneven sampling were already in hand, but the calculations were beyond the capabilities of the computers of the day. We needed to come up with about 3500 Fourier coefficients in the 3D transform of say a ribosome solved to ~3 nm resolution. We thus needed to invert a set of 3500 simultaneous, linear equations to obtain the 3500 unknown Fourier coefficients, given the many amplitudes and phases obtained from the images. On inverting the set of equations to get a least squares solution, we would determine from the eigen value spectrum whether the equations were ill-conditioned, that is, whether we had enough views at the right angles. Ill-conditioning arises when regions of reciprocal space are undersampled and thus when one is extrapolating from one region to fill in for the missing observations in an undersampled, neighboring region. A consequence of such extrapolation is that errors in the observations can be amplified in the values obtained for the unknowns. Sadly, the inversion of this large set of equations seemed beyond the capabilities of the computers of the day.

Our first thought was to try the method on an icosahedral virus, where each view was 60 views given the symmetry of the virus. Thus, only a few particles would be needed to generate a 3D map of the structure. One idea was to use the icosahedral harmonics since they embody the symmetry and would reduce the number of unknowns and hence the number of equations by a factor of 60 to about 60 simultaneous equations. But then, how could one get to a higher resolution? The use of icosahedral harmonics was not the answer. There was one good piece of news, though. Since each image of a virus particle corresponded to 60 symmetry-related images, the transforms of these 60 images intersected in "common" lines where the Fourier coefficients must be identical. Hence, the notion of common lines could be used to determine the orientation of a virus particle with respect to the symmetry axes.

The gloom remained. How could the set of equations be solved? Hugh Huxley suggested a form of back projection in which we would use a bunch of slide projectors and cigar smoke. Each projector would send out beams of light from transparencies of the images along the directions of view for the images. The cigar smoke would allow the intersecting beams to be visualized in 3D. The problem with such a solution was that one would not know whether the equations were ill-conditioned, meaning that there was an insufficient set of views. Since 3D reconstruction was a new method, it was essential to establish ways of proving that the answer was the right one. Simply reprojecting the 3D answer to generate images that could be compared to the images used in the reconstruction is a circular line of reasoning. It only tells one whether something is wrong with the program and not with the completeness of the data set.

The solution to the problem came from out of the blue and from our intimate relationship with Fourier Bessel transforms. The solution of one

3D set of 3500 simultaneous equations could be changed to many small 1D sets of equations. Each set corresponded to a ring of Fourier–Bessel coefficients with fixed values of R (radius) and Z (height above the equator); the input values occurred at irregular angular points. The solution was a set of Fourier–Bessel coefficients that could be used in a Fourier–Bessel inverse transform. The only penalty to reducing the dimensionality was that some of the data were thrown away but the method was tractable: the largest set of equations could be inverted with existing computers. The inversion of the equations would provide eigen values whose magnitude would uncover any ill conditioning due to lack of data in regions of reciprocal space, and the result would be a least squares solution. Moreover, the eigen values for each set of equations would reveal which rings in Fourier space were undersampled. Such rings might be eliminated or additional views could be found to fill in the missing data (Fig. 12).

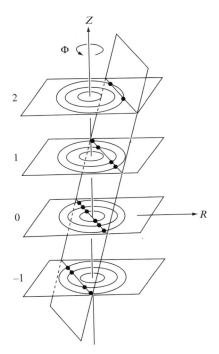

Figure 12 By sampling the Fourier transforms obtained from images on a set of concentric rings on a stack of equally spaced planes, one can set up a set of linear equations for each ring. These equations, when solved, will yield a set of Fourier–Bessel coefficients that describe the Fourier transform in 3D. There will be one small set of equations to invert for each ring, making the problem of solving the equations tractable. The image is taken from the work of Crowther (1970) with permission from the Royal Society of London.

In August 1967, after Aaron and I had figured how to do reconstruction of particles with any or no symmetry and to know the solution was tractable, we wrote up the principle of the method with the example from the helical T4 tail structure. We puzzled over what name should be given to the method. We considered 3D electron microscopy, but that suggested that the microscope itself generated 3D images rather than needing to reconstruct the original 3D structure given a bunch of different views. We decided on 3D reconstruction. We gave the paper to Hugh Huxley and Max Perutz, who not only read it thoroughly but also did a masterful job of editing and helping with the verbiage. We mailed the paper to *Nature* at the end of December. It was received on January 3, and to our surprise, it was accepted immediately. The editor wanted to read the proofs over the phone in order to get it into the next issue. We said we would like to see the proofs, and so the publication was "delayed" until January 13, 1968 (DeRosier and Klug, 1968).

We were not the only ones thinking about and having solutions for generating 3D maps from electron micrographs. Walter Hoppe also published his ideas in 1968 (Hoppe et al., 1968) and Michael Moody had thought along our lines all unknown to us. Later in 1968, Roger Hart published his reconstruction method, which he called polytropic montage (Hart, 1968). It was aimed at extracting 2D z-sections from tilted views of microtome-sectioned material. In his method, each tilted image is stretched by the inverse of the cosine of the angle of tilt. The stretching compensates for the foreshortening due to tilting. The stretched views are then shifted perpendicular to the tilt axis and added up to extract the image of a particular z-section within the 3D structure. Different shifts extract different z-sections. Thus, Hart could stack up the z-sections to get a 3D reconstruction. The resolution perpendicular in the z-direction was determined by the largest tilt angle in analogy, with the angle of illumination gathered by the objective lens in a light microscope. Even though the method seems quite different from the method that we described, it is simply the real space equivalent of the reciprocal space method that we described. To prove the equivalence of the methods, consider Hart's manipulations in Fourier space.

In 1968, Peter Moore, who worked with Hugh Huxley, joined the programming effort in order to produce a 3D map of actin decorated with the motor end of myosin. The programming goal was to automate many of the procedures and make them faster. By this time, the computing had moved from London to a faster IBM 360 operated by the astronomers at Cambridge University. Although we thought everything would be straightforward, there were some surprises. The amplitude map looked almost exactly like the optical transform of the actomyosin images, but the phases did not obey the expected phase relationships for helically symmetric objects. We had enforced the helical phase relationships by moving the origin of the transform to lie on the helix axis. Of course, we did all these calculations initially by hand. The phases were still not perfect. We wondered if the helical axis was tilted out of the plane of the projection. We worked out the math and figured out the corrections needed and applied them. Lo and behold, the phases of almost all the strong peaks were nearly perfect except those from one image. Annoyingly in that image, the phases of one very strong pair of symmetric peaks on the transform were disturbingly far from the expected values. How was this possible for just one pair of strong reflections and not the rest, some of which were weaker? Was it some bizarre distortion of the particle? It did not seem so, for the image looked perfectly good, except that the particle axis in the densitometer was slightly tilted relative scanning grid and hence to the edges of the mask we used to window the particle. We noticed that the masked edges of the boxed area gave rise to a strong spike of intensity in the Fourier transform. Because the image is digitized, this strong feature is convoluted with the transform of the sampling grid and is thus effectively folding back into the transform at the transform boundaries. The strong spike of intensity fell across one of the strong pair of reflections, thus perturbing its phases. This effect is known as aliasing, and we had discovered the need for floating the image to remove abrupt changes in density at the edges of the boxed image. We floated the image and, voila, the phases now obeyed helical symmetry. The effects of aliasing were known—just not to us—when we started. This and other "mysteries" made the programming of reconstructing an engrossing puzzle as addicting as a crossword puzzle. We spent time writing, debugging, and rewriting programs to make them better, faster, and more automated, obviating the need to do corrections by hand. When we had finished our final suite of programs, we celebrated by treating ourselves to an expensive and outstanding meal at a local French restaurant. We enjoyed a bottle of Chateau Haut Brion vintage 1936 (not a typo—the wine was older than we were). Our protocols (DeRosier and Moore, 1970) and the actomyosin map (Moore et al., 1970) were published in early 1970.

In 1969, Tony Crowther returned from a stint in Scotland, and with Linda Amos, took up the implementation of our algorithms to solve the first icosahedral virus. The work involved the use of those handy common lines to determine the orientation of each particle with regard to the symmetry axes. They applied the method to a tomato bushy stunt virus and to a human wart virus. No assumptions were made as to the T number of either virus and indeed for human wart virus, the results clearly showed that there were 72 capsomers (Crowther et al., 1970). Thus, my story has come full circle (Fig. 13).

Before ending, I want to add that when the papers on T4 tail, viruses, and actomyosin came out at the end of the 1960s and the start of the 1970s, I never dreamed that anyone would ever be able to do better than about 2 nm resolution. The reason is that a negative stain was needed to produce contrast and to provide a radiation-resistant casting of the specimen. The resolution would be limited by the ability of the stain to make a faithful

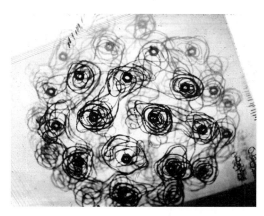

Figure 13 A contoured 3D map of the human wart virus. Each blob in this low-resolution map corresponds to a capsomer. The $T = 7$ design is easily verified. The image is taken from the work of Crowther *et al.* (1970) and is reproduced with permission from Elsevier.

casting of the features of the particle. It is a marvel that with the advances in methods, machinery, and image analysis in the intervening years, workers in the field have eliminated the requirement for stain and are now able to see details in viruses and other macromolecular structures to better than 0.4 nm. It is possible to do chain tracing with such maps. It is also possible to see features on viruses that do not obey the icosahedral symmetry such as specialized, tiny portals and even to detect and image conformational variations. An equal marvel are the 4–5 nm resolution tomograms of frozen hydrated whole cells, of sections of frozen hydrated cells, and of organelles. Averaging 3D particle images cut out of tomograms has lowered the resolution to 2–3 nm.

What will the future bring? One of our biggest needs is for a clonable marker that is the equivalent of the clonable markers for light microscopy so that we might identify the macromolecules we are visualizing. There are multiple groups working on such markers. We need better cameras that capture every electron in a digital format, and there are multiple groups working on them. We need phase plates to generate contrast without the need for defocus, and there are several groups working on that. We need more automated software and there are groups working on that. I imagine that one day, electron cryomicroscopy will be like light microscopy, a tool that anyone can use without any specialized training. I imagine a day when any undergraduate will be able to follow a recipe for sample preparation and labeling, insert the specimens in the microscope, and have 3D maps with identifying labels come out from the other end. Will this come to pass in the future? I think so, but to repeat a quote attributed to Yogi Berra, "It is hard to make predictions—especially about the future."

REFERENCES

Caspar, D. L. (1966). An anaolgue for negative staining. *J. Mol. Biol.* **15**, 365–371.

Caspar, D. L., and Klug, A. (1962). Physical principles in the construction of regular viruses. *Cold Spring Harb. Symp. Quant. Biol.* **27**, 1–24.

Crowther, R. A. (1970). Procedures for three-dimensional reconstruction of spherical viruses by Fourier synthesis from electron micrographs. *Philos. Trans. R. Soc. Lond.* **B261**, 221–230.

Crowther, R. A., Amos, L. A., Finch, J. T., De Rosier, D. J., and Klug, A. (1970). Three dimensional reconstructions of spherical viruses by Fourier synthesis from electron micrographs. *Nature* **226**, 421–425.

DeRosier, D. J., and Klug, A. (1968). Three-dimensional reconstruction from electron micrographs. *Nature* **217**, 130–134.

DeRosier, D. J., and Moore, P. B. (1970). Reconstruction of three-dimensional images from electron micrographs of structures with helical symmetry. *J. Mol. Biol.* **52**, 355–369.

De Rosier, D. J., and Klug, A. (1972). Structure of the tubular variants of the head of bacteriophage T4 (polyheads). I. Arrangement of subunits in some classes of polyheads. *J. Mol. Biol.* **14**, 469–488.

Finch, J. T., and Klug, A. (1965). The structure of viruses of the papilloma–polyoma type 3. Structure of rabbit papilloma virus, with an appendix on the topography of contrast in negative-staining for electron-microscopy. *J. Mol. Biol.* **13**, 1–12.

Hart, R. G. (1968). Electron microscopy of unstained biological material: The polytropic montage. *Science* **159**, 1464–1467.

Hitchborn, J. H., and Hills, G. J. (1967). Tubular structures associated with turnip yellow mosaic virus in vivo. *Science* **157**, 705–706.

Holmes, K. C., and Leberman, R. (1963). The attachment of a complex uranylfluoride to tobacco mosaic virus. *J. Mol. Biol.* **6**, 439–441.

Hoppe, W., Langer, R., Knesch, G., and Poppe, C. (1968). Protein crystal structure analysis with electron radiation. *Naturwissenschaften* **55**, 333–336.

Klug, A., and Berger, J. E. (1964). An optical method for the analysis of periodicities in electron micrographs, and some observations on the mechanism of negative staining. *J. Mol. Biol.* **10**, 565–569.

Klug, A., and De Rosier, D. J. (1966). Optical filtering of electron micrographs: Reconstruction of one-sided images. *Nature* **212**, 29–32.

Klug, A., and Finch, J. T. (1965). Structure of viruses of the papilloma–polyoma type. I. Human wart virus. *J. Mol. Biol.* **11**, 403–423.

Klug, A., and Finch, J. T. (1968). Structure of viruses of the papilloma–polyoma type. IV. Analysis of tilting experiments in the electron microscope. *J. Mol. Biol.* **31**, 1–12.

Markham, R., Frey, S., and Hills, G. J. (1963). Methods for the enhancement of image detail and accentuation of structure in electron microscopy. *Virology* **20**, 88–102.

Markham, R., Hitchborn, J. H., Hills, G. J., and Frey, S. (1964). The anatomy of the tobacco mosaic virus. *Virology* **22**, 342–359.

Matthews, R. M. C. (1972). Image Processing of Electron Micrographs Using Fourier Transform Holograms. University of Texas at Austin, Massachusetts.

Moore, P. B., Huxley, H. E., and DeRosier, D. J. (1970). Three-dimensional reconstruction of F-actin, thin filaments and decorated thin filaments. *J. Mol. Biol.* **50**, 279–295.

Reed, L. J., and Cox, D. J. (1970). Multienzyme complexes. *The Enzymes* **1**, 213–240.

Yanagida, M., DeRosier, D. J., and Klug, A. (1972). Structure of the tubular variants of the head of bacteriophage T4 (polyheads). II. Structural transition from a hexamer to a 6 + 1 morphological unit. *J. Mol. Biol.* **65**, 489–499.

CHAPTER ONE

PREPARATION OF 2D CRYSTALS OF MEMBRANE PROTEINS FOR HIGH-RESOLUTION ELECTRON CRYSTALLOGRAPHY DATA COLLECTION

Priyanka D. Abeyrathne, Mohamed Chami, Radosav S. Pantelic, Kenneth N. Goldie, *and* Henning Stahlberg

Contents

Abstract

Electron crystallography is a powerful technique for the structure determination of membrane proteins as well as soluble proteins. Sample preparation for 2D membrane protein crystals is a crucial step, as proteins have to be prepared for electron microscopy at close to native conditions. In this review, we discuss the factors of sample preparation that are key to elucidating the atomic structure of membrane proteins using electron crystallography.

C-CINA, Biozentrum, University of Basel, Basel, Switzerland

Methods in Enzymology, Volume 481
ISSN 0076-6879, DOI: 10.1016/S0076-6879(10)81001-8

1. INTRODUCTION TO ELECTRON CRYSTALLOGRAPHY

Membrane proteins are crucial to a wide variety of cellular processes and include the families of ion channels, transporters, as well as G-protein coupled receptors. Given their central role, malfunction of membrane proteins can lead to numerous pathological conditions. Hence, an understanding of their structure and function is essential to the development of new therapeutics.

X-ray diffraction (XRD), nuclear magnetic resonance (NMR), and electron crystallography of two-dimensional (2D) crystals have been used in the determination of membrane protein structures. Used in conjunction, each method may complement the other. For example, X-ray crystallography arranges the protein in a three-dimensional (3D) crystal that may restrict the conformational space of the protein. NMR has in the past been restricted to membrane proteins of smaller size. Electron crystallography images the membrane proteins in the form of 2D crystals, the formation of which requires smaller concentrations of protein, and establishes biological conformation by their arrangement in a lipid bilayer (Glaeser *et al.*, 2007; Renault *et al.*, 2006).

2. PURIFICATION OF MEMBRANE PROTEINS

The purification of membrane proteins close to homogeneity is one of the most important hurdles in structural studies. Prior to purification, the protein must be removed from the lipid bilayer without disturbing its structural integrity and maintaining its functional characteristics. Since there are many thousands of soluble proteins in each cell, the easiest way to exclude them is to isolate the membranes as a first purification step. The isolated membranes can then be used as a starting material for the solubilization process to free the proteins from the membrane, a prerequisite for further purification.

The following is an example of a simple method for preparing membranes from *Escherichia coli*.

1. Treat the *E. coli* cell suspension with a 1% lysozyme solution to strip away the outer membranes.
2. Harvest the cells and resuspend in a hypotonic solution (50 mM NaCl, 20 mM Tris pH 8.0) containing DNAase and protease inhibitors according to the manufacturer's recommendations.
3. Sonicate or French press the cells to lyse them.
4. Use low-speed centrifugation to remove unbroken cells.

5. Harvest the membranes by centrifugation at $100,000 \times g$. The membranes can be stored at $-80\ °C$ until further use.

The next step involves the solubilization of the target protein from the lipid bilayers. Once removed from the lipid bilayers, membrane protein would denature or aggregate. It is therefore necessary to replace the lipids that surround the protein with other amphiphilic molecules (usually detergents) to shield the protein's hydrophobic belt from water, which can then stabilize the native conformation of the membrane proteins. Several different detergents are commonly used to solubilize membrane proteins. Detergents vary in both the nature of the hydrophilic head group (sugar-based phospholipid-like) and the length and composition of the hydrophobic alkyl tail. At low concentrations, the detergents exist as monomers in solution. At a specific concentration, called the critical micellar concentration (CMC), the detergent molecules form micelles because of the hydrophobic effect. The CMC is inversely related to the length of the alkyl chain. Detergents with longer alkyl chains usually have lower CMC values. At concentrations above the CMC, a detergent can effectively disrupt the interactions between a membrane protein and the lipid bilayers, thereby solubilizing the membranes. The CMC value is an important feature of the detergent, since this value can have major implications on the solubilized protein. It is also important to keep in mind that the CMC varies with different salt concentrations and temperature. The detergent concentration has to be kept above the CMC during the entire protein purification process. Some detergents may be better suited to extract a certain membrane protein from its membrane than others, so that it is necessary to screen a number of detergents to optimize recovery of soluble material.

After establishing the solubilization condition for the protein of interest, purification procedures can be elaborated. Purification of membrane proteins is generally carried out by column chromatography. Recent advances in molecular biology allow engineering either the gene coding for the protein of interest or a suitable expression vector, to add affinity tags to the protein. The most common affinity tag is a poly-histidine tag (His-tag), which is formed by a series of histidine residues, which then have a high affinity to Ni^{2+} or Co^{2+} ions that can be immobilized by NTA chelators. The length of the His-tag can determine the specificity of the interaction and thus ultimately the purity of the eluted protein. Original studies with His-tags utilized six His-residues, but up to ten His-residues have been successfully used. A higher number of His-residues should ensure a stronger interaction with the resin, allowing washes with buffer containing higher concentrations of imidazole to remove more of unspecific bound protein to the column. However, there are some disadvantages of His-tag purification systems: The His-tag can have a negative influence on the expression level, His-tagged proteins can undergo nonspecific cleavage during expression, or a low affinity of the His-tag to the column material or binding of other non-His-tagged proteins to the column may sometimes hinder the use of this tag. To overcome these problems, alternative tag systems such as a Streptavidin-binding sequence (Strep-tag) (Schmidt and Skerra, 1994) or a hemagglutinin tag have been developed.

It is necessary to maintain a homogeneously dispersed state of the purified membrane protein. If partly unfolded or not sufficiently solubilized by detergents, membrane proteins are prone to aggregation to protect their hydrophobic region from the aqueous solution. Although denaturation and aggregation occur slowly over time for most solubilized membrane proteins, it can happen fast for some protein–detergent combinations, which then precipitate almost immediately after solubilization. Inclusion of glycerol in the buffers may help reduce this problem, as glycerol reduces the hydrophobic effect of the aqueous solution. Protein aggregates can be detected by dynamic light scattering (DLS), size exclusion chromatography, or single-particle transmission electron microscopy. DLS is of limited use with membrane proteins, since free detergent may also scatter light, making it difficult to interpret the data.

Owing to their hydrophobic nature, membrane proteins associate with a larger number of SDS molecules during SDS-PAGE analysis than soluble proteins of the same weight. This causes membrane proteins to migrate faster in SDS-PAGE than their molecular mass would suggest, so that their molecular weight appears to be lower than it is. However, many membrane proteins are also heavily glycosylated, which then lets them move slower on SDS-PAGE, resulting in a shift to heavier weight, which appears in a gel as a smeared rather than sharp band.

Most 2D crystallization protocols for membrane proteins require protein concentrations of 0.5–1.0 mg/ml. Purification methods and crystallization protocols vary between different membrane proteins. So far, there has been no one method or condition that has been optimal for all membrane proteins.

3. 2D Crystallization of Membrane Proteins

2D crystals of membrane proteins can be formed by various methods. As some proteins have a natural tendency to form regular arrays, 2D crystals can sometimes be found in their native membrane environment. Since in these cases the membrane proteins do not have to be dissociated from the lipid bilayers, it may be possible to prepare these for cryo-electron micros-copy (cryo-EM) without the need to use detergents, so that the native conformations can be maintained. However, spontaneously formed 2D crystals are limited to membrane proteins that occur in high density and rarely have high crystal order, as the incidental inclusion of other protein elements perturbs crystal growth. Some examples of natively forming 2D crystals found in both eukaryotic and prokaryotic organisms include bacteriorhodopsin (bR) of *Halicobacterium salinarium* (Henderson and Unwin, 1975), the gap junction channels (Oshima *et al.*, 2007; Unger *et al.*, 1999), and water channels (Gonen *et al.*, 2005). An interesting example is the native structure of *H. salinarium* purple membranes, first solved by electron crystallography to a resolution of 7 Å. During the crystallization process, octyl-β-glucopyranoside (OG) and dodecyl triammonium chloride were added, fusing smaller crystal patches into larger 2D crystals of $\sim$2 μm diameter. Later, a similar detergent-driven approach was used to improve the crystallinity of other samples such as the cardiac gap junctions (a hexagonal crystal of connexons (Yeager, 1994)), photosystem-1 (Böttcher *et al.*, 1992), and 2D crystals of mitochondrial porin in the presence of phospholipase A2 (Mannella, 1986). Aside from the inclusion of low-concentration detergents, 2D crystallization can be induced by the inclusion of protein analogues. One such case was the Ca^{2+} ATPase from the sarcoplasmic reticulum crystallized through induction by vanadate (Taylor *et al.*, 1988).

Given the limited number of membrane proteins that natively form 2D lattices, such 2D crystals can also be prepared artificially. This method consists of reconstitution of detergent-solubilized and purified membrane proteins at high density into lipid bilayers, while fine-tuning of the reconstitution conditions is used to induce 2D crystal formation (Jap *et al.*, 1992; Kühlbrandt, 1992). The protein is purified from other proteins and contaminants by solubilization of the original membrane with the detergent. After subsequent purification steps (assuming that the protein remains in its native and properly folded state), the protein–detergent solution often contains residual lipids. These lipids are thought to contribute to the stability of the membrane protein and are essential for successful 2D crystallization. The choice of the detergent for both protein solubilization/purification and subsequent crystallization trials is critical. The reconstitution of the membrane protein into bilayers is achieved by mixing lipids and proteins, both solubilized in detergents, to form a homogeneous solution of mixed protein–detergent and lipid–detergent micelles (so-called mixed micelles). The incorporation of membrane proteins into the lipid bilayer (reconstitution process) occurs when the detergent concentration falls below the CMC. Removal of the detergent induces the progressive formation of vesicles, tubes, and/or sheets with 2D crystalline regions or failed protein aggregates.

The following protocols describe the reconstitution of membrane proteins into lipid bilayers.

Important parameters in this process are the choice of lipids, the lipid to protein ratio, and the method and rate of detergent removal. The optimal lipid to protein ratio should be screened systematically. Detergent removal can be achieved by several methods. The following are short protocols for each of the most frequently used methods:

3.1. Materials

1. Lipid stock (Avanti Polar Lipids).
2. Vacuum evaporator.
3. Glass Hamilton syringe (10 or 100 μl).
4. 5 ml glass vials.
5. Protein of interest (0.5–1 mg/ml).
6. Lipids (5–10 mg/ml; presolubilized in desired detergent).
7. Dilution buffer.
8. Pretreated dialysis membrane (according to the manufacturer's instructions).
9. Dialysis buttons.
10. Dialysis buffer.

3.2. Methods

Preparation of lipids for 2D crystallization

- Wash the syringe and glass vials with chloroform/methanol (1:1 (v/v)), then with pure chloroform.
- Evaporate the organic solvent and dry the lipid with a vacuum evaporator at room temperature.
- Solubilize the lipid at 10 times the final concentration in buffer solution or water-containing detergent at a concentration of the CMC plus 2× concentration of lipids (mol/l). Sonication and mild heating can be used to get a clear solution of lipids.
- Dilute to the desired final concentration with the buffer or water devoid of detergent.

Then, use one of the following detergent removal methods:

Detergent removal using dilution

- Mix the protein and lipid to various lipid to protein ratios (0.1–1.0, w/w).
- Dilute the resulting solutions with the detergent-free dilution buffer slowly, so that the detergent concentration drops below its CMC, and incubate at desired temperatures (Remigy *et al.*, 2003).
- Monitor the solution by taking 2–3 μl samples at different time intervals and examine under an electron microscope using negative staining.

Detergent removal using dialysis

- Mix the lipid and protein to different ratios as aforementioned.
- Pipet the lipid and protein mixture into the dialysis buttons (Jap *et al.*, 1992).

- Cut a 2.5 × 2.5 cm piece of dialysis membrane and place it centrally over the top of the dialysis button. Carefully seal the buttons with rubber O-rings.
- Immerse the dialysis buttons in detergent-free dialysis buffer and dialyze for an appropriate time at a chosen temperature or temperature profile. This can also be done on 96-well plates, using a robotic setup (Vink *et al.*, 2007).
- After completed dialysis, punch the dialysis membrane with a pipette tip and harvest the samples.

Detergent removal using cyclodextrin

- Mix the lipid and protein to different ratios as aforementioned, and set up crystallization volumes of 40 μl each. This can be done in 96-well plates.
- Calculate the required amount of cyclodextrin needed to bring the detergent concentration below the CMC (Signorell *et al.*, 2007).
- Add this amount of cyclodextrin (e.g., 100 μl of 10% methyl-β-cyclodextrin (MBCD) solution) over a longer period of time (e.g., 48 h). This can be done with the help of a robotic setup (Iacovache *et al.*, 2010).
- During this process, measure the sample volume and add water as needed, to compensate for volume loss from evaporation.
- During this process, it is also helpful to measure sample turbidity by DLS to detect membrane or 2D crystal formation in the samples (Dolder *et al.*, 1996).
- After complete detergent removal by cyclodextrin, harvest the samples.

Detergent removal using Bio-Beads

- Wash the Bio-Beads with methanol and let sediment. Then wash with the detergent-free buffer used for the crystallization. Washed beads can be stored at 4 °C in water with 0.02% azide.
- Pipet the approximate quantity of Bio-Beads slurry needed. Remove the excess solution by blotting briefly with filter paper and weigh the precise quantity of wet beads.
- Add the wet Bio-Beads directly to the protein/lipid/detergent solution and place the samples on a rotating device (10 rpm) or agitate slowly with a very small magnetic stirrer (Rigaud *et al.*, 1997).
- At various time points take small aliquots from the supernatant to monitor the concentration of the remaining detergent using appropriate methods.

After the detergent concentration has dropped below the CMC, screen for the formation of 2D crystals by electron microscopy of negatively stained preparations. This can also be done using a robotic setup (Hu *et al.*, 2010; Iacovache *et al.*, 2010).

Astonishing results were obtained from the 2D crystallization of AQP0, extracted from native membranes of sheep lenses and reconstituted into a dimyristoyl phosphatidylcholine (DMPC) bilayer. The crystals allowed the determination of AQP0 to a resolution of 1.9 Å (Gonen *et al.*, 2005), thereby revealing the location of specific lipid molecules. AQP0 2D crystals have since been used as a model system to investigate interaction between lipid molecules and membrane proteins (Chapter 4). In recent work, the influence of *E. coli* polar lipids (EPLs) on the structure of AQP0 tetramers has been investigated (Hite *et al.*, 2010). This work has pioneered a new method to study the lipid–membrane protein interaction.

Another method involves the reconstitution of the membrane proteins at the water–air interface by attaching solubilized membrane proteins to a lipid monolayer spanning a water surface and then removing detergent (Levy *et al.*, 2001). In this process, membrane proteins are forced into 2D arrangement at the air–water interface, facilitating 2D crystal formation. This approach is applicable to membrane proteins that are purified in small amounts and are stably solubilized into low CMC detergent (Garavito and Ferguson-Miller, 2001). The lipid layer strategy can also be used to produce large, planar reconstituted membranes for the incorporation of proteins into a preferred orientation.

In the past, 2D crystallography had also been applied to the structural analysis of soluble proteins (e.g., Ganser-Pornillos *et al.*, 2007; Nogales *et al.*, 1998). Since there are few cases where periodic arrays of soluble proteins occur *in vivo*, crystallization must be artificially induced by lipid monolayer, mica, or mercury substrates (Ellis and Hebert, 2001). Some examples of soluble proteins determined by electron crystallography include several members of the annexin family (studied using lipid monolayer crystallization techniques; Ellis and Hebert, 2001) and annexin VI (p68) (using negatively charged phospholipids as an interfacial layer; Newman *et al.*, 1991).

4. REQUIREMENTS FOR ELECTRON CRYSTALLOGRAPHY DATA COLLECTION

High-resolution electron crystallography is often suffering from an inefficient collection of electron micrographs containing sufficiently high-resolution data. In addition to the quality of the 2D crystals themselves, several factors were identified that degrade image quality and limit resolution. Some of these factors include embedding media, support film (upon which the 2D crystals are imaged), crystal flatness, and beam-induced resolution loss on tilted crystal samples. As with all biological specimens, resolution of the recorded images is also affected by electron beam damage on the sample (Chapter 15). To attain the structure of 2D crystals at atomic and near-atomic resolution, several methods have been developed to address these issues. Some of these are discussed subsequently.

4.1. Electron microscopy grid preparation for 2D crystals

Electron microscopy analysis of biological protein samples requires the specimens to be imaged under high-vacuum conditions. However, the dehydration of the specimens caused by the vacuum in the electron microscope results in severe structural collapse. Hence, it is important to preserve the native 3D structure of the specimen and its molecular components. There are several methods that have been developed to overcome this problem, such as vitrification of the specimen by quick freezing in its aqueous buffer solution, which converts the liquid bulk water into amorphous vitrified water (Adrian et al., 1984; Dubochet et al., 1988; Chapter 3), or the embedding of the protein with a less volatile medium such as sugars.

Sugar embedding for high-resolution electron microscopy imaging was first successfully applied to 2D crystals of bR (Unwin and Henderson, 1975). Later, glucose embedding together with low-dose imaging helped to resolve the seven membrane spanning α-helices of bR, which became famous as the first integral membrane protein structure solved by electron crystallography (Henderson et al., 1990; Henderson and Unwin, 1975). Another embedding medium, tannin, is a glucose derivative introduced during investigation of the light-harvesting chlorophyll a/b–protein complex II (LHC-II). This new embedding medium allowed recording of high-resolution electron diffraction patterns and micrographs, which were then used to build an atomic model for LHC-II, resulting in a 3D density map at 3.4 Å resolution (Wang and Kühlbrandt, 1991). Later studies have shown that tannin in combination with glucose could be applied to determine the structure of the α,β tubulin dimer at 3.7 Å resolution (Nogales et al., 1995, 1998). Subsequently, trehalose (a nonreducing disaccharide consisting of two D-glucose molecules) was introduced as an embedding sugar for 2D crystals. Upon drying, trehalose does not crystallize and forms a temperature-stable vitreous material similar to amorphous ice that also stabilizes the 3D structure of proteins. Trehalose has since become the most frequently used embedding medium in the preparation of 2D crystals for cryo-EM (Jap et al., 1992; Kimura et al., 1997). Hirai et al. (1999) compared trehalose and glucose as an embedding medium for the preparation of 2D crystals of bR (Hirai et al., 1999) as partially hydrated and vacuum dried samples. Their results indicated that partially hydrated trehalose-embedded specimens were best preserved. Trehalose has since been used with several water channel 2D crystals (Braun et al., 2000; Gonen et al., 2005; Hite et al., 2010), prokaryotic V-ATPase (Toei et al., 2007), and human connexin26 gap junction (Oshima et al., 2007), further attesting to the excellent preservation capabilities of trehalose.

4.2. Preparation of flat support films

Current image processing algorithms still require perfectly flat 2D crystal arrays. Imperfect specimen flatness is a major problem when recording images and electron diffraction patterns at higher tilt angles (Glaeser *et al.*, 1991). The quality and surface properties of a support film are key factors affecting the resolution in electron crystallography. Carbon films are regularly used as support for specimens in electron microscopy because of their high transparency, good conductivity, and mechanical stability under the electron beam. Carbon films are commonly prepared by evaporation of carbon onto freshly cleaved mica surfaces using high-grade carbon and freshly cleaved (and therefore pristine) mica. Strong and flat carbon films were prepared with a carbon evaporator that has a very high vacuum and using an evaporation process that avoided spark formation by careful and slow heating of the carbon source (Vonck, 2000). A "preevaporation" of carbon from the carbon source before exposure of the mica was also found to improve the flatness of the evaporated films (Fujiyoshi, 1998). Freshly evaporated carbon is often hydrophilic, but becomes progressively hydrophobic as it ages over days and weeks. This changes the interaction between specimen and carbon, consequently influencing the number of crystals retained by the grids after blotting, and their flatness. Aging of carbon has been used to obtain favorable results, which tend to vary between different crystal specimens and evaporation methods. For example, carbon films after 10 days were used to prepare purple membrane specimens (Ceska and Henderson, 1990) and after 1–2 weeks used with tubulin 2D crystals (Wang and Downing, 2010).

Although copper grids are most commonly used in electron microscopy, "cryo-crinkling" of carbon films across copper grids is observed at liquid nitrogen temperatures due to the different thermal expansion rates of either material (Vonck, 2000). Cryo-crinkling can be significantly reduced using gold or molybdenum grids. In particular, molybdenum has about one-third the expansion coefficient of copper and is thus a better match for carbon. It has been shown that carbon films on molybdenum grids demonstrated little or no crinkling after cooling (Booy and Pawley, 1993) and provide a higher percentage of flat crystals (Vonck, 2000). Hence, molybdenum grids are most commonly used to prepare 2D crystals for imaging at cryo-temperatures. However, the surface of some commercially available molybdenum grids is rather rough (imposing wrinkles in the deposited carbon film), which can be avoided by the use of molybdenum grids manufactured by photochemical etching (Fujiyoshi, 1998). Such grids are available from EM supplies vendors (Pacific GridTech, San Francisco). Recently, other varieties of grids with near–atomically flat surfaces have been introduced, such as silicon nitride films (Rhinow and Kühlbrandt, 2008), which may present useful alternative supports for 2D crystals.

4.3. 2D crystal grid preparation with the back injection method

The "back injection method" is the most commonly used method for preparing cryo-grids for electron crystallography (Wall *et al.*, 1985). This method was first applied with purple membrane and then with LHC (Kühlbrandt and Downing, 1989), and has now been adopted as the most frequently used method for preparing 2D crystals for electron microscopy.

4.3.1. Back injection method

Materials

1. Carbon evaporator (oil-free high vacuum).
2. High grade mica sheets.
3. High grade carbon.
4. Anticapillary self-closing tweezers.
5. Molybdenum grids.
6. Whatman #2 filter paper.
7. Liquid nitrogen.
8. Embedding buffer, containing trehalose, glucose, sucrose, or tanning.
9. 2D crystal suspension.

Steps

- Evaporate carbon onto freshly cleaved mica.
- Float a small piece of carbon film (3 × 3 mm) onto the surface of an embedding buffer solution.
- Pick the floated carbon film up with a grid held by anticapillary self-closing tweezer.
- Turn the grid over and add 1–3 μl of crystal suspension on the carbon film through the grid bars.
- Gently mix the crystal solution on the grid with a pipette. Take care not to break the carbon film.
- Incubate crystals on the grid for an appropriate time in a humid atmosphere (1–5 min).
- Turn the grid over again and blot by placing onto two layers of filter paper.
- Let the grid air-dry for a few seconds.
- Plunge-freeze in liquid nitrogen.
- Cryo-grids are now ready for immediate observation, but can also be stored in liquid nitrogen for future use.

One of the major challenges for specimen preparation is to control the thickness of the ice of the specimen. This thickness is dependent on the blotting time, drying time between blotting and freezing, room temperature, and humidity. Most 2D crystals will suffer and become disordered when dried too much. However, too thick ice on the grids will contribute too much background scattering, thereby weakening the contrast of the recorded images or diffraction spots. It is therefore important to optimize the embedding medium, incubation time, blotting time, and air-drying time for each crystal.

Although offering a significant improvement in flatness and reproducibility over other methods, the back injection method promotes an inherently asymmetric specimen that may be susceptible to various forces causing specimen movement. To overcome this, the "carbon sandwich technique" was developed (Gyobu *et al.*, 2004; Koning *et al.*, 2003). This technique extends the back injection method by floating a second carbon film across the exposed side of the sample prior to blotting and demonstrates significant improvements in images recorded from tilted specimens (Gonen *et al.*, 2005; Hite *et al.*, 2010):

4.4. 2D crystal grid preparation with the carbon sandwich method

Materials

Material needed for the carbon sandwich method is same as for the back injection method described earlier, except for a 4 mm diameter platinum wire loop.

Steps

- Evaporate carbon onto freshly cleaved mica.
- Float a small piece of carbon film (3 × 3 mm) onto the surface of an embedding buffer solution.
- Pick the floated carbon film up with a grid held by anticapillary self-closing tweezer.
- Turn the grid over and add 1–3 μl of crystal suspension on the carbon film through the grid bars.
- Gently mix the crystal solution on the grid with a pipette. Take care not to break the carbon film.
- Float the second carbon film (2 × 2 mm) onto the surface of an embedding buffer solution.
- Pick up the floated second carbon film with a platinum wire loop and place it on the top of the grid. The grid and the 2D crystals are now sandwiched between the two carbon films.
- Blot the grid from the edge carefully with a 2 mm blotting paper. This step can also be done in a cold room to slow down the drying process.

- Plunge-freeze in liquid nitrogen.
- A movie of this process can be seen online at http://2dx.org/download/movies/Gyobu-Sandwich.mov/view.

4.5. 2D crystal grid preparation in continuous versus holey carbon film support

In electron crystallography, 2D crystals are adsorbed to thin carbon films that minimize background noise by reducing the degree of inelastic scattering of electrons within the film itself. For cryo-EM imaging, the films primarily have to provide a stable and flat support for the 2D crystals, so that a somewhat thicker film may be preferable. Electron diffraction is insensitive to specimen movement, so that a thinner film can be used, which then will result in diffraction patterns with a lower noise background level. Figure 1.1A and B shows examples of images and electron diffraction patterns collected on 2D crystals embedded in 7% Trehalose.

High-resolution electron crystallography imaging is mostly done with frozen hydrated crystal samples on continuous carbon, but can also be done with crystals spanning the holes of holey carbon films (e.g., Quantifoil (Quantifoil Micro Tools GmbH, Jena, Germany) or C-flat (Protochips, Inc. Raleigh, USA)). This can be done by glow-discharging the holey carbon film and adsorbing 2D crystals in suspension prior to blotting and plunge freezing (Cyrklaff and Kühlbrandt, 1994). The crystals are found suspended in the thin vitreous ice spanning the holes, free from background noise. However, these crystals are less likely to be perfectly flat in the absence of a 2D support. Nevertheless, this method also avoids the use of sugars, which may better suit certain 2D crystals that critically depend on the maintenance of a precise hydration state, or that have fragile extra-membranous domains that are easily distorted by contact with a supporting carbon film. Figure 1.1C and D shows examples of frozen hydrated 2D crystals spanning holey carbon film supports. This approach was successfully used on Aerolysin 2D crystals that had been prepared using the robotic 2D crystallization setup (Iacovache *et al.*, 2010). In this case, the crystals vitrified in buffer over holey carbon film showed better electron diffraction results than crystals prepared by trehalose embedding between a sandwich of two continuous carbon films (Fig. 1.1).

4.6. Reducing beam-induced resolution loss on tilted specimens

When imaging tilted cryo-EM samples under low-dose conditions, a severe resolution loss in the direction perpendicular to the tilt axis is often observed that cannot be explained by the lack of specimen flatness. This resolution

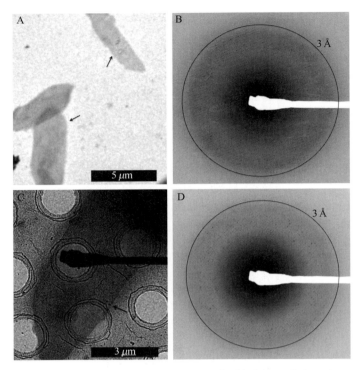

Figure 1.1 2D crystals of HoloHasA-HasR embedded in trehalose between two identical layers of continuous carbon film (A) or adsorbed frozen hydrated onto holey carbon film (C) and electron diffraction patterns of the same preparations (B on continuous and D on holey carbon film). Arrows indicate the 2D crystals.

loss can be observed by the disappearance of Thon rings in the Fourier transformations of the images. It is not observed when continuous illumination of the tilted specimen is employed, but then the high electron dose and consequent beam damage would prevent any recording of high-resolution data of the protein. The beam-induced resolution loss is also not, or only to a very minor degree, affecting the recording of images from *nontilted* specimens. The resolution loss is also not affecting the recording of *electron diffraction* patterns either at zero or at high specimen tilt angles.

The electron beam-induced resolution loss observed under low-dose illumination conditions was attributed to beam-induced charging of the sample (Brink *et al.*, 1998; Gyobu *et al.*, 2004). For thin samples, (less than 1 µm) charging may occur due to the accumulation of net positive charge, as the inelastic interaction of primary electrons within the sample releases Auger and secondary electrons from the sample during the exposure. This effect is stronger on vitrified aqueous samples at liquid helium temperature, due to the low electrical conductivity of vitreous water at those low temperatures, so that the induced charge fails to diffuse and consequently accumulates locally.

An alternative explanation was provided by Glaeser et al. (2007), who speculate that beam-induced physical movement of the specimens is the cause of the resolution loss, while the movement is normally in a direction perpendicular to the carbon film. Irradiation of sample with a 200 keV electron beam can cut covalent bonds in the specimen. The interatom distance would then increase from that of a covalent bond to that of an ionic bond or other bond types, resulting in physical expansion of the material, or the creation of significant pressures (Glaeser, 2008). The different layers of a typical nonsymmetrical cryo-EM grid are carbon film, vitreous water, a possible 2D crystal, and a possible water contamination layer on the surface of the sample. As these layers have different beam–related expansion coefficients, illumination results in different lateral pressures within the sample, which similarly to a bimetal thermostat that bends under heat, leads to bending of the sample, thereby inducing a vertical movement of the film. This model is referred to as "drum effect," where the carbon film support moves perpendicular to the sample plane under the electron beam (Glaeser, 2008).

A number of approaches have been developed to minimize this effect. Spot scanning has been introduced, which systematically illuminates only a small area of the sample at one time (Downing, 1991). In the charging model, this would reduce the lateral diameter of the charge-accumulating specimen area, so that its electric field would be limited in size and strength. In the drum-effect model, the spot scanning data collection scheme would restrict the size of the expanding area, thereby reducing the force that would lead to movement.

Other more direct approaches relate to sample preparation such as the aforementioned carbon sandwich technique (Gyobu et al., 2004), in which the 2D crystals across an amorphous carbon substrate are covered by a second layer of carbon, which should be identical to the first one. The symmetric carbon film sandwich then can either serve as conductive embedding, reducing the charge accumulation, or can provide symmetric forces during the expansion under electron irradiation, resulting in no net force and thereby no movement.

More recently, thin glass metallic substrates were produced by the evaporation of Ti88Si12 pellets across freshly cleaved mica. The substrates demonstrate an electrical resistance approximately six orders of magnitude lower than that of amorphous carbon (of the same thickness) at room temperature, and performed even better at liquid nitrogen temperature (77 K), where the electrical resistance for traditional amorphous carbon substrates climbs to almost immeasurably high values. Given their increased mechanical strength, the TiSi substrates can also be made thinner, thus decreasing imposed background in the image data. Tilting the substrate to

45°, beam-induced movement was reported to be reduced by ~50% compared to amorphous carbon imaged under the same conditions. It has also been suggested that the use of highly conductive substrates, such as TiSi, could avoid the need for additional conductive films across the sample (e.g., carbon sandwich), also minimizing the additional complication and attenuation of contrast (Rhinow and Kühlbrandt, 2008).

4.7. Low-dose data collection at helium temperatures

To avoid imaging the beam-damaged protein structure, data collection under a low-dose scheme is imperative. The sample is kept at cryo-temperatures, because the lower temperatures make it less likely (i.e., slower) for a protein to fall into a different conformation after the destruction of covalent bonds in the protein structure.

For samples that contain vitreous water, as often used in single-particle cryo-EM and electron tomography, some recent reports suggested that image recording at liquid helium temperature of 4.2 K may not be ideal, and samples kept at liquid nitrogen temperature (77 K) or slightly lower temperatures (25 or 42 K) might be better suited for cryo-EM data collection (e.g., Bammes *et al.*, 2010; Comolli and Downing, 2005; Iancu *et al.*, 2006; Wright *et al.*, 2006). Vitrified water appears to be less suited for cryo-EM imaging at ultra-low temperatures of 4 K, and the handling of samples becomes technically more difficult at these temperatures, due to the risk of increased contamination onto the sample from freezing gases like nitrogen and oxygen.

However, for samples that do not contain vitreous water, as is the case for sugar-embedded 2D membrane protein crystals on carbon film, the lower sample temperature of a helium cooled microscope has a strong advantage for recording high-resolution data at elevated electron dose levels (Fujiyoshi, 1998). The majority of the atomic resolution membrane protein structures from cryo-EM were determined from specimens kept at a temperature of 4.2 K, where a dose of up to 20 electrons per square Ångstrom still allows to record data in the 3 Å resolution range.

5. Conclusions

Structure determination of membrane and soluble proteins has been achieved by the complementary use of NMR, X-ray, and electron crystallographies. Electron crystallography is unique in that it facilitates the crystallization of membrane proteins in a near-native state, preserving function (Walz *et al.*, 1994). With refined data collection and data processing methods, there is no fundamental limit to acquiring the atomic structure of a membrane protein by electron crystallography, once suitable crystals are at hand. However, with the currently available software methods, the production of flat, highly ordered crystals is still a requirement. Highly flat carbon films prepared on molybdenum grids reduce crinkling of the substrate at low temperature. Carbon sandwich sample preparation reduces the beam-induced resolution loss when imaging tilted samples. Low-dose data collection of specimens kept at helium temperature is strongly advantageous for electron crystallography data collection of sugar-embedded specimens. New sample support films other than carbon may allow a further improvement in signal-to-noise ratio and success rate for imaging tilted membrane protein 2D crystals. The combination of the aforementioned methods with high-resolution electron microscopy data collection and high-throughput data processing methods (Chapter 16 of Vol. 482) will allow the determination of the atomic resolution structure of many membrane-embedded membrane proteins.

REFERENCES

Adrian, M., et al. (1984). Cryo-electron microscopy of viruses. *Nature* **308,** 32–36.

Bammes, B. E., Jakana, J., Schmid, M. F., and Chiu, W. (2010). Radiation damage effects at four specimen temperatures from 4 to 100 K. *J. Struct. Biol.* **169,** 331–341.

Booy, F. P., and Pawley, J. B. (1993). Cryo-crinkling: What happens to carbon films on copper grids at low temperature. *Ultramicroscopy* **48,** 273–280.

Böttcher, B., et al. (1992). The structure of photosystem I from the thermophilic cyanobacterium *Synechococcus* sp. determined by electron microscopy of two-dimensional crystals. *Biochim. Biophys. Acta* **1100,** 125–136.

Braun, T., Philippsen, A., Wirtz, S., Borgnia, M. J., Agre, P., Kuhlbrandt, W., Engel, A., and Stahlberg, H. (2000). The 3.7 A projection map of the glycerol facilitator GlpF: A variant of the aquaporin tetramer. *EMBO Rep.* **1,** 183–189.

Brink, J., Gross, H., Tittmann, P., Sherman, M. B., and Chiu, W. (1998). Reduction of charging in protein electron cryomicroscopy. *J. Microsc.* **191**(Pt 1), 67–73.

Ceska, T. A., and Henderson, R. (1990). Analysis of high-resolution electron diffraction patterns from purple membrane labelled with heavy-atoms. *J. Mol. Biol.* **213,** 539–560.

Comolli, L. R., and Downing, K. H. (2005). Dose tolerance at helium and nitrogen temperatures for whole cell electron tomography. *J. Struct. Biol.* **152,** 149–156.

Cyrklaff, M., and Kühlbrandt, W. (1994). High-resolution electron microscopy of biological specimens in cubic ice. *Ultramicroscopy* **55,** 141–153.

Dolder, M., et al. (1996). The micelle to vesicle transition of lipids and detergents in the presence of a membrane protein: Towards a rationale for 2D crystallization. *FEBS Lett.* **382,** 203–208.

Downing, K. H. (1991). Spot-scan imaging in transmission electron microscopy. *Science* **251,** 53–59.

Dubochet, J., et al. (1988). Cryo-electron microscopy of vitrified specimens. *Q. Rev. Biophys.* **21,** 129–228.

Ellis, M. J., and Hebert, H. (2001). Structure analysis of soluble proteins using electron crystallography. *Micron* **32,** 541–550.

Fujiyoshi, Y. (1998). The structural study of membrane proteins by electron crystallography. *Adv. Biophys.* **35,** 25–80.

Ganser-Pornillos, B. K., Cheng, A., and Yeager, M. (2007). Structure of full-length HIV-1 CA: A model for the mature capsid lattice. *Cell* **131,** 70–79.

Garavito, R. M., and Ferguson-Miller, S. (2001). Detergents as tools in membrane biochemistry. *J. Biol. Chem.* **276,** 32403–32406.

Glaeser, R. M. (2008). Retrospective: Radiation damage and its associated "information limitations". *J. Struct. Biol.* **163,** 271–276.

Glaeser, R. M., *et al.* (1991). Interfacial energies and surface-tension forces involved in the preparation of thin, flat crystals of biological macromolecules for high-resolution electron microscopy. *J. Microsc.* **161,** 21–45.

Glaeser, R., *et al.* (2007). Electron Crystallography of Biological Macromolecules. Oxford University Press, USA.

Gonen, T., *et al.* (2005). Lipid–protein interactions in double-layered two-dimensional AQP0 crystals. *Nature* **438,** 633–638.

Gyobu, N., *et al.* (2004). Improved specimen preparation for cryo-electron microscopy using a symmetric carbon sandwich technique. *J. Struct. Biol.* **146,** 325–333.

Henderson, R., and Unwin, P. N. (1975). Three-dimensional model of purple membrane obtained by electron microscopy. *Nature* **257,** 28–32.

Henderson, R., Baldwin, J. M., Ceska, T. A., Zemlin, F., Beckmann, E., and Downing, K. H. (1990). Model for the structure of bacteriorhodopsin based on high-resolution electron cryo-microscopy. *J. Mol. Biol.* **213,** 899–929.

Hirai, T., *et al.* (1999). Trehalose embedding technique for high-resolution electron cristallography: Application to structural study on bacteriorhodopsin. *J. Electron. Microsc.* **48,** 653–685.

Hite, R. K., *et al.* (2010). Principles of membrane protein interactions with annular lipids deduced from aquaporin-0 2D crystals. *EMBO J.* **29,** 1652–1658.

Hu, M., *et al.* (2010). Automated electron microscopy for evaluating two-dimensional crystallization of membrane proteins. *J. Struct. Biol.* (in press).

Iacovache, I., *et al.* (2010). The 2DX robot: A membrane protein 2D crystallization Swiss Army knife. *J. Struct. Biol.* **169,** 370–378.

Iancu, C. V., Wright, E. R., Heymann, J. B., and Jensen, G. J. (2006). A comparison of liquid nitrogen and liquid helium as cryogens for electron cryotomography. *J. Struct. Biol.* **153,** 231–240.

Jap, B. K., *et al.* (1992). 2D crystallization: From art to science. *Ultramicroscopy* **46,** 45–84.

Kimura, Y., Vassylyev, D. G., Miyazawa, A., Kidera, A., Matsushima, M., Mitsuoka, K., Murata, K., Hirai, T., and Fujiyoshi, Y. (1997). Surface of bacteriorhodopsin revealed by high-resolution electron crystallography. *Nature* **389,** 206–211.

Koning, R. I., Oostergetel, G. T., and Brisson, A. (2003). Preparation of flat carbon support films. *Ultramicroscopy* **94,** 183–191.

Kühlbrandt, W. (1992). Two-dimensional crystallization of membrane proteins. *Q. Rev. Biophys.* **25,** 1–49.

Kühlbrandt, W., and Downing, K. H. (1989). Two-dimensional structure of plant light-harvesting complex at 3.7 A [corrected] resolution by electron crystallography. *J. Mol. Biol.* **207,** 823–828.

Levy, D., *et al.* (2001). Two-dimensional crystallization of membrane proteins: The lipid layer strategy. *FEBS Lett.* **504,** 187–193.

Mannella, A. C. (1986). Mitochondrial outer membrane channel (VDAC, Porin) two-dimensional crystals from Neurospora. (S. Fleischer and B. Fleischer, eds.), Vol. 125, pp. 595–610. Academic press, London.

Newman, R. H., *et al.* (1991). 2D crystal forms of annexin IV on lipid monolayers. *FEBS Lett.* **279,** 21–24.

Nogales, E., Wolf, S. G., and Downing, K. H. (1998). Structure of the alpha beta tubulin dimer by electron crystallography. *Nature* **391,** 199–203.

Nogales, E., Wolf, S. G., Khan, I. A., Luduena, R. F., and Downing, K. H. (1995). Structure of tubulin at 6.5 A and location of the taxol-binding site. *Nature* **375**, 424–427.

Oshima, A., Tani, K., Hiroaki, Y., Fujiyoshi, Y., and Sosinsky, G. E. (2007). Three-dimensional structure of a human connexin26 gap junction channel reveals a plug in the vestibule. *Proc. Natl. Acad. Sci. USA* **104**, 10034–10039.

Remigy, H. W., *et al.* (2003). Membrane protein reconstitution and crystallization by controlled dilution. *FEBS Lett.* **555**, 160–169.

Renault, L., *et al.* (2006). Milestones in electron crystallography. *J. Comput. Aided Mol. Des.* **20**, 519–527.

Rhinow, D., and Kühlbrandt, W. (2008). Electron cryo-microscopy of biological specimens on conductive titanium-silicon metal glass films. *Ultramicroscopy* **108**, 698–705.

Rigaud, J.-L., *et al.* (1997). Bio-Beads: An efficient strategy for two-dimensional crystallization of membrane proteins. *J. Struct. Biol.* **118**, 226–235.

Schmidt, T. G., and Skerra, A. (1994). One-step affinity purification of bacterially produced proteins by means of the "Strep tag" and immobilized recombinant core streptavidin. *J. Chromatogr. A.* **676**, 337–345.

Signorell, G. A., *et al.* (2007). Controlled 2D crystallization of membrane proteins using methyl-beta-cyclodextrin. *J. Struct. Biol.* **157**, 321–328.

Taylor, K. A., *et al.* (1988). Analysis of two-dimensional crystals of Ca2+-ATPase in sarcoplasmic reticulum. *Methods Enzymol.* **157**, 271–289.

Toei, M., *et al.* (2007). Dodecamer rotor ring defines H+/ATP ratio for ATP synthesis of prokaryotic V-ATPase from Thermus thermophilus. *Proc. Natl. Acad. Sci. USA* **104**, 20256–20261.

Unger, V. M., Kumar, N. M., Gilula, N. B., and Yeager, M. (1999). Three-dimensional structure of a recombinant gap junction membrane channel. *Science* **283**, 1176–1180.

Unwin, P. N., and Henderson, R. (1975). Molecular structure determination by electron microscopy of unstained crystalline specimens. *J. Mol. Biol.* **94**, 425–440.

Vink, M., *et al.* (2007). A high-throughput strategy to screen 2D crystallization trials of membrane proteins. *J. Struct. Biol.* **160**, 295–304.

Vonck, J. (2000). Parameters affecting specimen flatness of two-dimensional crystals for electron crystallography. *Ultramicroscopy* **85**, 123–129.

Wall, J. S., *et al.* (1985). *Films that Wet without Glow Discharge. 35th EMSA Meeting* San Francisco Press, Louisville.

Walz, T., *et al.* (1994). Biologically active two-dimensional crystals of aquaporin CHIP. *J. Biol. Chem.* **269**, 1583–1586.

Wang, H., and Downing, K. H. (2010). Specimen preparation for electron diffraction of thin crystals. *Micron* (in press).

Wang, D. N., and Kühlbrandt, W. (1991). High-resolution electron crystallography of light-harvesting chlorophyll a/b–protein complex in three different media. *J. Mol. Biol.* **217**, 691–699.

Wright, E. R., Iancu, C. V., Tivol, W. F., and Jensen, G. J. (2006). Observations on the behavior of vitreous ice at approximately 82 and approximately 12 K. *J. Struct. Biol.* **153**, 241–252.

Yeager, M. (1994). In situ two-dimensional crystallization of a polytopic membrane protein: The cardiac gap junction channel. *Acta Crystallogr. D Biol. Crystallogr.* **50**, 632–638.

CHAPTER TWO

HELICAL CRYSTALLIZATION OF SOLUBLE AND MEMBRANE BINDING PROTEINS

Elizabeth M. Wilson-Kubalek,* Joshua S. Chappie,[†] *and* Christopher P. Arthur*

Contents

Abstract

Helical protein arrays offer unique advantages for structure determination by cryo-electron microscopy (cryo-EM). A single image of such an array contains a complete range of equally spaced molecular views of the underlying protein subunits, which allows a low-resolution, isotropic three-dimensional (3D) map to be generated from a single helical tube without tilting the sample in the

* The Scripps Research Institute, La Jolla, California, USA
[†] Laboratory of Molecular Biology, National Institute of Diabetes and Digestive and Kidney Diseases, National Institutes of Health, Bethesda, Maryland, USA

Methods in Enzymology, Volume 481
ISSN 0076-6879, DOI: 10.1016/S0076-6879(10)81002-X

electron beam as is required for two-dimensional (2D) crystals. Averaging many unit cells from a number of similar tubes can improve the signal-to-noise ratio and consequently, the quality of the 3D map. This approach has yielded reconstructions that approach atomic resolution [Miyazawa *et al.*, 1999, 2003; Sachse *et al.*, 2007; Unwin, 2005; Yonekura *et al.*, 2005]. Proteins that naturally adopt helical protein arrays, such as actin and microtubules, have been studied for decades. The wealth of information on how proteins bind and move along these cytoskeletal tracks, provide cross-talk between tracks, and integrate into the cellular machinery is due, in part, to multiple EM studies of the helical assemblies. Since the majority of proteins do not spontaneously form helical arrays, the power of helical image analysis has only been realized for a small number of proteins. This chapter describes the use of functionalized lipid nanotubes and liposomes as substrates to bind and form helical arrays of soluble and membrane-associated proteins.

1. INTRODUCTION

1.1. Lipid nanotubes as substrates for helical crystallization

The crystallization of proteins on lipid surfaces relies on the characteristics of the lipid head groups and the lateral mobility of the lipids. Submicrogram amounts of proteins and macromolecular complexes can be adsorbed, concentrated, and crystallized on lipid surfaces using specific ligand–derivatized head groups or by nonspecific, electrostatic interactions with charged lipids. Since His-tags can be inserted into recombinant proteins, nickel–chelating lipids provide a general means to concentrate such proteins on lipid substrates. By mixing lipids that have charged or derivatized head groups with the tube-forming properties of galactosylceramide (GalCer; Kulkarni *et al.*, 1999), several proteins have been shown to form helical arrays on lipid tubules (Dang *et al.*, 2005a,b; Melia *et al.*, 1999; Ringler *et al.*, 1997; Wilson-Kubalek, 2000; Wilson-Kubalek *et al.*, 1998). In this section, we describe methods for obtaining helical arrays of two model proteins— streptavidin (SA) and perfringolysin O (PFO)—to illustrate the utility of functionalized lipid nanotubes.

1.2. Streptavidin and perfringolysin O: Model proteins for helical crystallization on lipid nanotubes

SA molecules possess a high affinity for biotin ($K_D = 10^{-15}$ M), and two-dimensional (2D) crystals of SA bound to lipid monolayers via biotin have been thoroughly studied (Blankenburg *et al.*, 1989; Darst *et al.*, 1991; Ku *et al.*, 1993; Wang *et al.*, 1999a,b). Sigworth and colleagues have taken advantage of SA monolayers as a substrate for binding proteoliposomes and have used this method (for details, see Chapter 7 in this volume) to obtain structural details of an integral membrane protein (Wang and Sigworth, 2009; Wang et al., 2008). Using the methods previously described (Wilson-Kubalek, 2000; Wilson-Kubalek et al., 1998), wild-type and mutant forms of SA were used to form helical arrays with unique packing arrangements on lipid nanotubes composed of GalCer and biotinylated lipids. This work confirmed that lipid tubes produced by mixtures of derivatized lipids and GalCer can serve as an alternative substrate to study protein ordering and that a single surface residue can significantly affect intermolecular packing during helical crystallization (Dang et al., 2005a).

PFO is a member of the cholesterol-dependent cytolysin family of oligomerizing, pore-forming toxins found in a variety of Gram-positive bacterial pathogens. Both wild-type and a mutant version of C-terminally His-tagged PFO form helical arrays on three different kinds of nickel-lipid nanotubes. The packing arrangement of wild-type PFO on the nickel-lipid nanotubes differs from that of the mutant PFO. Cryo-EM (cryo-electron microscope) and image analysis of wild-type PFO arrays were used to reconstruct a three-dimensional (3D) density map. The results of this study demonstrate that nickel-lipid nanotubes provide a specific substrate for structural studies of a variety of His-tagged proteins and extend the applicability of lipid nanotubes as a method for helical crystallization of soluble proteins and membrane-associated proteins for EM structural studies (Dang et al., 2005b).

1.3. Liposomes as substrates for helical crystallization

Many proteins—especially those involved in intracellular signaling and trafficking—naturally associate with various cellular membranes as part of their normal biological function. In these molecules, membrane interactions are mediated by a recurring set of structural motifs that include pleckstrin homology (PH) domains (Lemmon, 2003, 2008), Bin-amphiphysin-Rvs domains (Itoh and De Camilli, 2006), phox homology domains (Ellson et al., 2002; Xu et al., 2001), epsin N-terminal homology domains (Ford et al., 2002; Legendre-Guillemin et al., 2004), AP180 N-terminal homology domains (Ford et al., 2001; Itoh and De Camilli, 2006; Mao et al., 2001), and amphipathic helices (Drin and Antonny, 2009). Each motif displays different lipid specificities and binding affinities yet shares the ability to sense and induce membrane curvature (McMahon and Gallop, 2005). In vivo, these properties spatially and temporally regulate the distribution of membrane-associating proteins throughout the cell; in vitro, they can be utilized to dissect protein structure and function.

Liposomes are spherical, unilamellar vesicles of varying lipid composition. Unlike lipid nanotubes, liposomes do not contain GalCer and are thus less rigid and more malleable as synthetic membrane mimics. The diameter of the liposomes can easily be varied from 0.03 to 1 μm, providing a simple way to sample different degrees of membrane curvature. As a result, proteins that promote membrane deformation can easily bind and often tubulate liposomes, producing ordered helical arrays that are suitable for cryo-EM structural studies. Such arrays have been observed with dynamins (Chen *et al.*, 2004; Mears and Hinshaw, 2008; Sweitzer and Hinshaw, 1998; Zhang and Hinshaw, 2001), Snx9 (Yarar *et al.*, 2008), amphiphysin (Peter *et al.*, 2004), endophilin (Gallop *et al.*, 2006), epsin (Ford *et al.*, 2002), epsin homology domain proteins (Daumke *et al.*, 2007), and *Escherichia coli* RNA polymerase (Opalka *et al.*, 2000; Polyakov *et al.*, 1998). In the next section, we describe procedures for generating helical crystals of dynamin and botulinum toxin from liposome templates and briefly discuss general approaches for extending this methodology to other membrane-associating proteins.

1.4. Dynamin and botulinum toxin as prototypes for liposome-mediated helical crystallization

Dynamin is a large, multidomain GTPase that functions as a key regulator of clathrin-mediated endocytosis (CME; Mettlen *et al.*, 2009). *In vivo*, dynamin tetramers are initially recruited to clathrin-coated pits by effector proteins and subsequently bind the plasma membrane via PH domain interactions with PIP_2 phospholipids. During the later stages of CME, these tetramers self-assemble into collar-like structures around the necks of deeply invaginated pits and catalyze membrane scission in a GTPase-dependent manner (Damke *et al.*, 1994, 2001; Pucadyil and Schmid, 2008; Song *et al.*, 2004; Stowell *et al.*, 1999). *In vitro*, dynamin forms rings and spirals—either under low-salt conditions (Hinshaw and Schmid, 1995) or in the presence of transition-state mimics (Carr and Hinshaw, 1997)—that resemble the collar-like structures observed *in vivo*. Further structural studies have demonstrated that dynamin adopts a similar architecture when incubated with negatively charged lipid templates (Stowell *et al.*, 1999; Sweitzer and Hinshaw, 1998; Zhang and Hinshaw, 2001). Dynamin assemblies derived from liposomes show a distinct helical symmetry that has been exploited to generate 3D maps of a truncated version of the proline-rich domain (ΔPRD) of dynamin in different nucleotide states (Chen *et al.*, 2004; Zhang and Hinshaw, 2001). Methods described subsequently were used to generate suitable helical samples of ΔPRD in the presence of the nonhydrolyzable GTP analog GMPPCP.

Botulinum neurotoxin (BoNT) is one of the most deadly bacterial neurotoxins known, with an LD_{50} of 0.1 ng/kg body weight; its seven serotypes (A–G) are secreted by the bacterium *Clostridium botulinum* (Binz and Rummel, 2009). BoNT B is a Zn^{2+}-endoprotease that acts on the exocytotic machinery at the neuromuscular synapse by cleaving unique components of the soluble *N*-ethylmaleimide-sensitive factor attachment protein receptor (SNARE) complex, thus inhibiting synaptic vesicle (SV) fusion and subsequent neurotransmitter release (Montecucco and Schiavo, 1994; Prabakaran *et al.*, 2001). This leads to flaccid paralysis and, if untreated, death. BoNTs are produced as 150 kDa single polypeptide chains that are cleaved into a 100-kDa heavy chain (HC) and a 50-kDa light chain (LC) linked by a disulfide bond. The 100-kDa HC contains a 50-kDa translocation domain and a 50-kDa receptor binding domain. The 50-kDa LC is made up of a single catalytic domain (Brunger and Rummel, 2009).

BoNTs initially enter neuronal cells via receptor-mediated endocytosis, after which they are translocated into the cytoplasm from an endosome-like compartment in a pH-dependent manner. BoNT/B requires the ganglioside GT1b and the synapse protein synaptotagmin I for neuronal cell intoxication (Dong *et al.*, 2003). The structural basis of the high specificity of binding of BoNT to the cell surface was recently accounted for with the solution of the crystal structure of BoNT/B in complex with the luminal domain of synaptotagmin II (Chai *et al.*, 2006; Jin *et al.*, 2006). While the mechanism of BoNT binding and cleavage of SNARE proteins is well characterized, the intermediate step in which the catalytic domain is translocated into the cytosol is not. Previous EM structural studies of BoNT have utilized the phospholipid monolayer technique (Flicker *et al.*, 1999) and lipid vesicle formation technique (Schmid *et al.*, 1993) and shown that BoNT in the presence of lipid will form ordered helical arrays. 3D helical reconstruction was used to show that BoNT binds to the surface of the lipid bilayer potentially penetrating at low pH to form a protein-conducting channel. The methods described in this chapter were utilized to form ordered helical arrays of BoNT at low pH in order to study the toxin's pore-forming capability.

2. Materials and Methods

2.1. Lipids

Synthetic D-galactosyl-*β*1-1'-*N*-nervonoyl-D-erythro-sphingosine (Gal-Cer), 1,2-dioleoyl-*sn*-glycero-3-phosphoethanolamine-*N*-(Cap Biotinyl) (*N*-Cap Biotinyl-PE), nickel-chelating lipid; 1,2-dioleoyl-*sn*-glycero-3-(*N*-(5-amino-1-carboxypentyl)iminodiacetic acid)succinyl, (DOGS-NTA-Ni); 1,2-dioleoyl-*sn*-glycero-3-(*N*-(5-amino-1-carboxypentyl)iminodiacetic acid) suberinyl, nickel salt (18:1-SUB-NTA (Ni)); 1,2 dimyristoyl-*sn*-glycero-3-(*N*-(5-amino-1-carboxypentyl)iminodiacetic acid), suberinyl nickel salt (14:0-SUB-NTA (Ni)), synthetic 1,2-dioleoyl-*sn*-glycero-3-phospho-L-serine (18:1 DOPS), 1,2-dioleoyl-*sn*-glycero-3-phosphocholine (18:1

DOPC), porcine brain L-α-phosphatidylinositol-4,5-bisphosphate (PIP2) 1,2-dioleoyl-*sn*-glycero-3-phosphoethanolamine (18:1 DOPE), and 1, 2-diheptanoyl-*sn*-glycero-3-phosphocholine (7:0 DHPC) were purchased from Avanti Polar Lipids (Alabaster, AL). Biotinylated lipid, *N*-((6-(biotinoyl)amino)hexanoyl)-1,2-dihexadecanoyl-*sn*-glycero-3-phosphoethanolamine (Biotin-X-DHPE), was purchased from Molecular Probes (Eugene, OR) biotinylated lipid, *N*-((6-(biotinoyl)amino)hexanoyl)-1,2,distearoyl-*sn*-glycero-3-phosphoethanolamine (Biotin-X-DSPE) was purchased from Northern Lipids (Vancouver, Canada). All lipids were stored at − 20 °C.

2.2. Nanotube preparation for streptavidin

Three biotinylated lipids were used to obtain biotinylated nanotubes as previously described (Wilson-Kubalek, 2000; Wilson-Kubalek *et al.*, 1998). Biotin lipids, Biotin-X-DHPE, Biotin-X-DSPE, and *N*-Cap Biotinyl-PE, were mixed with GalCer at varying ratios (5:95, 10:90, 20:80, 30:70, 40:60, and 50:50, w/w) in organic solvents. The organic solvents were removed by drying under a steady stream of argon or nitrogen. After rehydration in 50 mM Tris/HCl, 200 mM NaCl, pH 7.0 (final lipid concentration ∼0.5 mg/ml), the aqueous suspensions were vortexed for ∼30 s and then sonicated for 2 min in a water bath at room temperature to promote nanotube formation. The best ratio of biotinylated lipid to GalCer was determined empirically. Five microliters of each of the nanotube preparations was applied to a carbon–coated EM grid. After blotting and negatively staining with 1% uranyl acetate, the grids were observed with an FEI CM208 EM. The quality of the preparations was assessed on the basis of a high abundance of nanotubes of uniform diameter, lengths generally exceeding 1 μm, and a low abundance of other lipid structures such as liposomes and amorphous aggregates. Nanotube preparations were also compared based on their propensity to yield ordered arrays of SA.

2.3. Helical crystallization of streptavidin

Lipid nanotubes containing any of the three biotinylated lipids tested (Biotin-X-DHPE, Biotin-X-DSPE, or *N*-Cap Biotinyl-PE) permitted the formation of SA arrays. Biotin-X-DHPE:GalCer, which consistently yielded the most uniform nanotubes, was used to optimize the crystallization conditions. Crystallization trials were carried out using all of the aforementioned ratios of Biotin-X-DHPE:GalCer (in 50 mM Tris/HCl, 200 mM NaCl, pH 7.0, at a final lipid concentration ∼0.45 mg/ml).

Twenty microliters aliquots of the aqueous nanotube suspensions were mixed with different protein concentrations (0.05–0.20 mg/ml) of both wild-type and mutant SA. The protein/nanotube mixtures were incubated at room temperature or at 4 °C for various time intervals (5 min–24 h). Helical array formation was observed by negative stain EM. SA bound and began ordering ∼5 min after incubation on all biotinylated lipid nanotubes tested in this study. The lipid mix of 20:80 Biotin-X-DHPE lipid:GalCer (w/w) provided the most uniform biotinylated lipid nanotubes and were subsequently used for helical crystallization and structure determination. Well-ordered arrays were formed by incubating either wild-type or mutant SA (final protein concentration ∼0.15 mg/ml) with the Biotin-X-DHPE:GalCer nanotubes (20:80, w/w, final lipid concentration ∼0.45 mg/ml) at room temperature for 1 h. Negatively stained images were analyzed using both Phoelix, a Fourier-based helical image analysis program package (Carragher *et al.*, 1996; Whittaker *et al.*, 1995, for details on this method, see Chapter 5 in Vol. 482), and an iterative helical real space reconstruction (IHRSR) method (Egelman, 2000, for details on this method, see Chapter 6 in Vol. 482). The resulting 3D EM density maps are depicted in Fig. 2.1.

2.4. Nanotube preparation for perfringolysin O

Three nickel-lipids were used in this study; the DOGS-NTA (Ni) lipid that was previously used to obtain helical arrays of two Fab fragments (Wilson-Kubalek *et al.*, 1998), a 18:1-SUB-NTA (Ni) lipid with an identical alkyl chain but a longer (suberimide) linker between the alkyl

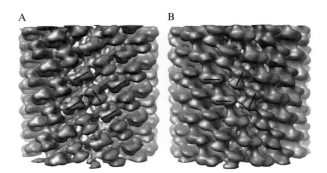

Figure 2.1   Helical crystallization of SA. (A, B) Side views of the surface rendered 3D maps of wild-type SA (A) and mutant SA (B). Black lines indicate organization of molecules on biotinylated lipid nanotube surface.

chain and the NTA head-group, and 14:0-SUB-NTA lipid that has the same longer (suberimide) linker but a shorter, saturated alkyl chain. Nanotubes were prepared by mixing either DOGS-NTA (Ni), 18:1-SUB-NTA (Ni), or 14:0-SUB-NTA (Ni) at 20 or 50 wt% with GalCer. The lipid mixtures in chloroform were then dried under argon or nitrogen and rehydrated in 20 mM HEPES, pH 7.0, by vortexing for ∼30 s or sonication for 1–2 min (Wilson-Kubalek, 2000; Wilson-Kubalek *et al.*, 1998). Addition of imidazole (200 mM) or NaCl (200 mM), filler lipid, or detergent improved the solubility of all three nickel-lipids when mixed with GalCer. The addition of imidazole not only improved solubility, but also improved the quality of nanotubes. Nanotubes prepared in the presence of imidazole were more uniform, aggregated less, and tended to be longer. Preparations of nickel-lipid nanotubes prepared with imidazole can simply be diluted (to 25–50 mM), before incubation with His-tagged proteins, to permit protein binding and crystal formation. The addition of DOPC improved the solubility of the lipid mixtures; however, it did not improve the amount of well-ordered PFO crystals. Overall, the DOGS-NTA (Ni) lipid produced the highest yield of uniform nanotubes that allowed helical crystallization of both wild-type and mutant PFO (Dang *et al.*, 2005b).

2.5. Helical crystallization of wild-type and mutant PFO

Crystallization trials of wild-type PFO with GalCer:DOGS-NTA (Ni) nanotubes were prepared under a wide range of buffer conditions, including TRIS, HEPES, and MES (20–100 mM), pH ranges (6–8), and with various NaCl concentrations (50–200 mM). Wild-type PFO helical arrays formed under all the conditions tested. However, buffers with 100–200 mM NaCl tended to yield multilayered crystalline arrays. Single-layered PFO arrays were consistently obtained in a low-salt buffer, 20 mM HEPES, pH 7.0, at a final protein concentration of 50–100 μg/ml and a final lipid concentration of 60–250 μg/ml. Lipid nanotubes containing DOGS-NTA (Ni) or 18:1-SUB-NTA (Ni) permitted the formation of both wild-type and mutant PFO helical arrays, whereas lipid nanotubes containing 14:0-SUB-NTA (Ni) needed the addition of the filler-lipid DOPC for array formation. There was also an increase in the number of crystals when the 14:0-SUB-NTA (Ni):DOPC:GalCer nanotubes were incubated with PFO at room temperature rather than at 4 °C. A lipid mix of DOGS-NTA (Ni):GalCer (20:80, w/w) yielded the most uniform nickel-lipid nanotubes with the lowest amount of undesired lipid structures, such as liposomes or aggregates, and were subsequently used as substrates upon which to form arrays of both wild-type and mutant PFO.

2.6. Helical crystallization of wild-type PFO for 3D map

The stock solution of wild-type PFO (1.8 mg/ml) was diluted with 20 mM HEPES, pH 7.0, to a concentration of 0.1 mg/ml and then mixed 1:1 with lipid nanotubes (DOGS–NTA (Ni):GalCer (20:80, ww) at 0.125 mg/ml). The final concentrations in the protein/nanotube/imidazole mixture were 50 μg/ml, 62.5 μg/ml, and 25 mM, respectively. The mixture was then incubated on ice for 1 h before cryopreservation.

For cryopreservation, 5 μl of sample was applied to a glow–discharged Quantifoil grid (2/4 Cu/Rh grid Quantifoil Micro Tools GmbH, Jena, Germany), blotted, and rapidly plunged into liquid ethane; (Dubochet *et al.*, 1988) for details on freezing specimens, see Chapter 3 in this volume. Images of the PFO arrays were recorded with a Philipis CM200 FEG EM at a nominal magnification of $\sim$ 38,000. An example of PFO helical arrays is shown in Fig. 2.2A. Images of well-ordered helical segments identified by strong diffraction (Fig. 2.2B) were analyzed using Phoelix (Carragher *et al.*, 1996; Whittaker *et al.*, 1995). Several different helical families were identified and indexed. Segments from one helical family were averaged and the

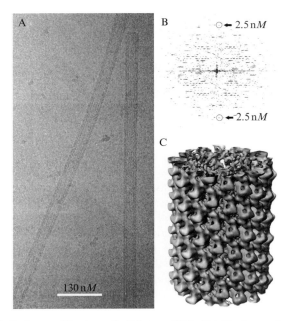

Figure 2.2 Helical crystallization of wild-type PFO. (A) Electron micrograph of ice-embedded wild-type PFO helical arrays on nickel-lipid nanotubes. Scale bar = 130 nM. (B) Computed diffraction pattern of a single wild-type PFO helical crystal (arrows point to 25 Å layer line spots). (C) Side view of the surface rendered 3D map of wild-type PFO.

resulting 3D map is shown in Fig. 2.2C. Since functional His–tags can be inserted at the N– or C–terminus or even within the sequence of a recombinant protein, nickel-chelating lipids provide a general means to attach such proteins to lipid tubular substrates.

2.7. Liposome preparation for dynamin

Liposomes were prepared as follows: Lipid mixtures containing either 100% DOPS or a combination of DOPC and PIP_2 (85:15, mol%) were prepared in chloroform and then dried using nitrogen gas. Dried lipids were rehydrated to a final concentration of 2.5 mM ($\sim$2 mg/ml) in 20 mM HEPES, pH 7.5. This suspension was warmed at 37 °C for 30 min to maximize lipid solubility, vortexed, and then sonicated for 2 min in a bath sonicator. Liposomes were then generated by extrusion ($>$11 passes) through prehydrated polycarbonate membranes of varying pore sizes (1, 0.4, and 0.1 μm) using an Avanti Mini-Extruder (Avanti Polar Lipids, catalog # 610000). Smaller liposomes were made in sequential extrusion steps, each with decreasing pore size. For comparison, PIP_2–containing lipid nanotubes (15% PIP2, 40% DOPC, and 40% GalCer) were prepared similarly but without extrusion. All lipid templates were diluted to 1 mg/ml in 20 mM HEPES, pH 7.5, prior to the addition of dynamin.

2.8. Helical crystallization of dynamin

Purified ΔPRD dynamin at 1 mg/ml in 20 mM HEPES, pH 7.5, 100 mM NaCl, 2 mM $MgCl_2$, 2 mM EGTA was mixed 1:1 (v/v) with each lipid template (DOPS liposomes, PIP_2 liposomes, and PIP_2 nanotubes) at 1 mg/ml in 20 mM HEPES, pH 7.5, and 1 mM GMPPCP. The mixture was incubated at room temperature for 2 h and then applied to glow-discharged, carbon-coated grids, washed with 20 mM HEPES, pH 7.5, and stained with 1% uranyl acetate. Samples were visualized in a Phillips Technai F20 EM operating at 120 kV and images were collected using Leginon (Potter *et al.*, 1999; Suloway *et al.*, 2005) at $\sim$2.0 μm underfocus with a 4K $\times$ 4K Gatan CCD camera at a nominal magnification of 50,000 (2.26 Å per pixel). Helical samples suitable for structure determination by cryo-EM were prepared as aforementioned, absorbed to C-flat holey grids (Electron Microscopy Sciences, Hatfield, PA), washed with 20 mM HEPES, pH 7.5, blotted, and frozen in liquid ethane. Grids were stored in liquid nitrogen prior to data collection at liquid nitrogen temperatures. Cryo images were obtained using the same parameters as aforementioned. Helical image processing was carried out using Phoelix (Carragher *et al.*, 1996; Whittaker *et al.*, 1995).

Because dynamin preferentially binds to PIP2 phospholipids, crystallization was initially carried out using PIP_2–containing liposomes (0.4 μm diameter) and lipid nanotubes. The lattice packing on these substrates was irregular and highly variable, which limited helical diffraction. This variability likely arose from incomplete incorporation of PIP_2 into both templates, resulting in heterogeneity among the underlying lipid scaffolds. To circumvent this problem, liposomes composed entirely of DOPS were incubated with ΔPRD dynamin and GMPPCP as described earlier. Three different liposome diameters (1, 0.4, and 0.1 μm)—representing increasing amounts of membrane curvature—were tested. Binding to each of these substrates was confirmed prior to crystallization using a sedimentation assay (Ramachandran *et al.*, 2007, data not shown). The two larger templates yielded well-ordered helical arrays of varying diameter (Fig. 2.3A and B), of which a subset exhibited strong helical diffraction extending to a resolution beyond 15 Å (Fig. 2.3C and D). Multiple tubulation events arising from a single vesicle were often observed with the 1-μm liposomes (Fig. 2.3A), whereas the 0.4-μm liposomes tended to produce a collection

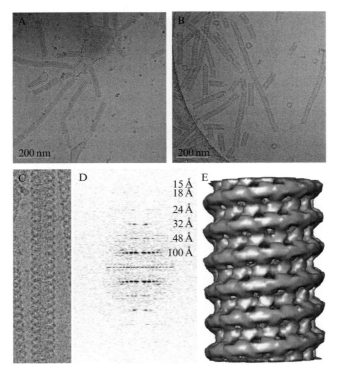

Figure 2.3 Helical crystallization of ΔPRD dynamin. (A, B) ΔPRD was crystallized using either 1 μm (A) or 0.4 μm DOPS liposomes (B). Note the greater propensity for multiple tubulation events with the larger, less curved lipid substrates. Scale bar = 200 nm. (C, D) Zoomed in view of a single, highly ordered ΔPRD tube (C) and its computed diffraction pattern (D). (E) Side view of the surface rendered 3D map of ΔPRD arrays in the presence of GMPPCP.

of single, ordered tubes (Fig. 2.3B). Assemblies generated from the 0.1-μm liposomes, in contrast, were extremely short and typically did not contain more than a single helical repeat (data not shown). Helical averaging of the best diffracting ΔPRD tubes yielded a 3D map at ∼18 Å resolution (Fig. 2.3E) that closely resembles the previously published helical reconstruction of dynamin in the constricted state (Zhang and Hinshaw, 2001).

2.9. Helical crystallization of BoNT

Expressed and purified BoNT was dialyzed against citrate-phosphate buffer (5 m*M* citric acid, 10 m*M* Na phosphate buffer, pH 6.0) for 12 h at 4 °C. DOPC, DOPE, DOPS, and DMPC (Avanti Polar Lipids) were dried down individually in 1 ml glass vials under nitrogen and resuspended in citrate-phosphate buffer containing 0.1 mg/ml BoNT. The solution was vortexed for 5 s and allowed to incubate at room temperature for 60 min. Samples were then applied to a glow-discharged carbon-coated EM grid and stained with 2% uranyl acetate. Screening of the samples revealed that BoNT bound only to DOPS and formed small (< 100 nm^2) ordered arrays on the surface of vesicles. Optimal array formation was obtained by incorporating DOPS and disialoganglioside-G_{D1a} (Sigma) in a 20:1 ratio (w/w). The lipid mixtures were dried down in a 1-ml glass vial under nitrogen and resuspended in citrate-phosphate buffer containing 0.1 mg/ml BoNT. The solution was vortexed for 5 s and allowed to incubate at room temperature for 60 min. Screening of the lipid/protein mixture revealed the formation of tubular helical arrays (Fig. 2.4A). BoNT/B crystal was prepared for

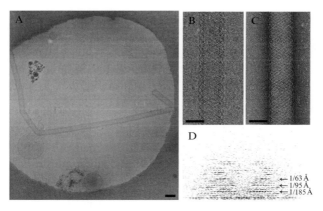

Figure 2.4 Helical crystallization of BoNT. (A) Electron micrograph of an ice-embedded helical array of BoNT. Scale bar = 70 n*M*. (B) Electron micrograph of ice-embedded BoNT helical crystal. (C) Electron micrograph of negatively stained BoNT helical crystal (scale bar for B & C = 70 n*M*). (D) Computed diffraction pattern of a single BoNT helical tube. Pattern shows strong layer lines at an interval of 1/185 Å^{-1}.

cryo-TEM imaging by applying 4 μl of crystallized BoNT/B solution to a 3-mm copper, holey-carbon-coated EM grid. The solution was then blotted, immediately plunged into liquid ethane and subsequently stored under liquid nitrogen. For cryo-EM, grids were loaded into a Gatan 626 cryotransfer specimen holder (Gatan, Inc.) and imaged using a Philips CM200 EM with a field emission gun (FEI, Eindhoven, Netherlands) at an accelerating voltage of 120 kV. All images were recorded under low-dose conditions ($\sim$10 electrons/Å^2) on Kodak SO163 film at a nominal magnification of 38,000 and a defocus range of 0.7–1.4 μm underfocus. The tubes have an inner diameter of $\sim$400 Å and an outer diameter of $\sim$700 Å. The power spectra of the tubular arrays preserved in vitreous ice reveal a helical periodicity of 1/185 Å^{-1} (Fig. 2.4D).

2.10. Alternative strategies and future outlooks

The number of protein structures determined by EM is small compared to those determined by X-ray crystallography. It is therefore essential to develop new methods to obtain well-ordered specimens so that the potential of EM as a tool for protein structure determination and structure–function analysis can be fully realized. Methods described in this chapter present a means for studying a variety of protein structures and structure–function relationships. Improving existing lipid substrates and producing novel substrates would extend the applicability of the two approaches described. His-tagged PFO was used to demonstrate the utility of preformed nickel-lipid nanotubes as a tool for the structural analysis of recombinant His-tagged proteins. It is foreseeable that glutathione and maltose functionalized lipids could be used to bind glutathione-S-transferase or maltose-binding recombinant fusion proteins. One can also take advantage of nonspecific electrostatic interactions between charged lipid head groups and accessible charged surfaces on proteins. A number of proteins have formed helical arrays on lipid nanotubes derived from mixing GalCer with positively or negatively charged lipids (Wilson-Kubalek et al., 1998). Replacement of the nervonyl ($24:1^{\Delta 15 \ (cis)}$) alkyl chain of GalCer with ($22:1^{\Delta 13 \ (cis)}$) alkyl chain yields nanotubes of larger diameter (30–35 nm vs. 25–30 nm; Kulkarni et al., 1999). It may be possible to produce functionalized nanotubes of consistently larger diameter, which would be useful for larger protein complexes. Nickel-lipid nanotubes form in the presence of a number of different detergents; with the addition of n-hexyl-β-D-glucopyranoside, the length and uniformity of the nickel-lipid nanotubes are improved (Dang et al., 2005b). Forming lipid nanotube substrates in the presence of detergent may prevent different helical families being produced under the same crystallization conditions. For some proteins, the addition of detergent or a filler lipid may aid the crystallization process.

The ability to form functionalized tubes in the presence of detergents is exciting, as it raises the possibility of using preformed tubes as a scaffold for recruitment of detergent-solubilized membrane proteins and subsequent reconstitution of a lipid protein membrane encircling the scaffold. An alternative approach would be to take advantage of the properties of functionalized lipids, filler lipids, and tube-forming lipids and reconstitute the membrane protein of choice with more conventional methods such as dialysis or the use of bio-beads.

Lipid compositions, which were uniquely determined to form the helical arrays of dynamin and BoNT, can be modified for use with other membrane-associating proteins. Importantly, it is necessary to determine the ideal lipid composition and the curvature preference for each protein of interest. Although we have described the use of sonication and vortexing, liposomes can also be generated by multiple freeze–thaw cycles that alternate between liquid nitrogen temperature and 37 °C prior to extrusion. Others have also reported that helical crystals can be improved by a slow annealing protocol that gradually reduces the temperature after the initial mixing of the protein and liposomes (Frost *et al.*, 2008). It is possible that this procedure could also improve long-range order of the arrays formed on lipid nanotube substrates.

Liposomes can also be used to examine binding affinities to different charged lipid head groups by sedimentation (Ramachandran *et al.*, 2007), floatation (Buser and Mclaughlin, 1998), fluorescence (Ramachandran and Schmid, 2008), and light scattering (Bigay and Antonny, 2005) and to determine how membrane binding affects activity in enzymatic assays (Leonard *et al.*, 2005). These templates therefore provide a general means to study the structural and functional properties of membrane-associating proteins simultaneously. By investigating crystallization on the surface of either preformed lipid nanotubes or liposomes, the chances of obtaining ordered arrays of the protein of interest can only be increased.

ACKNOWLEDGMENTS

The original work described here was supported in part by grants from the National Institute of Health (GM75820, GM52468 to Ronald A. Milligan, GM61941 to Nigel P. Unwin, and GM61938 to Elizabeth M. Wilson-Kubalek).

REFERENCES

Bigay, J., and Antonny, B. (2005). Real-time assays for the assembly–disassembly cycle of COP coats on liposomes of defined size. *Methods Enzymol.* **404,** 95–107.
Binz, T., and Rummel, A. (2009). Cell entry strategy of clostridial neurotoxins. *J. Neurochem.* **109,** 1584–1595.

Blankenburg, R., Meller, P., Ringsdorf, H., and Salesse, C. (1989). Interaction between biotin lipids and streptavidin in monolayers: formation of oriented two-dimensional protein domains induced by surface recognition. *Biochemistry* **28,** 8214–8221.

Brunger, A. T., and Rummel, A. (2009). Receptor and substrate interactions of clostridial neurotoxins. *Toxicon* **54,** 550–560.

Buser, C. A., and McLaughlin, S. (1998). Ultracentrifugation technique for measuring the binding of peptides and proteins to sucrose-loaded phospholipid vesicles. *Methods Mol. Biol.* **84,** 267–281.

Carr, J. F., and Hinshaw, J. E. (1997). Dynamin assembles into spirals under physiological salt conditions upon the addition of GDP and gamma-phosphate analogues. *J. Biol. Chem.* **272,** 28030–28035.

Carragher, B., Whittaker, M., and Milligan, R. A. (1996). Helical processing using PHOELIX. *J. Struct. Biol.* **116,** 107–112.

Chai, Q., Arndt, J. W., Dong, M., Tepp, W. H., Johnson, E. A., Chapman, E. R., and Stevens, R. C. (2006). Structural basis of cell surface receptor recognition by botulinum neurotoxin B. *Nature* **444,** 1096–1100.

Chen, Y. J., Zhang, P., Egelman, E. H., and Hinshaw, J. E. (2004). The stalk region of dynamin drives the constriction of dynamin tubes. *Nat. Struct. Mol. Biol.* **11,** 574–575.

Damke, H., Baba, T., Warnock, D. E., and Schmid, S. L. (1994). Induction of mutant dynamin specifically blocks endocytic coated vesicle formation. *J. Cell Biol.* **127,** 915–934.

Damke, H., Binns, D. D., Ueda, H., Schmid, S. L., and Baba, T. (2001). Dynamin GTPase domain mutants block endocytic vesicle formation at morphologically distinct stages. *Mol. Biol. Cell* **12,** 2578–2589.

Dang, T. X., Farah, S. J., Gast, A., Robertson, C., Carragher, B., Egelman, E., and Wilson-Kubalek, E. M. (2005a). Helical crystallization on lipid nanotubes: Streptavidin as a model protein. *J. Struct. Biol.* **150,** 90–99.

Dang, T. X., Milligan, R. A., Tweten, R. K., and Wilson-Kubalek, E. M. (2005b). Helical crystallization on nickel-lipid nanotubes: Perfringolysin O as a model protein. *J. Struct. Biol.* **152,** 129–139.

Darst, S. A., Ahlers, M., Meller, P. H., Kubalek, E. W., Blankenburg, R., Ribi, H. O., Ringsdorf, H., and Kornberg, R. D. (1991). Two-dimensional crystals of streptavidin on biotinylated lipid layers and their interactions with biotinylated macromolecules. *Biophys. J.* **59,** 387–396.

Daumke, O., Lundmark, R., Vallis, Y., Martens, S., Butler, P. J., and McMahon, H. M. (2007). Architectural and mechanistic insights into an EHD ATPase involved in membrane remodelling. *Nature* **449,** 923–927.

Dong, M., Richards, D. A., Goodnough, M. C., Tepp, W. H., Johnson, E. A., and Chapman, E. R. (2003). Synaptotagmins I and II mediate entry of botulinum neurotoxin B into cells. *J. Cell Biol.* **162,** 1293–1303.

Drin, G., and Antonny, B. (2010). Amphipathic helices and membrane curvature. *FEBS Lett.* **584,** 1840–1847.

Dubochet, J., Adrian, M., Chang, J. J., Homo, J. C., Lepault, J., McDowall, A. W., and Schultz, P. (1988). Cryo-electron microscopy of vitrified specimens. *Q. Rev. Biophys.* **21,** 129–228.

Egelman, E. H. (2000). A robust algorithm for the reconstruction of helical filaments using single-particle methods. *Ultramicroscopy* **85,** 225–234.

Ellson, C. D., Andrews, S., Stephens, L. R., and Hawkins, P. T. (2002). The PX domain: A new phosphinositide-binding module. *J. Cell Sci.* **115,** 1099–1105.

Flicker, P. F., Robinson, J. P., and DasGupta, B. R. (1999). Is formation of visible channels in a phospholipid bilayer by botulinum neurotoxin type B sensitive to its disulfide? *J. Struct. Biol.* **128,** 297–304.

Ford, M. G., Pearse, B. M., Higgins, M. K., Vallis, Y., Owen, D. J., Gibson, A., Hopkins, C. R., Evans, P. R., and McMahon, H. T. (2001). Simultaneous binding of PtdIns(4,5)P2 and clathrin by AP180 in the nucleation of clathrin lattices on membranes. *Science* **291**, 1051–1055.

Ford, M. G., Mills, I. G., Peter, B. J., Vallis, Y., Praefcke, G. J., Evans, P. R., and McMahon, H. T. (2002). Curvature of clathrin-coated pits driven by epsin. *Nature* **419**, 361–366.

Frost, A., Perera, R., Roux, A., Spasov, K., Destaing, O., Egelman, E. H., De Camilli, P., and Unger, V. M. (2008). Structural basis of membrane invagination by F-BAR domains. *Cell* **132**, 807–817.

Gallop, J. L., Jao, C. C., Kent, H. M., Butler, P. J., Langen, R., and McMahon, H. T. (2006). Mechanism of endophilin N-BAR domain-mediated membrane curvature. *EMBO J.* **25**, 2898–2910.

Hinshaw, J. E., and Schmid, S. L. (1995). Dynamin self-assembles into rings suggesting a mechanism for coated vesicle budding. *Nature* **374**, 190–192.

Itoh, T., and De Camilli, P. (2006). BAR, F-BAR (EFC) and ENTH/ANTH domains in the regulation of membrane–cytosol interfaces and membrane curvature. *Biochim. Biophys. Acta* **1761**, 897–912.

Jin, R., Rummel, A., Binz, T., and Brunger, A. T. (2006). Botulinum neurotoxin B recognizes its protein receptor with high affinity and specificity. *Nature* **444**, 1092–1095.

Ku, A. C., Darst, S. A., Robertson, C., Gast, A., and Kornberg, R. D. (1993). Molecular Analysis of Two-dimensional protein crystallization. *J. Phys. Chem.* **97**, 3013–3016.

Kulkarni, V. S., Boggs, J. M., and Brown, R. E. (1999). Modulation of nanotube formation by structural modifications of sphingolipids. *Biophys. J.* **77**, 319–330.

Legendre-Guillemin, V., Wasiak, S., Hussain, N. K., Angers, A., and McPherson, P. S. (2004). ENTH/ANTH proteins and clathrin-mediated membrane budding. *J. Cell Sci.* **117**, 9–18.

Lemmon, M. A. (2003). Phosphoinositide recognition domains. *Traffic* **4**, 201–213.

Lemmon, M. A. (2008). Membrane recognition by phospholipid-binding domains. *Nat. Rev. Mol. Cell Biol.* **9**, 99–111.

Leonard, M., Song, B. D., Ramachandra, R., and Schmid, S. L. (2005). Robust colorimetric assays for dynamin's basal and stimulated GTPase activities. *Methods Enzymol.* **404**, 490–503.

Mao, Y., Chen, J., Maynard, J. A., Zhang, B., and Quiocho, F. A. (2001). A novel all helix fold of the AP180 amino-terminal domain for phosphoinositide binding and clathrin assembly in synaptic vesicle endocytosis. *Cell* **104**, 433–440.

McMahon, H. T., and Gallop, J. L. (2005). Membrane curvature and mechanisms of dynamic cell membrane remodeling. *Nature* **438**, 590–596.

Mears, J. A., and Hinshaw, J. E. (2008). Visualization of dynamins. *Methods Cell Biol.* **88**, 237–256.

Melia, T. J., Sowa, M. E., Schutze, L., and Wensel, T. G. (1999). Formation of helical protein assemblies of IgG and transducin on varied lipid tubules. *J. Struct. Biol.* **128**, 119–130.

Mettlen, M., Pucadyil, T. J., Ramachandran, R., and Schmid, S. L. (2009). Dissecting dynamin's role in clathrin-mediated endocytosis. *Biochem. Soc. Trans.* **37**, 1022–1026.

Miyazawa, A., Fujiyoshi, Y., Stowell, M., and Unwin, N. (1999). Nicotinic acetylcholine receptor at 4.6 Å resolution: Transverse tunnels in the channel wall. *J. Mol. Biol.* **288**, 765–786.

Miyazawa, A., Fujiyoshi, Y., and Unwin, N. (2003). Structure and gating mechanism of the acetylcholine receptor pore. *Nature* **423**, 949–955.

Montecucco, C., and Schiavo, G. (1994). Mechanism of action of tetanus and botulinum neurotoxins. *Mol. Microbiol.* **13**, 1–8.

Opalka, N., Mooney, R. A., Richter, C., Severinov, K., Landick, R., and Darst, S. A. (2000). Direct localization of a beta-subunit domain on the three-dimensional structure of *Escherichia coli* RNA polymerase. *Proc. Natl. Acad. Sci. USA* **97,** 617–622.

Peter, B. J., Kent, H. M., Mills, I. G., Vallis, Y., Butler, P. J., Evans, P. R., and McMahon, H. T. (2004). BAR domains as sensors of membrane curvature: The amphiphysin BAR structure. *Science* **303,** 495–499.

Polyakov, A., Richter, C., Malhotra, A., Koulich, D., Borukhov, S., and Darst, S. A. (1998). Visualization of the binding site for the transcript cleavage factor GreB on *Escherichia coli* RNA polymerase. *J. Mol. Biol.* **281,** 465–473.

Potter, C. S., Chu, H., Frey, B., Green, C., Kisseberth, N., Madden, T. J., Miller, K. L., Nahrstedt, K., Pulokas, J., Reilein, A., *et al.* (1999). Leginon: a system for fully automated acquisition of 1000 electron micrographs a day. *Ultramicroscopy* **77,** 153–161.

Prabakaran, S., Tepp, W., and DasGupta, B. R. (2001). Botulinum neurotoxin types B and E: Purification, limited proteolysis by endoproteinase Glu-C and pepsin, and comparison of their identified cleaved sites relative to the three-dimensional structure of type A neurotoxin. *Toxicon* **39,** 1515–1531.

Pucadyil, T. J., and Schmid, S. L. (2008). Real-time visualization of dynamin-catalyzed membrane fission and vesicle release. *Cell* **135,** 1263–1275.

Ramachandran, R., Surka, M., Chappie, J. S., Fowler, D. M., Foss, T. R., Song, B. D., and Schmid, S. L. (2007). The dynamin middle domain is critical for tetramerization and higher-order self-assembly. *EMBO J.* **26,** 559–566.

Ramachandran, R., and Schmid, S. L. (2008). Real-time detection reveals that effectors couple dynamin's GTP-dependent conformational changes to the membrane. *EMBO J.* **27,** 27–37.

Ringler, P., Wolfgang, M., Helmut, R., and Brisson, A. (1997). Functionalized lipid tubes as tools for helical crystallization of proteins. *Chem. Eur. J.* **3–4,** 620–625.

Sachse, C., Chen, J. Z., Coureux, P. D., Stroupe, M. E., Fandrich, M., and Grigorieff, N. (2007). High-resolution electron microscopy of helical specimens: A fresh look at tobacco mosaic virus. *J. Mol. Biol.* **371,** 812–835.

Schmid, M. F., Robinson, J. P., and DasGupta, B. R. (1993). Direct visualization of botulinum neurotoxin-induced channels in phospholipid vesicles. *Nature* **364,** 827–830.

Song, B. D., Leonard, M., and Schmid, S. L. (2004). Dynamin GTPase domain mutants that differentially affect GTP binding, GTP hydrolysis, and clathrin-mediated endocytosis. *J. Biol. Chem.* **279,** 40431–40436.

Stowell, M. H., Marks, B., Wigge, P., and McMahon, H. T. (1999). Nucleotide-dependent conformational changes in dynamin: Evidence for a mechanochemical molecular spring. *Nat. Cell Biol.* **1,** 27–32.

Suloway, C., Pulokas, J., Fellmann, D., Cheng, A., Guerra, F., Quispe, J., Stagg, S., Potter, C. S., and Carragher, B. (2005). Automated molecular microscopy: the new Leginon system. *J. Struct. Biol.* **151,** 41–60.

Sweitzer, S. M., and Hinshaw, J. E. (1998). Dynamin undergoes a GTP-dependent conformational change causing vesiculation. *Cell* **93,** 1021–1029.

Unwin, N. P. (2005). Refined structure of the nicotinic acetylcholine receptor at 4 Å resolution. *J. Mol. Biol.* **346,** 967–989.

Wang, L., and Sigworth, F. J. (2009). Structure of the BK potassium channel in a lipid membrane from electron cryomicroscopy. *Nature* **461,** 292–295.

Wang, L., Ounjai, P., and Sigworth, F. J. (2008). Streptavidin crystals as nanostructured supports and image-calibration references for cryo-EM data collection. *J. Struct. Biol.* **164,** 190–198.

Wang, S. W., Robertson, C. R., and Gast, A. P. (1999a). Molecular arrangement in two-dimensional streptavidin crystals. *Langmuir* **15,** 1541–1548.

Wang, S. W., Robertson, C. R., and Gast, A. P. (1999b). Two-dimensional crystallization of streptavidin mutants. *J. Phys. Chem. B.* **103,** 7751–7761.

Whittaker, M., Carragher, B. O., and Milligan, R. A. (1995). PHOELIX: A package for semi-automated helical reconstruction. *Ultramicroscopy* **58,** 245–259.

Wilson-Kubalek, E. M. (2000). Preparation of functionalized lipid tubules for electron crystallography of macromolecules. *Methods Enzymol.* **312,** 515–519.

Wilson-Kubalek, E. M., Brown, R. E., Celia, H., and Milligan, R. A. (1998). Lipid nanotubes as substrates for helical crystallization of macromolecules. *Proc. Natl. Acad. Sci. USA* **95,** 8040.

Xu, Y., Seet, L. F., Hanson, B., and Hong, W. (2001). The Phox homology (PX) domain, a new player in phosphoinositide signaling. *Biochem. J.* **360,** 513–530.

Yarar, D., Surka, M. C., Leonard, M. C., and Schmid, S. L. (2008). Snx9 activities are regulated by multiple phosphoinositides through both PX and BAR domains. *Traffic* **9,** 133–146.

Yonekura, K., Maki-Yonekura, S., and Namba, K. (2005). Building the atomic model for the bacterial flagellar filament by electron cryomicroscopy and image analysis. *Structure* **13,** 407–412.

Zhang, P., and Hinshaw, J. E. (2001). Three-dimensional reconstruction of dynamin in the constricted state. *Nat. Cell Biol.* **3,** 922–926.

CHAPTER THREE

PLUNGE FREEZING FOR ELECTRON CRYOMICROSCOPY

Megan J. Dobro,* Linda A. Melanson,† Grant J. Jensen,*,‡ *and* Alasdair W. McDowall*

Contents

Abstract

Aqueous biological samples must be "preserved" (stabilized) before they can be placed in the high vacuum of an electron microscope. Among the various approaches that have been developed, plunge freezing maintains the sample in the most native state and is therefore the method of choice when possible. Plunge freezing for standard electron cryomicroscopy applications proceeds by spreading the sample into a thin film across an EM grid and then rapidly submerging it in a cryogen (usually liquid ethane), but success depends critically on the properties of the grid and sample, the production of a uniformly thin film, the temperature and nature of the cryogen, and the plunging conditions. This chapter reviews plunge-freezing principles, techniques, instrumentation, common problems, and safety considerations.

* Division of Biology, California Institute of Technology, Pasadena, California, USA
† Gatan, Inc., Pleasanton, California, USA
‡ Howard Hughes Medical Institute, California Institute of Technology, Pasadena, California, USA

Methods in Enzymology, Volume 481
ISSN 0076-6879, DOI: 10.1016/S0076-6879(10)81003-1

1. INTRODUCTION

Because electrons have such high scattering cross-sections, their path through the electron microscope must be kept at extremely high vacuum. Aqueous biological samples must therefore be stabilized or "preserved" before they can be imaged. The first set of methods that were developed to preserve biological samples for EM involved dehydration: protein and viruses were negatively stained; tissues and cells were first chemically fixed, then dehydrated, plastic embedded, sectioned, and then stained. Dehydration perturbs structure, however, so methods were sought to preserve samples in their naturally hydrated state through freezing. The basic problem is, of course, that when frozen gradually, water crystallizes and expands, again denaturing macromolecules and perturbing cellular structures. One approach to solving this problem is to apply high pressures and cryoprotectants ("high pressure freezing"), which inhibit the nucleation and growth of ice crystals (Chapter 8, this Volume).

It was wondered, however, whether water or biological molecules could instead be cooled so rapidly that molecular rearrangements would simply stop before ice crystals had time to form. In the early 1970s, Taylor and Glaeser plunged hydrated catalase crystals into liquid nitrogen and showed that the crystals still diffracted to 3.4 Å (proving that the structure of the proteins had been preserved to at least that resolution; Taylor and Glaeser, 1973, 1974). Then in 1981, the Dubochet group showed that pure water could be frozen in a noncrystalline, liquid-like ("vitreous") state by spreading it into a thin layer across a standard carbon-coated EM grid and plunging it into liquid ethane (Dubochet and McDowall, 1981). At first, this claim was met with skepticism, but the impact the advance would have on structural biology became clear when macromolecular complexes were later added to the water and shown to be preserved in a native, "frozen-hydrated" state (Adrian *et al.*, 1984). The development of dedicated cryo-EM instrumentation (anticontaminators; low-dose kits, and tools to insert and hold frozen grids) and complementary advances in software, computational power, and other aspects of the work have now fully capitalized on this advance, producing reconstructions of specimens in their native states that are interpretable at the atomic level (Chapter 11, this volume; Chapters 9 and 15, Vol. 482).

Today, plunge freezing is being used to study macromolecules, drug delivery vehicles, 2D protein crystals, cell fractionations, vesicle suspensions, filaments, virus particles, thin bacteria, polymers, matrices, colloids, nanoparticulate catalysts, and even emulsion paints (Cerritelli *et al.*, 2009; Finnigan *et al.*, 2006). A variety of plunge freezers and protocols have been optimized for different applications (Grassucci *et al.*, 2007; Iancu *et al.*, 2006). The process of preparing samples for electron cryomicroscopy can be arduous, however, and often requires extensive troubleshooting to determine the best freezing conditions for each sample. This chapter describes the plunge-freezing protocol in detail and will enable the novice to recognize the potential rewards and challenges that await them.

2. GRIDS AND SUPPORTS

The first step in preparing samples for electron cryomicroscopy is to choose the right grid and support film. The grid itself can be made from a variety of metals. Copper is the most common, but if cells are to be grown on or in the presence of the grid, gold is a better choice because it is less toxic. Molybdenum has the advantage that it has a similar coefficient of thermal contraction as carbon, so that when frozen, the grid and the carbon support shrink more similarly, preventing "crinkling" (Booy and Pawley, 1993). Larger mesh sizes (smaller squares between grid bars) provide more support, but the grid bars block more area on the grid, especially at high tilt angles. Specialized "finder" grids are decorated with symbols to help mark particular locations on the grid, which can be critical, for instance, in correlative light and electron microscopy (Chapter 13, this Volume).

Grids for cryo-EM applications are almost invariably coated with a thin carbon film, although just recently, a new silicon ceramic combination called Cryomesh^{TM} has been introduced, which shows promise in providing greater strength and stability (Quispe et al., 2007; Yoshioka et al., 2010). The carbon film can be either "continuous" (no holes) or "holey." Continuous-carbon support films can be better for 2D crystals, for instance, where maintaining a perfectly flat crystal is more important than reducing background noise (Chapter 4, Vol. 482). Holey carbon films allow background noise to be reduced, as samples can be imaged suspended in vitreous ice alone across the holes. While "lacey" grids (prepared either in the lab or purchased) have an irregular array of varying hole sizes (Fig. 3.1A), commercially available Quantifoil® (Fig. 3.1B) and C-flat^{TM} films have a regular pattern of holes to facilitate automatic image acquisition. Typical hole sizes are around 1 μm. Larger holes maximize the sample imaging area, but it is helpful to have at least some carbon film (and maybe the full periphery surrounding a hole) in each image to reveal the defocus more precisely (Chapter 9, Vol. 482) and to reduce charging (Chapter 10, this volume).

Unfortunately, in our experience, the surface properties and integrity of the support film vary from batch to batch. We therefore recommend that a few grids from each batch be tested before use. The integrity of the carbon film can be checked easily in a light microscope, both before and after plasma cleaning (see Section 3). The surface properties can be checked by plunge-freezing pure water on a grid with standard plasma-cleaning and plunge-freezing protocols to make sure that uniformly thin vitreous ice is formed. Commercial suppliers can

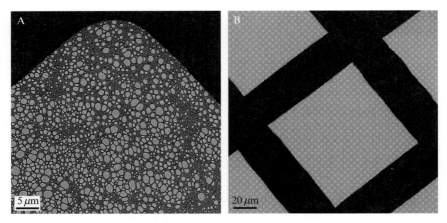

Figure 3.1   EM grid types. (A) Lacey carbon support film with an array of various hole sizes (grid bar 5 μm) and (B) Quantifoil® grid showing regular pattern of holes in carbon (grid bar 20 μm).

customize their support films to particular needs to reduce substrate bubbling, for instance, or improve film stability through the use of extra thick layers.

 ## 3. Cleaning the Grids

Freshly prepared carbon films are hydrophilic, but they become pro-gressively more hydrophobic over time. It is therefore usually necessary to restore their hydrophilicity so that the liquid sample will spread evenly over their surface. Before the advent of electron cryomicroscopy, plasma cleaning, also called "glow discharging," was used to modify the adhesive properties of a variety of substrates for room temperature microscopy (Dubochet *et al.*, 1971). The plasma is created from the ionization of a gas, such as air, argon, oxygen, and hydrogen, or combinations thereof, such as argon/oxygen or hydrogen/oxygen, under low vacuum. Radicals within the plasma react with the surface of the substrate. As a result, the surface of the grid typically becomes hydrophilic. When liquid samples are then placed on the grid, they spread evenly across the surface and can be blotted to form films as thin as just tens of nanometers thick (Gan *et al.*, 2008). On a properly cleaned grid, in the chamber of a humidity-controlled plunge freezer, this thin sample film is remarkably stable and can remain suspended across holes in the grid for many seconds. If the plasma cleaning does not make the whole surface of the film uniform, the liquid will not spread over the grid or blot evenly off the grid, causing denser ice in some areas. One result can be, for instance, a bulge of ice in the center of each grid square (Fig. 3.2C).

Figure 3.2 The effect of plasma cleaning on the ice. (A) Montaged serial EM atlas of a grid showing uniform ice thickness, image courtesy of Dr. Guenter Resch. (B) A Solarus^{TM} plasma-cleaned Quantifoil® film with thin, uniform ice in the holes, image courtesy of Dr. Chen Xu, Rosentiel Basic Medical Sciences Research Center, Waltham, MA (grid bar 1 μm). (C) When plasma cleaning fails, one result can be a dense core of ice in the center of each grid square (grid bar 100 nm).

Plasma-cleaning parameters, such as the chamber pressure, radio frequency (RF) power, the gas mixture used to form the plasma, and the overall system geometry, should all be explored and optimized. The system settings can vary for each machine and application, but in our lab at Caltech, we use a platform height of 35 mm, a glow time of 60 s and an electrical current of 15 mA. If the fields are too strong or the glow time is too long, bombardment by the highly energetic ions can break the carbon film. The carbon film can also break if the vacuum is vented too quickly. Small organic molecules like amylamine and polylysine can be introduced as vapors during the ion discharge and subsequently affect how purified macromolecules partition in the ice over the carbon film and the holes. Once cleaned, grids can be stored in their original storage grid box and sealed in an air-tight bag or chamber for later use. There are several diagrams available that show the proper setup for a plasma-cleaning system (Aebi and Pollard, 1987; Kumar *et al.*, 2007).

Building homemade plasma cleaners can be dangerous because of the high currents and voltages used and the specialized gases required to create the plasma. Commercial instruments are widely available. Some are designed solely for plasma cleaning and others also offer carbon coating. The Cressington 208 plasma–cleaning module attached to the Cressington carbon coater has a fixed 40-mm grid platform height and programmable time and power values. The Emitech K100X free standing unit uses a programmable protocol sequence, an adjustable height platform, and options for introducing alternate gases, making this a versatile unit. The smaller Harrick PDC-32 unit has fewer control settings, a fixed glow tube diameter, and may be easily transported to more remote research locations. Further development by Gatan, Inc. has produced the autotuning Solarus^{TM} 950 Advanced Plasma Cleaning System, with a chamber to accommodate grid cleaning as well as two ports for cleaning microscope specimen holders. The Solarus^{TM} 950 is configured to use a hydrogen/oxygen gas mixture that cleans with minimal sputter damage, making it especially suitable for cleaning fragile carbon support substrates. Because of the efficiency of the hydrogen/oxygen plasma, the cleaning time for carbon substrates is very short, typically 15–30 s using a hydrogen/oxygen gas mixture and a RF setting of 50 W. This has produced carbon films that are uniformly hydrophilic and can remain so for several weeks (Melanson, 2009b).

Though plasma cleaning is the preferred method to clean grids, when plasma cleaners are not available, the grids can also be dipped into ethanol, acetone, or chloroform or be recoated with a fresh carbon layer (Quispe *et al.*, 2007). Grids can also be coated with polylysine or other organic molecules to promote the adherence of cells, for instance. We and others have found that more extreme grid treatments (such as overnight "preirradiation" in an EM) can cause certain macromolecular complexes to partition into the holes in the carbon.

4. Preparing the Cryogen

For water to vitrify, the temperature has to drop faster than $\sim 10^5$ K/s (Dubochet and McDowall, 1981). The reason why samples have to be thin is that the heat conductivity of the water in the sample is the limiting factor. The cryogen that the sample is plunged into has to have a high thermal conductivity in order to transfer heat out of the specimen quickly, a freezing point below the temperature needed to vitrify the sample, and both a high boiling point and a large heat capacity to prevent a layer of vapor forming between the sample and the cryogen (Bellare *et al.*, 1999). While the temperature of liquid nitrogen at ambient pressures is very low (77 K), it is readily available, and it is relatively inexpensive, unfortunately its thermal conductivity is only about 400 K/s and so frequently produces crystalline ice. The most commonly used cryogens are therefore ethane and propane, primarily because their thermal conductivity is 300–400 times higher (in excess of 13–15 kK/s). The freezing point of ethane is 90 K, its boiling point is 184 K, and it has a high heat capacity (68.5 J/mol K at 94 K). Liquid nitrogen is used instead as the primary coolant to first liquefy the ethane or propane and then keep it cold during the procedure.

An inconvenience arises, however, because the freezing points of both ethane and propane are higher than the temperature of the nitrogen, so they slowly solidify during the experiment. Some plunge freezers have therefore been constructed with built-in heating elements or special designs that limit the heat transfer between the nitrogen and ethane/propane cups to maintain the cryogen just above its melting point. Tivol *et al.* (2008) found that a mixture of 37% ethane and 63% propane remains liquid even when in direct contact with liquid nitrogen. This mixture produces consistently thin vitrified layers and facilitates long plunge-freezing sessions without heaters or special cup configurations.

After the grid is plunged into the cryogen and then transferred into liquid nitrogen for storage, excess ethane (or propane) on the grid will freeze, forming a solid crust. Usually, this crust falls off the grid in subsequent handling, but if not, it will sublime rapidly when the grid is inserted into the high vacuum of the microscope. Impurities will remain, however, so it is important to use very pure cryogen. Lower grade "camping gas propane" is too full of contaminants to produce clean samples.

4.1. Condensing the cryogen

Cryogens come as compressed gases and therefore need to be liquefied. This is done by releasing the gas slowly into a cup cooled by liquid nitrogen. The flow of the gas can be controlled with a 2-stage regulator fitted with a needle valve on the second stage and narrow-bore Tygon tubing on the nozzle. A pipette tip is usually inserted into the end of the Tygon tubing to further restrict and better direct the gas flow. The following is a typical protocol for condensing the cryogen:

1. Work in a fume hood and wear a lab coat and goggles.
2. Pour liquid nitrogen into the space around the cryogen cup. When the cup has reached at least − 175 °C, the liquid nitrogen will stop bubbling violently (the "Leidenfrost point"). Depending on the plunge freezer design, this can take 5–15 min, and the procedure is usually outlined in the instruction manual specific to the instrument being used.
3. Before starting the condensation process, check to make sure that the cryogen cup is free of any residual liquid nitrogen.
4. With the needle and main tank valves on the 2-stage regulator closed at this point, adjust the gas outlet pressure on the second stage to

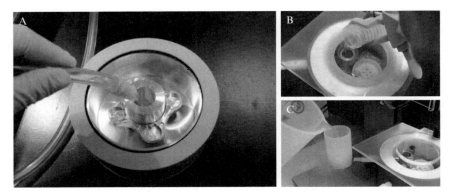

Figure 3.3 (A) Condensing the cryogen by flowing cryogen gas into a precooled cup surrounded by liquid nitrogen. (B) Pouring liquid ethane into the cold cryogen cup after condensing ethane gas in a separate container. (C) Refilling the liquid nitrogen through an external port maintains a clean, cold nitrogen gas environment. (B) and (C) Images courtesy of Gatan, Inc., Pleasanton, CA.

approximately 0.14–0.28 bar. Use low pressure to avoid unnecessary venting of the gas into the fume hood or splashing of condensed cryogen.
5. Place the tubing attached to the gas tank regulator into the bottom of the precooled cryogen cup (as in Fig. 3.3A).
6. Open the main tank and needle valves to allow delivery of the gas at the preset pressure.
7. You will start to notice the liquid filling the cup. When the liquid reaches the top, decrease the flow of gas and slowly pull the tip of the tubing out. Quickly turn the gas off. If you turn the gas off while the tip is still submerged, the liquid will aspirate back into the tubing.
8. Remember to close the main tank valve on the cylinder and bleed the line of any residual gas. Always leave the gas cylinder in a safe configuration as defined by the safety procedures for your laboratory.

Alternatively, the cryogen can be condensed in a separate container cooled by liquid nitrogen and then poured into the precooled cryogen cup (Fig. 3.3B).

4.2. Safety considerations

Before handling cryogens, read about them thoroughly in the latest Materials Safety Data Sheets. Ethane and propane are highly flammable and are even more so when condensed, so do not condense these gases in the presence of an open flame. Only condense the smallest volume necessary to fill the cryogen cup (usually less than 10 ml). Rather than having one large tank of cryogen gas, try to limit the size and keep reserve tanks in flameproof cabinets. Two refillable cylinders containing 67 lb of gas last 2–3 months in a busy laboratory. Ethane or propane gas cylinders, and their associated 2-stage regulators, should be ordered and installed in consultation with the on-site laboratory safety officer.

The liquid nitrogen that is used to maintain the low temperature of the condensed cryogen will evaporate over time and must be continually refilled during a freezing session. Replenishing the liquid nitrogen also serves to maintain a layer of cold, dry nitrogen gas surrounding the condensed cryogen. This helps to minimize condensation of atmospheric moisture into ice that will contaminate the cryogen and the sample, and provide a protective interface for transferring the frozen specimen grid. However, try not to splash liquid nitrogen into the cryogen. The surface of the cryogen can freeze solid, entrapping an underlying volume of warmer, liquid cryogen that can explode through the frozen layer. The Cryoplunge^{TM} 3 has a shield over the workstation to prevent splashing, as well as an external funnel for refilling the liquid nitrogen (Fig. 3.3C). Physical exposure to these low temperature cryogens can produce severe frostbite. Always wear adequate eye and face protection when working with these cryogens. Also, exercise caution when handling any materials that come in contact with the condensed cryogen, since these surfaces can also freeze skin and underlying tissues.

While small volumes of liquid nitrogen can safely be poured over a large ventilated surface, such as a floor, to dispose of it, it is recommended that propane and ethane be allowed to evaporate in a dedicated fume hood for several reasons. First, they are highly flammable. As the cryogens evaporate, they will expand rapidly by factors in excess of 700 times. Since the cryogens are odorless and colorless, there is also a risk of asphyxiation as atmospheric oxygen is displaced. Even at low concentrations, ethane gas can cause narcotic effects with symptoms of dizziness, headache, nausea, and loss of coordination. The plunge-freezing area should be well ventilated, and labs handling large volumes of cryogens can be equipped with oxygen displacement sensors to warn people when oxygen gets low. In addition to educating the staff on the risks from cryogens, always provide plenty of protective cryo gloves and eye shields and post signs alerting visitors and emergency responders to the location of cryogens.

5. PLUNGING THE GRID

5.1. Basic procedure

The process of plunge freezing generally involves three main steps: a small liquid droplet containing the specimen is applied to the carbon surface of an EM grid, the liquid droplet is blotted with filter paper until only a very thin film of fluid remains, and then the grid is plunged into the cryogen. The grid is then stored in liquid nitrogen in a custom-made grid box until it is finally loaded into the electron cryomicroscope for imaging. Blotting can be done from either one or two sides. Unilateral blotting can be particularly helpful in reducing the direct contact of fibers in the blotting paper with cells, for instance, growing on the other side (Lepper *et al.*, 2010). The best ice thickness depends on several factors, including the size and shape of the specimen and the accelerating voltage of the electron cryomicroscope that will be used. Thicker ice may provide more stability, but if the fluid sample to be vitrified is too thick, the ice may not vitrify. If too much fluid is blotted away, the cells can become dehydrated. The thinness of the ice will also effect how particles distribute across the holes: large particles may be displaced to deeper regions of the film, such as the edge of a hole. Particles may also be oriented preferentially in very thin layers in part because of surface charges at the air/liquid interface (Glaeser *et al.*, 2007). The temperature, humidity, blotting pressure, and blotting duration should be optimized for each specimen. It has recently been shown that blotting can damage and even kill large cells (Lepper *et al.*, 2010). Such samples should therefore be blotted gently for longer times.

As an example protocol for plunge-freezing protein or bacteria,

1. Suspend the sample in an aqueous medium (e.g., water or low ionic buffer solution to reduce background noise during imaging) at a concentration of 1–3 mg/ml for protein complexes or an OD_{600} of 0.5 for bacteria.
2. Plasma clean EM grids, following the instructions provided in the user manual for your particular machine.
3. Secure an EM grid with the tweezers provided with your plunge freezer and attach the tweezers to the machine.
4. If the plunge freezer has a humidity-controlled chamber, set the humidity to 100%.
5. Apply 3–5 μl of the sample to the carbon side of the grid (see the manufacturer's instructions on the grid box).
6. Blot the EM grid with #1 grade filter paper for 1–3 s to produce an aqueous film less than 1 μm in thickness.
7. Plunge into liquid cryogen to produce a thin glass–like solid.
8. Transfer the grid into a labeled four-grid-slot box in liquid nitrogen, being careful not to expose the grid to atmospheric moisture.
9. Grid boxes are stored within a 50-ml conical tube placed in a large nitrogen cryostorage dewar.

Part of the skill of plunge freezing is knowing when the cryogen is at the right temperature. When gaseous cryogens are first liquefied, they are still warmer than the surrounding liquid nitrogen, and it takes time for them to cool further. The best indication for when the liquid ethane reaches the right temperature for plunge freezing is when the bottom of the cup freezes, but enough liquid remains at the surface for plunging the grid. This state does not last very long before the rest of the volume freezes, however, so unless a mixture of ethane and propane is used or the freezing device somehow keeps the cryogen temperature just above its freezing point (see earlier), the cryogen will have to be melted periodically. This can be done by inserting a warm metal rod or adding more (room temperature) cryogen gas, but neither strategy is ideal, since rods can introduce contamination and adding more gas can cause the cryogen to overflow the cup. One must also wait again until the cryogen has recooled to its freezing point before the next grid is frozen.

After a grid has been plunged, it should be handled very carefully to avoid damage. The grid should never be bent, because it will then fail to seat securely in the holder, causing drift and instability, so try to avoid touching the grid to any walls of the freezing cup during manipulation. When transferring the frozen grid from the cryogen cup to the storage holder, the grid may need to be lifted out of the cryogen very quickly in a space filled with cold dry nitrogen gas to prevent exposing it to moisture in the air. Floating cylindrical barriers and purpose-built covers are also to be used to trap more dry nitrogen gas and protect the specimen. For more details on the plunge-freezing procedure, see Iancu *et al.* (2006). Training courses are frequently available from vendors and the NIH-funded National Research Resource Centers.

5.2. Controlling humidity

Atmospheric moisture in the cryolab is undesirable. If precautions are not taken, moisture will form ice crystals on liquid nitrogen storage containers and subsequently on the grid sample. In humid regions of the world, cryolabs employ complex ventilation systems and dehumidifiers for reducing relative humidity to less than 25%. Additionally, instruments may be entirely enclosed within a humidity-controlled chamber. (A cautionary tip: low humidity may increase static electricity. Certain floor coverings and clothes can reduce these discharges.) The regulations at some institutions require that plunge freezers be operated in a fume hood. The strong air currents within fume hoods can introduce ice contamination and air-drying artifacts. A shield around the plunging area is recommended. Automated plunge freezers now provide covers that facilitate a dry nitrogen gas flow over the cryogen container to reduce contamination. In addition, all liquid nitrogen dewars must be kept dry between freezing sessions and fitted with a loose lid to reduce water vapor condensation. Always invert portable tanks and dewars to dry, since moisture will collect on their cold surfaces. A large drying incubator at 30 °C is useful to ensure that all components remain moisture free. Workstations with heat blocks set at 50 °C can be used to dry small tools, and a source of low pressure "lab air" can be used to dry fixtures.

While atmospheric moisture must be controlled to avoid contamination, a higher relative humidity in the immediate area of the sample grid is preferred for preventing desiccation of the sample prior to freezing. Many commercial plunge-freezing instruments offer an environmental chamber for controlling temperature and humidity. This controlled environment is especially impor- · tant when blotting, as evaporation and surface air/liquid interfaces play important roles in electrostatic forces and on how macromolecules organize at the surface. In general, the humidity in the chamber before plunging should be greater than 80%, as demonstrated in the preservation of liposomes. Humidity values in the 40% range will create osmotic imbalances as water evaporates from the film, causing liposome inversions (Frederik and Hubert, 2005). Other effects are seen in the preferential organization of viral capsid complexes as they align because of surface interactions (Dubochet et al., 1985).

5.3. Instrumentation

The designs of plunge freezers used in pioneering experiments contributed to the design of modern-day instruments (Fernandez-Moran, 1960; Handley et al., 1981; McDowall et al., 1984). Early laboratory prototypes for plunge freezers were often a basic construction of makeshift stands and Styrofoam boxes (Dubochet et al., 1983; McDowall et al., 1983; McDowall, 1984). One of the first plunge freezers was a simple pivoting fine-forcep holding a carbon-coated grid. The grid fell in a gravity arc past a vaporized mist sprayed through an aperture slit. The continuous-carbon substrate collected microfine droplets, which were vitrified in a pot of viscous ethane. The first image of vitreous water was prepared in this way (Dubochet and McDowall, 1981). Eventually, the carbon substrate was removed and the "bare grid" method was the precursor of the unsupported liquid film (Adrian et al., 1984). The Dubochet group's 1980s plunge freezer design was an elastic-driven rod supporting gold electronic circuitry pins for freezing suspensions and filaments. A water-driven magnetic stir bar kept the ethane fluid and successfully vitrified samples for early cryosectioning experiments (Dubochet et al., 1988). The plunge freezers today are much more sophisticated but owe their design to these early prototypes and years of experience in many laboratories.

A key requirement for obtaining good frozen-hydrated specimens is the ability to produce the uniformly thin vitrified ice layer. In the early days of cryospecimen preparation, many of the "homemade" plunge-freezing instruments required manual blotting of the specimen grid. Although, very high-quality results can be obtained in this manner, manual blotting of the liquid from the surface of a fragile EM grid is often variable, and success depends on the skill of the individual. Manual plungers rely on gravity to plunge the EM grid into a cryogen and have therefore been called "gravity plungers" (Fig. 3.4A). Because they are not automated they offer more user control over the blotting. For experiments involving cells growing on the EM grid, harsh blotting pressures from both sides in an automated plunger run the risk of "peeling" the adherent cells off the grid. In the case of gravity plungers, the user can blot from the back of the grid if the intent is not to disturb the cells. The liquid flows into the filter paper through the holes in the carbon. A typical practice is to blot with filter paper from one side until the liquid stops wicking into the filter paper (Fig. 3.4B), but a variation of blotting times should be tested for each sample. The gravity plungers are fairly mobile and are often used

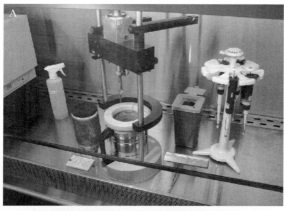

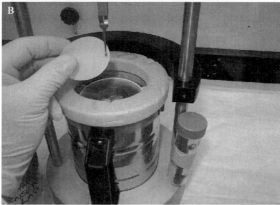

Figure 3.4 (A) Manual plunger (custom-made in the Department of Biochemistry, Max Planck Institute, Martinsried, Germany) in a biosafety cabinet. (B) Manually blotting the liquid from the back of the grid before plunging.

when traveling to laboratories where specimens are to be frozen on-site. However, the lack of an environment-controlled chamber means that the grids are exposed to atmospheric humidity and temperature, and the users are more exposed to the danger of having sharp tweezers and biological samples near their hands. Modern manual plungers have built-in lights to help visualize the blotting process and a foot pedal for dropping the tweezers.

Currently, there are a variety of automated plunge-freezing instruments designed to make the plunge-freezing process efficient and reproducible (Fig. 3.5). Automated plunge freezers provide precise control of several parameters such as humidity, blot pressure, and blot duration in order to eliminate variability in the thickness of the vitrified ice layer. They also provide the means to select, store, and recall a set of parameters. Finally, automated plunge freezers incorporate a variety of safety features for protecting the user.

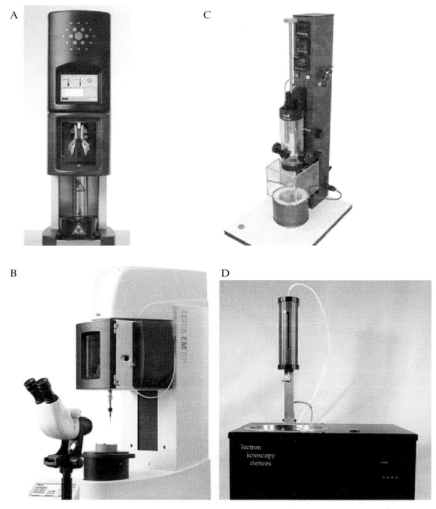

Figure 3.5 Variety of automated plunge freezers. (A) Vitrobot™ Mark IV, image courtesy of FEI, Inc., Hillsboro, OR. USA. (B) Leica EM GP, image courtesy of Leica Microsystems, Inc., Vienna, Austria. (C) Cryoplunge™ 3, image courtesy of Gatan, Inc., Pleasanton, CA. (D) EMS-002 Rapid Immersion Freezer, image courtesy of Electron Microscopy Sciences, Hatfield, PA, USA.

The first fully automated, computer-controlled plunge freezer was developed in the late 1990s by Dr. Peter Frederik and Paul Bomans (Frederik and Hubert, 2005). This machine, the Vitrobot™ Mark I, was the first in a series of "vitrification robots" now commercially offered through the FEI Company. The Vitrobot Mark IV is the latest version, and can dip the grid into a liquid sample or allow a sample to be pipetted onto the grid from an opening in the side. The Leica EM GP, developed in conjunction with Dr. Gunter Resch, provides one-sided blotting of cell monolayers grown on the specimen grid. An attached stereomicroscope allows the user to view the specimen grid to monitor the process. Cryoplunge™ 3 from Gatan, Inc. is a versatile, semiautomated plunge-freezing instrument that provides timed blotting functions, a removable humidity chamber, and the temperature of the liquid ethane can be held just above the melting point of the cryogen. A shield over the cryogenic workstation provides a protective environment for transferring the frozen-hydrated grid to its storage container to prevent the formation of contaminating ice (Melanson, 2009a). The Rapid Immersion Freezer from Electron Microscopy Sciences requires that the specimen be manually blotted, but it is a portable and economical plunge freezer that provides an environmental chamber, temperature control of the cryogen, and freeze-substitution capabilities.

Consistency of results, control parameters, and overall cost are some of the criteria that can be used to determine which type of plunge freezer best suits the needs of the individual investigator. Some laboratories use several different types of plungers based on the features required to prepare their diverse specimens.

5.4. Safety considerations

Many people with different samples typically share a plunge freezer. A good practice is to decontaminate surfaces and mechanical parts of the machine with 70% ethanol before and after every use. In order to keep track of all the samples that come in contact with the machine, an accurate sample history should be kept on a database and staff should be informed of the daily freezing schedule. Only microliter volumes should be cultured and used in freezing procedures, and aseptic technique should be followed in safety hoods. For biohazardous samples in automated plunge freezers, a heat cycle overnight will help to decontaminate the chamber. In manual plungers, the whole machine can be placed in a biosafety cabinet in an isolated room. When using manual plungers, be aware of the location of the foot pedal to avoid premature release of the tweezers and try to keep fingers from ever going beneath the sharp tweezers. The automated plunge freezers are usually controlled by compressed air and have enclosed chambers so the user is less likely to be injured by the tweezers. Training for new staff and yearly refresher lab safety courses should be offered.

6. Common Problems and Their Diagnoses

Ideally, the ice across a grid will be uniformly thin and vitreous (Fig. 3.2A). If the ice and embedded sample is too thick, or there is too much contamination, no electrons will penetrate through the grid. If the ice is at least thin enough for electron penetration, electron diffraction can be used to evaluate the quality of the ice. Vitreous ice is the absence of a detectable crystalline structure and most closely resembles the liquid state of water. It is the preferred form of "ice" for electron imaging and is thought to be the least damaging to structures. However, there are many points in the freezing, transfer, and loading procedures that can cause damaging hexagonal or cubic ice crystals to form: slow cooling of the sample because the cryogen was too warm or had a low thermal conductivity, contaminating ice floating in the cryogen or storage liquid nitrogen, or the frozen sample warming to a temperature higher than $-135\ ^\circ$C at any time during storage or transfer (Cavalier *et al.*, 2009; Dubochet *et al.*, 1988). Contaminating ice can adhere to the sample during the transfer of the grid from the cryogen to the storage container or during loading into the microscope. Care should be taken at these points to protect the EM grid by always keeping it in liquid nitrogen, limiting exposure to atmospheric moisture, and always cooling tools in liquid nitrogen before using them to manipulate the grid.

Hexagonal ice is the most common ice on earth and is formed when water molecules attach to each other at each point of their tetrahedral structure and extend indefinitely (Dubochet *et al.*, 1988). This proliferation of bonded molecules can severely damage the cellular ultrastructure. Hexagonal ice can form during slow freezing (streaks of crystalline ice, as seen in Fig. 3.6A), or could have condensed after freezing in the form of discrete spherical spots. Cubic ice is very similar to hexagonal ice, except that the bond angles of neighboring water molecules are rotated 180°, making it only stable below $-70\ ^\circ$C (Dubochet *et al.*, 1988). The dimensions of cubic ice crystals range from 30 nm to 1 μm and they usually look like fine grain spots (Fig. 3.6B). While cubic ice does not tend to cause extensive structural damage to the specimen, it can cause background noise that will disrupt the image. For reference, the three forms of ice and their diffraction patterns can be found in the landmark article by Dubochet *et al.* (1988).

There are many decisions to be made and conditions to be controlled during the plunge-freezing process in hopes of creating a well-preserved sample in a thin, vitreous ice. However, when the process works, high-resolution structural detail of a sample can be obtained and the reward can be great. Figure 3.7 demonstrates the level of detail that can be achieved in different samples.

ACKNOWLEDGMENTS

We wish to thank our colleagues, Steve Coyle and John Hunt of Gatan, Inc., Mark Ladinsky, Elitza Tocheva, Ariane Briegel, and Martin Pilhofer at Caltech, who proofread the manuscript. FEI, Inc.; Gatan, Inc.; Leica Microsystems, Inc.; and Electron Microscopy Sciences

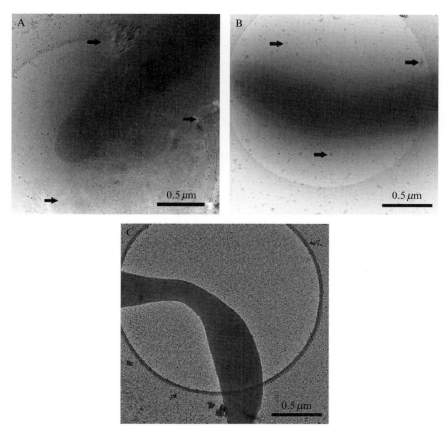

Figure 3.6 Examples of poor ice. (A) Hexagonal ice surrounding the bacterium, indicated by the arrows. (B) Cubic ice contamination, indicated by arrows. (C) The result of extreme rewarming: loss of water and structural detail. Grid bars = 0.5 μm.

supplied images and data. Guenter Resch, IMP/IMBA Electron Microscopy Facility, Stefan Westermann and Angela Pickl-Herk, at the Max F. Perutz Laboratories in Vienna, Austria, provided specimens, images, and protocols from their experience using the Leica EM GP. Chen Xu, Rosentiel Basic Medical Sciences Research Center, Waltham, MA provided the image for Fig. 3.2B. This work was supported in part by NIH grant P01 GM066521 to G. J. J., the Beckman Institute at Caltech, and gifts to Caltech from the Gordon and Betty Moore Foundation and Agouron Institute. Finally, a special thank you to Ted and Chris Pella, Pella, Inc., for their bequest to the Jensen laboratory tomography database project.

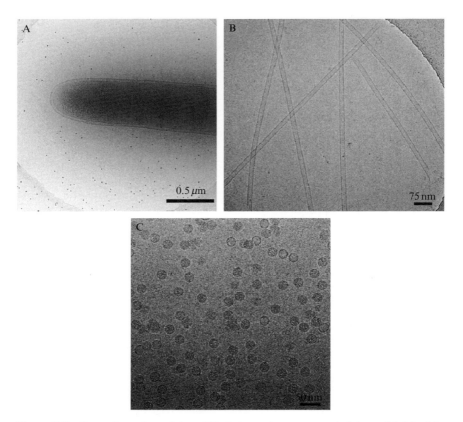

Figure 3.7 Examples of good ice. (A) A bacterium surrounded by gold fiducials, which are used to align tomographic tilt-series (grid bar 0.5 μm). (B) Microtubules, image courtesy of Dr. Guenter Resch (grid bar = 75 nm). (C) Rhinovirus particles labeled with Fab fragments, image courtesy of Angela Pickl-Herk (grid bar = 50 nm).

REFERENCES

Adrian, M., Dubochet, J., Lepault, J., and McDowall, A. W. (1984). Cryo-electron microscopy of viruses. *Nature* **308,** 32–36.

Aebi, U., and Pollard, T. D. (1987). A glow discharge unit to render electron microscope grids and other surfaces hydrophilic. *J. Electron Microsc. Tech.* **7,** 29–33.

Bellare, J. R., Haridas, M. M., and Li, X. J. (1999). Characterization of microemulsions using fast freeze-fracture and cryo-electron microscopy. *In* Handbook of Microemulsion Science and Technology pp. 411–436. Marcel Dekker, Inc., New York.

Booy, F. P., and Pawley, J. B. (1993). Cryo-crinkling: What happens to carbon films on copper grids at low temperature. *Ultramicroscopy* **48,** 273–280.

Cavalier, A., Spehner, D., and Humbel, B. M. (2009). Handbook of Cryo-Preparation Methods for Electron Microscopy. CRC Press, Boca Raton, FL.

Cerritelli, S., O'Neil, C. P., Velluto, D., Fontana, A., Adrian, M., Dubochet, J., and Hubbell, J. A. (2009). Aggregation behavior of poly(ethylene glycol-bl-propylene sulfide) di- and triblock copolymers in aqueous solution. *Langmuir* **25**, 11328–11335.

Dubochet, J., and McDowall, A. (1981). Vitrification of pure water for electron microscopy. *J. Microsc.* **124**, RP3–RP4.

Dubochet, J., Ducommun, M., Zollinger, M., and Kellenberger, E. (1971). A new preparation method for dark-field electron microscopy of biomacromolecules. *J. Ultrastruct. Res.* **35**, 147–167.

Dubochet, J., McDowall, A. W., Menge, B., Schmid, E. N., and Lickfeld, K. G. (1983). Electron microscopy of frozen-hydrated bacteria. *J. Bacteriol.* **155**, 381–390.

Dubochet, J., Adrian, M., Lepault, J. C., and McDowall, A. W. (1985). Emerging techniques: Cryo-electron microscopy of vitrified biological specimens. *Trends Biochem. Tech.* **10**, 143–146.

Dubochet, J., Adrian, M., Chang, J. J., Homo, J. C., Lepault, J., McDowall, A. W., and Schultz, P. (1988). Cryo-electron microscopy of vitrified specimens. *Q. Rev. Biophys.* **21**, 129–228.

Fernandez-Moran, H. (1960). Low-temperature preparation techniques for electron microscopy of biological specimens based on rapid freezing with liquid helium II. *Ann. NY Acad. Sci.* **85**, 689–713.

Finnigan, B., Halley, P., Jack, K., McDowall, A., Truss, R., Casey, P., Knott, R., and Martin, D. (2006). Effect of the average soft-segment length on the morphology and properties of segmented polyurethane nanocomposites. *J. Appl. Polym. Sci.* **102**, 128–139.

Frederik, P. M., and Hubert, D. H. (2005). Cryoelectron microscopy of liposomes. *Methods Enzymol.* **391**, 431–448.

Gan, L., Chen, S., and Jensen, G. (2008). Molecular organization of Gram-negative peptidoglycan. *Proc. Natl. Acad. Sci. USA* **105**, 18953–18957.

Glaeser, R., Downing, K., DeRosier, D., Chiu, W., and Frank, J. (2007). Electron Crystallography of Biological Macromolecules. Oxford University Press, New York, pp. 1–476.

Grassucci, R., Taylor, D. J., and Frank, J. (2007). Preparation of macromolecular complexes for cryo-electron microscopy. *Nat. Protoc.* **2**, 3239–3246.

Handley, D. A., Alexander, J. T., and Chien, S. (1981). The design and use of a simple device for rapid quench-freezing of biological samples. *J. Microsc.* **121**, 273–282.

Iancu, C. V., Tivol, W. F., Schooler, J., Dias, D. P., Henderson, G. P., Murphy, G. E., Wright, E., Li, Z., Yu, Z., Briegel, A., Gan, L., He, Y., *et al.* (2006). Electron cryotomography sample preparation using the Vitrobot. *Nat. Protoc.* **1**, 2813–2819.

Kumar, R., Singh, R. K., Kumar, M., and Barthwal, S. K. (2007). Effect of DC glow discharge treatment on the surface energy and surface resistivity of thin film of polypropylene. *J. Appl. Polym. Sci.* **104**, 767–772.

Lepper, S., Merkel, M., Sartori, A., Cyrklaff, M., and Frischknecht, F. (2010). Rapid quantification of the effects of blotting for correlation of light and cryo-light microscopy images. *J. Microsc.* **238**, 21–26.

McDowall, A. (1984). Ultracryotomy: An investigation of the cryotechnical problems involved in the preparation of frozen-hydrated cells and tissues for high resolution electron microscopy. PhD Thesis, Universite Pierre et Marie Curie, Paris, France.

McDowall, A. W., Chang, J. J., Freeman, R., Lepault, J., Walter, C. A., and Dubochet, J. (1983). Electron microscopy of frozen hydrated sections of vitreous ice and vitrified biological samples. *J. Microsc.* **131**, 1–9.

McDowall, A. W., Hofmann, W., Lepault, J., Adrian, M., and Dubochet, J. (1984). Cryo-electron microscopy of vitrified insect flight muscle. *J. Mol. Biol.* **178**, 105–111.

Melanson, L. (2009a). A versatile and affordable plunge freezing instrument for preparing frozen hydrated specimens for cryo transmission electron microscopy (CryoEM). *Microsc. Today* 14–17.

Melanson, L. (2009b). The importance of the specimen support film for cryo TEM. http://www.gatan.com/knowhow/knowhow_15/cryo.htm.

Quispe, J., Damiano, J., Mick, S. E., Nackashi, D. P., Fellmann, D., Ajero, T. G., Carragher, B., and Potter, C. S. (2007). An improved holey carbon film for cryo-electron microscopy. *Microsc. Microanal.* **13,** 365–371.

Taylor, K. A., and Glaeser, R. M. (1973). Hydrophilic support films of controlled thickness and composition. *Rev. Sci. Instrum.* **44,** 1546–1547.

Taylor, K. A., and Glaeser, R. M. (1974). Electron diffraction of frozen, hydrated protein crystals. *Science* **186,** 1036–1037.

Tivol, W. F., Briegel, A., and Jensen, G. (2008). An improved cryogen for plunge freezing. *Microsc. Microanal.* **14,** 375–379.

Yoshioka, Y., Carragher, B., and Potter, C. S. (2010). **Cryomesh**^{TM}: A new substrate for cryo-electron microscopy. *Microsc. Microanal.* **16,** 43–53.

CHAPTER FOUR

A Practical Guide to the Use of Monolayer Purification and Affinity Grids

Deborah F. Kelly,* Danijela Dukovski,*,† *and* Thomas Walz*,†

Contents

* Department of Cell Biology, Harvard Medical School, Boston, Massachusetts, USA
† Howard Hughes Medical Institute, Harvard Medical School, Boston, Massachusetts, USA

Methods in Enzymology, Volume 481
ISSN 0076-6879, DOI: 10.1016/S0076-6879(10)81004-3

Abstract

Lipid monolayers have traditionally been used in electron microscopy (EM) to form two-dimensional (2D) protein arrays for structural studies by electron crystallography. More recently, monolayers containing Nickel-nitrilotriacetic acid (Ni-NTA) lipids have been used to combine the purification and preparation of single-particle EM specimens of His-tagged proteins into a single, convenient step. This monolayer purification technique was further simplified by introducing the Affinity Grid, an EM grid that features a predeposited Ni-NTA lipid-containing monolayer. In this contribution, we provide a detailed description for the use of monolayer purification and Affinity Grids, discuss their advantages and limitations, and present examples to illustrate specific applications of the methods.

ABBREVIATIONS

2D	two-dimensional
AQP9	aquaporin-9
DLPC	1,2-dilauroyl-*sn*-glycero-3-phosphocholine
DMPC	1,2-dimyristoyl-*sn*-glycero-3-phosphocholine
DOPC	1,2-dioleoyl-*sn*-glycero-3-phosphocholine
EM	electron microscopy
GP1	glycoprotein subunit 1
IgG	immunoglobulin G
NECD	Notch extracellular domain
Ni-NTA	Nickel-nitrilotriacetic acid
RNAP II	RNA polymerase II
rpl3	ribosomal protein L3
STEM	scanning transmission electron microscopy
Tf–TfR	transferrin–transferrin receptor

1. A BRIEF HISTORY OF THE USE OF LIPID MONOLAYERS IN ELECTRON MICROSCOPY

In 1983, Uzgiris and Kornberg introduced the use of lipid monolayers to produce two-dimensional (2D) arrays of soluble proteins for structure determination by electron crystallography (Uzgiris and Kornberg, 1983). The principle underlying the approach is simple and elegant. When lipids (dissolved in an organic solvent) are added to an aqueous solution, they spontaneously form a monolayer at the air–water interface. If the protein of interest in the aqueous phase interacts with the polar head groups of the lipid monolayer, the protein becomes concentrated and partially oriented on the lipid monolayer. As the lipids diffuse laterally, the attached proteins rearrange and come in contact with one another. If the proteins make favorable interactions, they can form an ordered 2D array on the monolayer. Monolayer crystallization requires only nanogram quantities of purified protein and can be used to obtain 2D arrays under mild, near-physiological conditions. It has thus been used for structural studies on a wide variety of proteins (e.g., Asturias *et al.*, 1998; Brisson *et al.*, 1991; Darst *et al.*, 1989; Tang *et al.*, 2001; Taylor and Taylor, 1993).

While some proteins have an intrinsic affinity for natural lipid head groups that can be exploited for monolayer crystallization (e.g., Norville *et al.*, 2007; Olofsson *et al.*, 1990; Ward and Leonard, 1992), other proteins can interact with lipid head groups through unspecific, electrostatic complementation. Alternatively, proteins and lipids can be modified to engineer specific interactions. Lipids have been synthesized with head groups carrying functional groups, such as Nickel-nitrilotriacetic acid (Ni-NTA) or biotin. Such lipids can be incorporated into lipid monolayers to specifically adsorb proteins carrying the complementary tag, that is, a His or streptavidin tag. Kubalek *et al.* (1994) were the first to use a lipid monolayer containing Ni-NTA lipids to form 2D arrays of His-tagged recombinant HIV1 reverse transcriptase. Since then, a number of monolayer crystals were produced using both Ni-NTA lipids (e.g., Al-Kurdi *et al.*, 2004; Ganser *et al.*, 2003; Kelly *et al.*, 2006; Thess *et al.*, 2002; Venien-Bryan *et al.*, 1997) and biotinylated lipids (e.g., Avila-Sakar and Chiu, 1996; Qin *et al.*, 1995).

Monolayer crystallization requires very small quantities of purified protein, and crystals form within hours or days, but finding conditions that promote crystal formation is not trivial and can be very time-consuming. Furthermore, the order of monolayer crystals is often limited, which may be partially due to disorder introduced when the fragile crystals are transferred from the air–water interface to an EM grid. Monolayer crystals transferred to an EM grid are also often not flat enough to allow for high-resolution imaging of tilted specimens, causing severe problems for the calculation of 3D density maps.

Lipid monolayers are not only useful for growing 2D arrays but can also be used to prepare EM specimens of single particles. For example, using a positively charged lipid monolayer, it was possible to adsorb a sufficiently high number of large nuclear ribonucleoprotein particles to the monolayer to prepare vitrified specimens for cryo-EM imaging (Azubel *et al.*, 2004; Medalia *et al.*, 2002). In recent work from our laboratory, we further developed this idea and established monolayer purification (Kelly *et al.*, 2008a) and the Affinity Grid (Kelly *et al.*, 2008b). The idea behind these techniques is the use of monolayers containing Ni-NTA lipids to combine the purification of His-tagged proteins or complexes with the preparation of specimens for single-particle EM into a single, rapid step. In addition to the convenience of the monolayer purification and Affinity Grid techniques, the quick preparation of single-particle EM samples directly from cell extracts has proved beneficial for the visualization of complexes that cannot be produced in large quantities and labile complexes that do not survive lengthy biochemical purification procedures. In this contribution, we provide practical guidelines for the use of monolayer purification and Affinity Grids and discuss the advantages and limitations of these techniques as well as future prospects.

2. Preparation of Lipid Monolayer Specimens

Teflon blocks used to prepare lipid monolayer samples are currently not commercially available, but can easily be produced in workshops. The simplest design, typically used to prepare monolayer samples of soluble proteins, is a rectangular Teflon block containing circular wells (typically ~ 5 mm in diameter and ~ 1.3 mm deep), each accommodating a volume of ~ 25 μl depending on the exact dimensions of the well (Fig. 4.1A, panel 1). The wells are filled with a solution containing the protein of interest (Fig. 4.1A, panel 2). One microliter of a 1-mg/ml lipid mixture, consisting of a charged or functionalized lipid and a neutral filler lipid dissolved in organic solvent, is applied to the surface of the well with a Hamilton syringe. The lipids spontaneously form a monolayer, to which the proteins can adsorb (Fig. 4.1A, panel 3). In a slightly modified design, ports (~ 1 mm in diameter) are drilled into the Teflon block that connect to the wells at a $\sim 45°$ angle (Fig. 4.1B, panel 1; Levy *et al.*, 2001). In this case, the wells are filled with buffer solution, and the lipid mixture is added to form a monolayer (Fig. 4.1B, panel 2). Using a pipetman, 1–2 μl of protein solution is added to each well through the side ports, allowing the proteins to adsorb to the already formed monolayer (Fig. 4.1B, panel 3). Once a monolayer is formed at the air–water interface, it is somewhat less sensitive to detergents, which is the reason why this method is typically used when preparing monolayer samples of detergent-solubilized membrane proteins (Levy *et al.*, 2001).

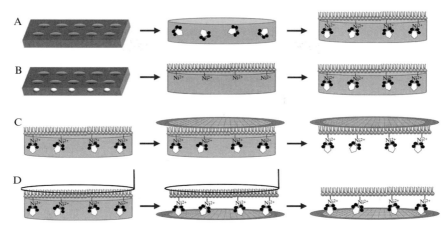

Figure 4.1 The lipid monolayer technique for binding macromolecular complexes and methods to transfer lipid monolayers to EM grids. (A, B) Different designs of Teflon blocks can be used to set up lipid monolayers. With the simplest design, a Teflon block with circular holes (A, panel 1), the wells are filled with solution containing the His-tagged target protein (A, panel 2), and a Ni-NTA lipid-containing monolayer is cast over the well for the proteins to adsorb (A, panel 3). In a different design, the Teflon block features side ports that connect to the wells (B, panel 1). In this case, the wells are filled with buffer solution and the Ni-NTA lipid-containing monolayer is cast (B, panel 2). The solution containing the His-tagged protein is finally injected through the side port, allowing it to adsorb to the lipid monolayer (B, panel 3). (C, D) Once prepared, two methods can be used to transfer monolayer samples to EM grids. In the direct method (C), a hydrophobic grid is placed on the lipid monolayer (C, panel 2), which attaches to the carbon film through the acyl chains of the lipids (C, panel 3). In the loop transfer (D), the monolayer is lifted off the well (D, panel 1) and placed on a hydrophilic grid (D, panel 2), causing the attached proteins to interact with the carbon film (D, panel 3).

Once set up, the Teflon blocks containing the monolayer samples are incubated for at least 20 min and up to several hours at 4 °C or at room temperature, depending on the stability of the protein sample. To avoid excessive evaporation of the sample solution during incubation, the Teflon block is placed in a Petri dish containing a wet filter paper, which is then sealed with parafilm.

After adsorption of the proteins or complexes, the monolayer sample is transferred to an EM grid using either the direct transfer method (Uzgiris and Kornberg, 1983; Fig. 4.1C) or the loop-transfer method (Asturias and Kornberg, 1995; Fig. 4.1D). In the direct transfer method, an EM grid covered with a hydrophobic (not glow-discharged) carbon film is placed on the well, with its carbon side touching the lipid monolayer (Fig. 4.1C, panel 2). After 30 s to 1 min, the grid (with the hydrophobic acyl chains of the lipid monolayer attached to the hydrophobic carbon film) is carefully lifted off with a pair of forceps (Fig. 4.1C, panel 3). For the loop-transfer method, a platinum wire loop with a diameter slightly larger than the diameter of the sample well is used. The loop is placed on top of the sample well (Fig. 4.1D, panel 1) and the lipid monolayer is lifted up from the Teflon block. The monolayer is carefully placed on the carbon side of an EM grid covered with a hydrophilic (glow-discharged) carbon film (Fig. 4.1D, panel 2) and the loop is carefully removed (Fig. 4.1D, panel 3).

To prepare monolayer samples by negative staining or cryo–negative staining, continuous carbon grids are commonly used. Holey carbon grids are typically used for the preparation of vitrified specimens. In this case, the direct transfer method is used, and the grid is placed on top of the lipid monolayer for 1 min, with the side of the grid not covered by the holey carbon film touching the lipid monolayer.

2.1. Practical considerations

2.1.1. Preparation of lipid solutions
We purchase lipids as lyophilized powders from Avanti Polar Lipids (Alabaster, Alabama), reconstitute them in chloroform at a stock concentration of 1 mg/ml in glass test tubes, and store the reconstituted lipid stocks at − 20 °C until use.

2.1.2. Filler lipids
1,2-dilauroyl-*sn*-glycero-3-phosphocholine (DLPC), 1,2-dimyristoyl-*sn*-glycero-3-phosphocholine (DMPC), and 1,2-dioleoyl-*sn*-glycero-3-phosphocholine (DOPC) as well as other lipids can be used as neutral filler lipids. The length and saturation of the acyl chains influence the phase transition temperature of the lipids. While maintaining the monolayer in a fluid state is necessary for array formation, it is not as important for producing single-particle specimens.

2.1.3. Active lipids
Lipids with positively or negatively charged head groups or with functionalized head groups, for example, Ni–NTA or biotinylated lipids, are used as active lipids to recruit the target protein to the lipid monolayer. The percentage of active lipid added to the monolayer determines how much protein can bind to the monolayer.

2.1.4. Sample level
A sufficiently high level of the sample in the well is essential to ensure that the grid can be properly lifted off from the well. It is therefore important to make sure that sufficient sample is added to each well and that the incubation is carried out in a humid environment to prevent excessive evaporation of the sample solution.

2.1.5. Lifting of the grid

In the direct transfer method, the grid should be lifted from the sample solution in one smooth motion perpendicular to the sample. Improper lifting of the EM grid can result in uneven transfer of the lipid monolayer to the grid. Baking the grid to render the carbon film more hydrophobic can help in accomplishing a more complete transfer of the lipid monolayer to the EM grid.

3. MONOLAYER PURIFICATION

Recently, we introduced monolayer purification as a method that exploits Ni-NTA lipid-containing monolayers to produce single-particle EM specimens of His-tagged proteins and complexes without prior biochemical purification (Kelly *et al.*, 2008a). The method was first established by adsorbing His-tagged transferrin–transferrin receptor (Tf–TfR) complexes that were added to cell extracts and conditioned media to Ni-NTA lipid-containing monolayers. Monolayer purification was then further tested by isolating ribosomal complexes from an extract of *Escherichia coli* cells expressing a His-tagged construct of the human ribosomal subunit rpl3. In both cases, highly pure samples could be prepared and imaged by negative stain and cryo-EM. A general protocol for preparing single-particle EM specimens by monolayer purification is provided in Section 3.1.

3.1. Protocol for monolayer purification

1. Centrifuge cell culture at $3500 \times g$ for 20 min at 4 °C and resuspend cell pellet in an appropriate (detergent-free) lysis buffer (1/10 of the original cell volume).
2. Remove insoluble cell debris by centrifugation at $100,000 \times g$ for 30 min at 4 °C.
3. Add imidazole to the cleared supernatant to a final concentration of 20–50 mM.
4. Add ~ 25 μl aliquots of the cell extract to sample wells in a Teflon block.
5. Apply 1 μl of a Ni-NTA lipid-containing lipid mixture (1 mg/ml in chloroform) to each sample well and incubate for 20 min at 4 °C, with the Teflon block placed in a Petri dish containing a wet filter paper and sealed with parafilm.
6. Apply an EM grid to the monolayer sample and allow the grid to incubate for approximately 30 s for continuous carbon grids or 1 min for holey carbon grids.
7. For negative staining, lift the grid off the sample well and blot from the side with Whatman #1 filter paper (Whatman International Ltd.,

Middlesex, England). Add a drop of negative stain solution (ideally
0.75% uranyl formate solution) to the grid and immediately blot away
with filter paper. Stain the sample with another drop of staining solution
for 30 s before blotting from the side with filter paper.
8. For vitrification, lift the grid off the sample well, blot from both sides,
and plunge into liquid ethane (see McDowall *et al.*, 2010).

3.2. Practical considerations

3.2.1. Particle density

The density of particles on the monolayer can be adjusted by varying the
incubation time and the percentage of Ni–NTA lipid in the monolayer.
Two percent Ni–NTA lipid and DLPC as filler lipid usually produces a
good particle distribution for negatively stained specimens, but vitrified
specimens tend to require a 10 times higher Ni–NTA lipid concentration
(20%) to obtain the same particle density. Potential reasons for the lower
number of particles seen in vitrified than in negatively stained specimens at
the same Ni–NTA lipid concentration may be that particles are removed by
the double-sided blotting of the grid or that particles are ripped off the lipid
monolayer as the grid is plunged into liquid ethane.

3.2.2. Unspecific background binding

Imidazole is added to the cell extract to suppress unspecific binding of
proteins to the Ni–NTA lipid-containing monolayer. The best imidazole
concentration has to be determined empirically. An imidazole concentration
that is too low increases background binding, while too high a concentration may interfere with the binding of the target protein. If imidazole does
not completely eliminate unspecific binding, potential solutions are to
dilute the cell extract used for monolayer purification or to partially purify
the His-tagged target protein using chromatographic methods.

3.3. Advantages of monolayer purification

Monolayer purification is fast and convenient, and allows preparation of
specimens suitable for single-particle EM without lengthy biochemical
purifications. In addition to its speed, making the method particularly
suitable for preparing labile and transient complexes directly from cell
extracts, monolayer purification has several additional advantages that may
make the use of Ni–NTA lipid-containing monolayers beneficial even for
the preparation of vitrified specimens from pure protein samples.

Since the Ni–NTA lipids specifically bind the His-tagged proteins, prolonged incubations can produce specimens with a sufficient number of particles
even if the protein is present at only very low concentrations. This feature

makes the technique very useful for the preparation of proteins or complexes that exist only in low abundance or that can only be expressed at low levels.

Furthermore, structural heterogeneity is a severe problem in the structure determination of multisubunit complexes by single-particle EM. Monolayer purification could potentially help to alleviate this problem by strategically selecting the protein that is being His-tagged. For example, one could tag subunits, substrates, or activators that are particularly prone to dissociate from the complex. As a result, all the complexes recruited to the monolayer and seen in EM images are guaranteed to contain the tagged subunit, substrate, or activator.

Finally, the lipid monolayer also has advantages for the vitrification of EM specimens. Due to the specific attachment of the His-tagged proteins to the lipid monolayer, monolayer samples can usually be blotted for longer time periods and still maintain thin ice and a useful particle concentration. The lipid monolayer also helps to obtain a more even ice thickness across holes, and once ideal freezing conditions have been established, the preparation of vitrified monolayer samples is very reproducible. Most importantly, the attachment of the proteins to the monolayer prevents them from coming into contact with the air–water interface, thus protecting the complexes from the disruptive surface tension forces that can cause complexes to dissociate and even proteins to denature (Taylor and Glaeser, 2008).

3.4. Limitations of monolayer purification

Monolayer purification can substantially simplify the preparation of labile complexes for EM imaging, but the method also has inherent limitations. In particular, the presence of a lipid monolayer will add noise to EM images. While noise is averaged out during image processing, it does reduce the contrast of the raw images, which can make it more difficult to see particles in vitrified specimens and potentially limit the achievable resolution.

Another unavoidable issue is that the Ni–NTA lipids in the monolayer will recruit any complex that contains the His-tagged subunit, including partial complexes as well as different complexes containing the same His-tagged subunit. The resulting particle population can provide biological insights as it may identify different complexes a particular protein can associate with or reveal intermediates on the assembly pathway to the fully assembled, biologically active complex. On the other hand, the structural heterogeneity of the complexes adsorbed on the monolayer can make structure analysis using vitrified specimens very difficult. One way to deal with this problem is to prepare monolayer purification samples by negative staining, which often causes the particles to adopt preferred orientations and is thus beneficial for heterogeneous specimens (Ohi *et al.*, 2004). By recording image pairs of tilted and untilted specimens, the particles selected from the images of the untilted specimen can be used to classify the particles into structurally homogeneous groups and the corresponding particles selected from the images of the tilted specimen can be used to calculate 3D reconstructions of the individual particle classes with the random conical tilt approach (Radermacher *et al.*, 1987). The density maps for the different complexes obtained with the negatively stained specimens can then be used as initial references to classify and align particles selected from vitrified specimens.

As the complexes are recruited to the monolayer through the His-tagged subunit, monolayer purification can potentially lead to preferred orientations of the adsorbed particles even in vitrified specimens. However, His tags are typically added to one of the termini of the protein and since these tend to be flexible and to allow for significant motional freedom, preferred orientations are not often observed if specimens prepared by monolayer purification are vitrified. If preferred orientations occur, this problem can be addressed by introducing long, flexible linker sequences between the His tag and the protein. Preferred orientations pose a more severe problem if a complex contains more than one copy of the His-tagged subunit. The introduction of long, flexible linkers may still overcome the problem, but if possible, subunits should be chosen for tagging that exist in only one copy in the target complex.

Lipid monolayers are not compatible with all buffer components. In particular, detergents, mainly used to keep membrane proteins stable in solution, and glycerol, often added to protein solutions to minimize aggregation, tend to have a detrimental effect on the integrity of a lipid monolayer and are thus not compatible with monolayer purification.

Finally, the lipids in the monolayer diffuse and attached proteins move along with the lipids. While this feature is the basis for the formation of ordered protein arrays on monolayers, it often causes proteins to cluster and is thus counterproductive for the preparation of specimens for single-particle EM, in which particles should be well separated from one another.

4. EXAMPLES OF MONOLAYER PURIFICATION APPLICATIONS

After establishing monolayer purification using the Tf–TfR complex, the technique was used to examine a variety of complexes. To illustrate the properties of monolayer purification, we describe here the preparations of ribosomal complexes from clarified cell extract and of human C complex from partially purified nuclear extract.

4.1. Ribosomal complexes — Purifying complexes from cell extract

To isolate ribosomal complexes, an N-terminally His-tagged construct of the human 60S ribosomal protein L3 (rpl3) was expressed in *E. coli*. The rpl3 protein is highly homologous to the *E. coli* subunit, rplC, but has 47 additional residues at its N terminus. After lysing the cells, the clarified extract was overlaid with a Ni–NTA lipid-containing monolayer, which was then transferred onto an EM grid and vitrified. EM images revealed large particles, suggesting that the human homolog of rplC was indeed incorporated into the *E. coli* large subunit and could be used to isolate ribosomal complexes (Fig. 4.2A). Averages identified both the 50S subunit and 70S ribosome to be present in the preparation. Using the previously determined X-ray structure of the *E. coli* ribosome (Klaholz *et al.*, 2003) as an initial model, it was possible to calculate a 3D reconstruction of the 50S ribosomal subunit at 22 Å resolution without purifying it by chromatographic means (Kelly *et al.*, 2008a; Fig. 4.2B).

4.2. C complex — Preparing labile complexes that usually cannot be vitrified

Many macromolecular complexes occur in low abundance or are labile. Such complexes are difficult to purify in concentrations suitable for vitrification. Furthermore, the surface tension at the air–water interface can

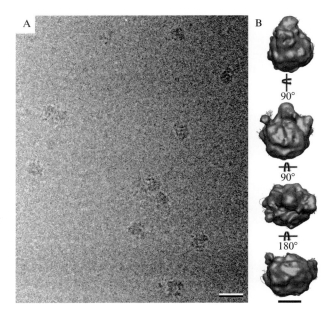

Figure 4.2 Ribosomal complexes prepared from *E. coli* cell extract using monolayer purification. (A) Image of vitrified ribosomal complexes prepared by monolayer purification by means of a His tag on the rpl3 subunit. Scale bar is 30 nm. (B) Different views of the 3D reconstruction of the 50S subunit (gold surface) with the fit crystal structure (red; Klaholz *et al.*, 2003). Scale bar is 10 nm. (See Color Insert.)

Figure 4.3 Image of vitrified human C complex prepared by monolayer purification from a partially purified nuclear extract. Vitrification of the complex using the traditional technique was unsuccessful. Scale bar is 40 nm.

disrupt labile complexes. Spliceosomes are a typical example of complexes that are very difficult to vitrify. One solution is to chemically fix spliceosomes before vitrification (e.g., Kastner *et al.*, 2008). Monolayer purification provides an alternative to chemical fixation and was used to vitrify human C complex. Using commercially available nuclear extracts and by adding the tandem His-tagged-MBP-MS2 RNA stem loop affinity tag (Stroupe *et al.*, 2009) to the arrested mRNA splicing complex (Jurica *et al.*, 2004), it was possible to use Ni-NTA lipid-containing monolayers to purify and to concentrate the splicing complexes, producing vitrified samples suitable for cryo-EM imaging (Fig. 4.3).

 ## 5. THE AFFINITY GRID

Monolayer purification requires that a fresh lipid monolayer is set up every time a specimen is being prepared for EM imaging. The Affinity Grid was conceived as a simplification of monolayer purification by fabricating an EM grid that features a predeposited Ni-NTA lipid-containing monolayer (Kelly *et al.*, 2008b). Since Affinity Grids can be dried and stored for several months under ambient conditions, they can be produced in large batches and used whenever needed. The Affinity Grid was tested by isolating ribosomal complexes from cell extracts of *E. coli* expressing a His-tagged version of the human rpl3 subunit (Kelly *et al.*, 2008b). In addition to simplifying monolayer purification, Affinity Grids turned out to have a number of further advantages that will be discussed in Section 5.5.

Affinity Grids are prepared in the same way as described for monolayer purification, using 25 μl of distilled water (or sample buffer specific for the target protein or complex) as the aqueous phase in the sample well. After addition of the Ni-NTA lipid-containing lipid mixture to the sample well, the monolayer is allowed to equilibrate for 20 min at 4 °C in a humid environment. A non-glow-discharged EM grid is applied to the monolayer and allowed to incubate for 30 s for continuous carbon grids (carbon touching the monolayer) or 1 min for holey carbon grids (grid-bar side touching the monolayer). After lifting off the grid, a Hamilton syringe is used to remove excess fluid from the grid (typically ~ 2 μl), and the grid is allowed to air-dry. For Affinity Grids prepared with a holey carbon film, a thin layer of carbon (~ 1–2 nm thick) is evaporated onto the grid (on the opposite side of where the sample will be applied to) for additional stability during storage.

5.1. Use of Affinity Grids for His-tagged complexes

Affinity Grids are very easy to use. A ~ 3-μl drop of sample solution is simply applied to the side of the grid that is coated with the lipid monolayer (the grid should not be glow-discharged). Because of the rapid binding of His-tagged complexes to the Ni-NTA lipids and the small sample volume, Affinity Grids typically have to be incubated with the sample for only 2–5 min. If His-tagged complexes are to be prepared directly from cell extracts, clarified extract should be prepared as described for monolayer purification, and imidazole should be added to the extract to a final concentration of 20–50 mM to suppress unspecific binding. After incubation, the grid can be prepared by negative staining or vitrification as described for monolayer purification. Although Affinity Grids are more resistant to glycerol, they do not withstand the high concentrations needed for cryonegative staining (de Carlo and Stark, 2010; Golas et al., 2003; Ohi et al., 2004). For a modified version of cryo-negative staining, after incubation, 2 μl of the sample solution is removed from the Affinity Grid with a Hamilton syringe, and a 3-μl drop of solution containing a 1:1 mixture of 2% (w/v) uranyl formate and 2% (w/v) trehalose is added to the grid. The grid is blotted from both sides and plunged into liquid ethane.

5.2. Use of Affinity Grids for complexes without a His tag

The functionality of Affinity Grids is based on Ni–NTA lipids, thus limiting their use to His-tagged proteins. His-tagged adaptor molecules make it possible, however, to use Affinity Grids for the preparation of proteins or complexes that do not carry a His tag themselves (Kelly et al., 2010a). The most general strategy is to use His-tagged protein A, which binds to the constant domain of mammalian immunoglobulin G (IgG) antibodies (Sjoquist and Stalenheim, 1969). Thus, as long as a specific antibody exists or can be raised against a protein or a particular tag on the protein, the antibody can be immobilized on the Affinity Grid through His-tagged protein A, and the antibody-coated Affinity Grid can then be used for the rapid purification of untagged or alternatively tagged target proteins or complexes from cell extracts. A general protocol for purifying complexes from cell extracts using Affinity Grids combined with His-tagged protein A and a specific antibody is provided in Section 5.3.

5.3. Protocol for the use of Affinity Grids to purify non-His-tagged proteins

1. Prepare clarified, imidazole-containing cell extract, as described in Section 3.1.
2. Prepare a 0.1-mg/ml solution of His-tagged protein A and several dilutions of the specific antibody in buffer compatible with the target protein or complex.
3. Add 3 μl of His-tagged protein A solution to an Affinity Grid and incubate for 1 min at room temperature.
4. Add a 3-μl aliquot of the specific antibody to the Affinity Grid and incubate for 1 min at room temperature (various dilutions of the antibody, typically from 0.01 to 0.2 mg/ml, may need to be tested).
5. Use a Hamilton syringe to remove most of the protein A and antibody solution ($\sim$5 μl), apply 3 μl of cell extract to the Affinity Grid, and incubate for 2–5 min.
6. Prepare the Affinity Grid by negative staining, cryo–negative staining, or vitrification, as described earlier.

5.4. Practical considerations

5.4.1. Particle density
Since Affinity Grids are prefabricated, it is not as straightforward to adjust the Ni-NTA lipid concentration in the monolayer as in the case of monolayer purification. It is therefore useful to prepare Affinity Grids with different Ni-NTA lipid concentrations and to test several of these Affinity Grids. Once the Affinity Grid with the most suitable Ni-NTA lipid concentration has been identified for the preparation of a particular specimen, the density of the particles adsorbed to the grid can be fine-tuned by varying the incubation time with the cell extract.

5.4.2. Antibody concentration and incubation time
Our experience suggests that for the use of Affinity Grids with His-tagged protein A, it is best to use an Affinity Grid with a Ni-NTA lipid content of 2% for negative staining and 20% for vitrification and to incubate it for

1 min with a 0.1 mg/ml protein A solution. On the other hand, the optimal concentrations of the antibody solution and the cell extract as well as the optimal incubation time with cell extract have to be determined empirically, as the resulting number of adsorbed particles will depend strongly on the affinity of the antibody for the target protein.

5.4.3. Protein background

The incubations with His-tagged protein A and antibody can lead to a high protein background. While this background will eventually be eliminated during image averaging, too high a background may complicate particle picking or prevent the determination of a high-resolution density map.

5.5. Advantages of the Affinity Grid

The option to prefabricate and store Affinity Grids until use makes the application of monolayer purification easier and more convenient, but Affinity Grids have additional advantages. Because the lipid monolayer is attached to a carbon film, the lipids no longer diffuse, and the particle clustering seen with monolayer purification does not occur with Affinity Grids. Without particle clustering and with the number of binding sites defined by the percentage of Ni-NTA lipids in the monolayer, it should be possible to use Affinity Grids for the preparation of single-particle EM specimens even from highly concentrated protein solutions without crowding the particles on the grid. Although not yet tested, this feature may be particularly useful for preparing EM specimens of complexes and protein oligomers that dissociate at low concentrations. Another advantage of the lipid monolayer being attached to a carbon film is that the monolayer becomes more resistant to detergents and glycerol, which makes it possible to use Affinity Grids for the preparation of detergent-solubilized membrane proteins from membrane extracts (Kelly *et al.*, 2008b).

Affinity Grids also appear to be milder for the preparation of complexes than monolayer purification. For example, many ribosomal complexes adsorbed to an Affinity Grid directly from *E. coli* cell extract were still attached to RNA strands, which was not observed when the ribosomal complexes were prepared using monolayer purification.

A particular advantage of Affinity Grids is the ease with which complexes can be assembled on the monolayer in a stepwise manner. It is thus possible to attach a His-tagged ligand or substrate to the Affinity Grid and then use the decorated Affinity Grid to fish out interacting proteins or complexes from a cell extract. The Affinity Grid may thus potentially be used as an alternative to immunoprecipitation to determine whether two molecules interact with each other. The only requirement is that one of the components must be His-tagged. Interaction between the two components, either tested by incubating Affinity Grids with purified components or by using Affinity Grids to recruit the complex directly from cell extracts, can be verified by EM imaging and/or by mass spectrometry (see below).

The possibility to assemble complexes on Affinity Grids in a step-wise fashion is the basis for using His-tagged adaptor proteins to isolate proteins or complexes that are either untagged or have a tag other than a His tag. His-tagged protein A may be the most universal adaptor, as it allows virtually any molecule to be recruited to the Affinity Grid provided a specific antibody exists or can be raised against the target protein or its tag. Because the specificity of the antibody defines what complexes are recruited, Affinity Grids decorated with different antibodies can potentially be used to recruit different complexes from the same cell extract. Furthermore, His-tagged adaptor proteins other than protein A may be useful. For example, His-tagged calmodulin could potentially be used to recruit proteins carrying a calmodulin-binding domain/protein A Tandem Affinity Purification (TAP) tag and His-tagged avidin could be used to recruit biotinylated targets.

5.6. Limitations of Affinity Grids

The Affinity Grid remedies some of the issues of the underlying monolayer purification technique, such as particle clustering and the incompatibility with detergents and glycerol. Furthermore, the introduction of His-tagged adaptor proteins removes the requirement for the target protein or complex to carry a His tag, thus extending the use of Affinity Grids to the preparation of native complexes that lack a His tag or to complexes that contain a different tag. The additional requirement for this approach to work is the availability of a protein that interacts with the target protein or complex and can thus be used as His-tagged adaptor. This requirement is substantially simplified when His-tagged protein A is used as adaptor, as it is usually possible to raise antibodies against the target protein, if such antibodies do not already exist. However, because the Affinity Grid is based on the same fundamental principles as those of monolayer purification, some limitations remain the same. Issues that persist are the added background noise caused by the presence of the lipid monolayer, the potential recruitment of more than one kind of complex to the Affinity Grid, and the potential that adsorbed particles show preferred orientations.

6. EXAMPLES OF AFFINITY GRID APPLICATIONS

Affinity Grids have already been used to prepare a number of different complexes. The following examples were chosen to illustrate the advantages of Affinity Grids, especially in comparison to the monolayer purification technique.

6.1. The Tf–TfR complex—No particle clustering

When Tf–TfR complex was added to Sf9 cell extract, prepared by mono-layer purification, and vitrified, most of the complexes were found in clusters (Kelly *et al.*, 2008a; Fig. 4.4A). This particle clustering presumably results from the diffusion of the Ni-NTA lipids in the monolayer, allowing the attached complexes to come in contact with one another and form clusters. In the case of Affinity Grids, the lipid monolayer is fixed to a carbon film, preventing diffusion of the lipids and thus clustering of the attached particles. Tf–TfR complex isolated from Sf9 cell extract using Affinity Grids showed much more homogeneously distributed particles, ideal for structure determination (Kelly *et al.*, 2008b; Fig. 4.4B).

6.2. The complex of Tf–TfR with GP1—Recruiting targets using tagged ligands

Often, interactions of ligands with their receptors are not strong enough for the complex to survive purification by gel filtration chromatography. Such weak interactions are particularly problematic for EM studies if the ligand is small, making it difficult, for example, to decide whether the observed particles are ligand-bound or unliganded receptors. By decorating the Affinity Grid with a small, His-tagged ligand, which is then used to recruit the receptor, the Affinity Grid can be used to ensure that all the receptors seen in EM images are bound to ligand, producing a homogeneous particle population. This strategy was tested using a His-tagged domain (residues 79–258) of the glycoprotein subunit 1 (GP1) from the Machupo virus envelope, which has recently been shown to bind to the transferrin receptor (Abraham *et al.*, 2010). As the GP1 construct has a molecular weight of only

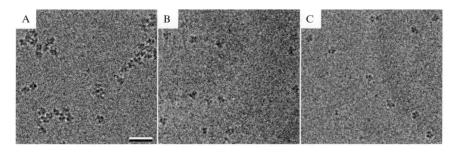

Figure 4.4 Tf–TfR complexes in vitrified ice. (A) Image of His-tagged Tf–TfR complexes isolated by monolayer purification, showing the clustering of particles often observed with this technique. (B) Image of His-tagged Tf–TfR complexes isolated using an Affinity Grid, which eliminates particle clustering. (C) Image of untagged Tf–TfR complex prepared using an Affinity Grid that was decorated with His-tagged GP1. Scale bar is 40 nm.

19 kDa, small compared to the ~300-kDa Tf–TfR complex, it is impossible from raw images to judge whether the Tf–TfR complexes have the GP1 domain bound or not. Decorating an Affinity Grid with His-tagged GP1 and then incubating the grid with untagged Tf–TfR complex produced vitrified EM specimens clearly showing individual Tf–TfR complexes, which by design have to be associated with GP1 (Fig. 4.4C). Images of Affinity Grids prepared without prior incubation with the His-tagged GP1 construct showed no Tf–TfR complexes (not shown).

6.3. Ribosomal complexes—Gentler purification of complexes

Ribosomal complexes containing the His-tagged rpl3 subunit were initially isolated from *E. coli* extract using monolayer purification (Kelly *et al.*, 2008a; Figs. 4.2 and 4.5A). Ribosomal complexes prepared with Affinity Grids from the same *E. coli* extract showed that now many of the complexes were attached to RNA molecules (Kelly *et al.*, 2008b; Fig. 4.5B). This result suggests that preparation of complexes with Affinity Grids is gentler than with monolayer purification, preserving even easily disrupted interactions.

6.4. Aquaporin-9—Purification of membrane proteins

Lipid monolayers cannot be formed at the air–water interface if the sample solution contains detergents. Attachment of the lipid monolayer to the carbon film of the Affinity Grid makes the monolayer more resistant to detergents (for details, see Kelly *et al.*, 2008b). It was thus possible to use Affinity Grids to isolate His-tagged aquaporin-9 (AQP9) directly from octyl glucoside-solubilized membranes prepared from AQP9-expressing Sf9 cells. The negatively stained samples obtained with Affinity Grids as well as the

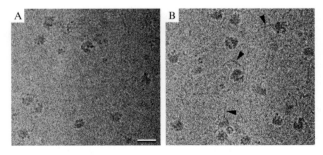

Figure 4.5 Ribosomal complexes in vitrified ice. (A) Image of His-tagged ribosomal complexes isolated by monolayer purification, showing the absence of RNA molecules. (B) Image of His-tagged ribosomal complexes isolated using an Affinity Grid, which preserves their interaction with RNA (indicated by arrows). Scale bar is 30 nm.

resulting class averages were virtually indistinguishable from those obtained by conventionally prepared EM specimens of chromatographically purified AQP9 (Kelly *et al.*, 2008b).

6.5. The Notch extracellular domain—Purification of proteins with low expression levels

Some types of proteins, such as membrane proteins, very large proteins, proteins containing many disulfide bridges, and heavily glycosylated proteins, can be difficult to express in large quantities. Low expression levels can make it impossible to purify recombinant proteins chromatographically, even if the proteins carry a tag that would allow affinity chromatography to be used for purification. The 186-kDa Notch extracellular domain (NECD), containing 36 EGF domains with up to three potential disulfide bonds each, is a typical example of a recombinant protein that can only be expressed at a low level (<0.05 mg/l cell culture). Despite its C-terminal tandem Flag-His$_6$ tag, chromatographic purification of the NECD was thus not successful. Using Affinity Grids, it was possible to prepare cryo-negatively stained EM specimens of the NECD directly from conditioned Sf9 cell medium (Fig. 4.6A) and to produce density maps of the NECD in different conformations using the random conical tilt approach (Kelly *et al.*, 2010b; Fig. 4.6B).

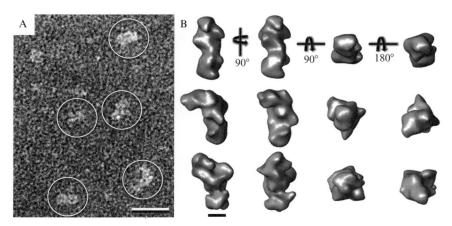

Figure 4.6 Isolation of NECD using an Affinity Grid. (A) Image of NECD particles prepared using an Affinity Grid from conditioned Sf9 cell medium and stained using a modified cryo–negative staining procedure. Scale bar is 25 nm. (B) Different views of the 3D reconstructions of the NECD dimer in three distinct conformations. Scale bar is 5 nm.

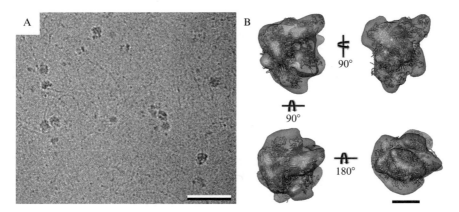

Figure 4.7 Isolation of mammalian RNAP II using an Affinity Grid in combination with His-tagged protein A and a specific antibody against the largest subunit. (A) Image of vitrified RNAP II particles attached to the Affinity Grid through His-tagged protein A and a specific IgG. Scale bar is 40 nm. (B) Different views of the 3D reconstruction (gray surface), which accommodates the crystal structures of the yeast RNAP II (green; Kettenberger *et al.*, 2004). Scale bar is 5 nm. (See Color Insert.)

6.6. RNA polymerase—Using His-tagged adaptors to purify untagged complexes

Potentially, the greatest advantage of Affinity Grids is the option to build complexes and to use His-tagged adaptors to prepare EM specimens of untagged complexes. This strategy was tested by isolating native, untagged RNA polymerase II (RNAP II) from mammalian cell extract. Affinity Grids were decorated with His-tagged protein A (from *Staphylococcus aureus* and expressed in *E. coli*), to which commercially available IgG antibodies against the C-terminal domain of the largest subunit of human RNAP II were bound. The Affinity Grid primed in this way was then used to adsorb native RNAP II particles from an extract of mammalian 293T cells. Images of the vitrified specimens (Fig. 4.7A) could be used to calculate a 3D density map that was consistent with the X-ray crystal structure of yeast RNAP II (pdb entry, 1Y77; Kelly *et al.*, 2010a; Kettenberger *et al.*, 2004; Fig. 4.7B).

7. CHARACTERIZATION OF MONOLAYER SPECIMENS

The use of monolayer purification and Affinity Grids makes it possible to prepare single-particle EM specimens directly from cell extracts. While very convenient, the lack of a biochemical purification scheme with its associated characterization of the purified protein or complex makes it essential to determine what proteins have actually been recruited to the lipid monolayer. Methods have thus been developed to analyze the adsorbed proteins by SDS-PAGE and mass spectrometry as well as by scanning transmission electron microscopy (STEM).

7.1. SDS-PAGE and mass spectrometry

To analyze the adsorbed proteins, they can be eluted from the lipid mono-layer by a high concentration of imidazole. Because of the small amount of protein adsorbed to an individual lipid monolayer sample, it is necessary to elute the protein from several monolayer specimens. Furthermore, the elution volume has to be kept small to obtain a sufficiently high protein concentration for analysis. To accomplish this task, a 20-μl drop of a 300 mM imidazole solution is added to a freshly prepared monolayer sample and incubated for 2 min to allow the proteins to be released from the Ni-NTA lipids. The 20-μl drop is removed from the grid with a Hamilton syringe and added to the next freshly prepared monolayer sample. This procedure is repeated until 20 or more monolayer samples have been eluted into the same 20-μl drop of 300 mM imidazole solution. To maximize the amount of protein that will be available for analysis, it is best to use monolayers with a high percentage of Ni-NTA lipids. The identity of the proteins released from the lipid monolayers can then be determined by SDS-PAGE, Western blot, and mass spectrometry analyses. This method was used, for example, to analyze both the Tf–TfR and ribosomal com-plexes, verifying their purity (Kelly et al., 2008a,b).

7.2. STEM analysis

STEM is a method that can be used to determine the molecular mass of protein complexes from images of unstained molecules (Muller and Engel, 2001; Thomas et al., 1994). The method can thus also be used to determine the mass of molecules bound to lipid monolayers. Mass determination at the STEM facility at Brookhaven National Laboratory requires the use of special grids. Ni-NTA lipid-containing monolayers can be deposited on these grids in the same way as described for conventional EM grids, thus creating a "STEM Affinity Grid." After adsorbing His-tagged molecules to a STEM Affinity Grid, the specimen is freeze-dried and used for STEM analysis according to the procedures described on the Web site of the Brookhaven STEM facility http://www.bnl.gov/biology/stem/.

Using STEM Affinity Grids to isolate NECD from conditioned Sf9 cell medium, it was possible to determine its molecular mass to be 381 ± 23 kDa, establishing that soluble NECD forms a dimer (expected mass ∼380 kDa; Kelly et al., 2010b).

8. CHALLENGES AND FUTURE DIRECTIONS

In their current form, monolayer purification and Affinity Grids have already proved very useful for preparing specimens for single-particle EM analysis, but challenges remain, and there are many directions for potential developments that could make lipid monolayers even more powerful for single-particle EM studies.

The greatest limitation may be that Affinity Grids can currently only be produced by hand, essentially one at a time. A method for the automated production of large quantities of Affinity Grids is lacking and would be highly desirable. Automated fabrication of large numbers of Affinity Grids would most likely make the quality of Affinity Grids more reproducible as well as open the way for their use in current efforts for high-throughput structure determination of macromolecular complexes by single-particle EM. While much progress has been made in the automation of data collection and image processing (e.g., Carragher *et al.*, 2010; Stagg *et al.*, 2006), the biochemical purification of macromolecular complexes suitable for structural studies remains a major bottleneck. The convenience and versatility of Affinity Grids, which make biochemical purification of complexes unnecessary, may enable the production of a large number of EM specimens in a short time, especially when combined with the already available libraries of tagged constructs.

Currently, DLPC is typically used as filler lipid and Ni–NTA lipid as the active lipid, but there are many directions to explore for improving the characteristics of the lipid monolayer. For instance, when DLPC and Ni–NTA lipid are mixed, they segregate into separate microdomains. As the two lipids stain differently, these microdomains can be seen in negatively stained specimens as a "leopard skin" pattern. It would be worthwhile to test other filler lipids for their ability to better mix with Ni–NTA lipid and thus to avoid the "leopard skin" staining. Furthermore, while Affinity Grids are stable in the presence of detergents to a certain degree, using fluorinated lipids to form the monolayer could make Affinity Grids even more resistant to detergents. Fluorinated lipids have already been shown to be more detergent-resistant (Lebeau *et al.*, 2001) and have been used for 2D crystallization of membrane proteins on lipid monolayers (Levy *et al.*, 2001). Although not yet commercially available, fluorinated lipids can be synthesized (Hussein *et al.*, 2009), and initial trials indicate that fluorinated lipids indeed make Affinity Grids even more resistant to detergents.

Not only could the acyl chains of the lipids in the monolayer be changed, but also the head groups could be modified. For example, efforts are underway to synthesize lipids with longer linkers between the Ni–NTA group and the lipid head group (Daniel Lévy, personal communication).

Longer linkers would substantially increase the motional freedom of His-tagged molecules bound to the monolayer, thus reducing the likelihood that the adsorbed molecules adopt preferred orientations. Furthermore, Ni-NTA is not the only functional group that can be attached to lipid head groups. For example, lipids are commercially available with biotinylated head groups, which could be used for purification schemes, exploiting the high affinity of avidin for biotin ($K_D \sim 10^{-15}$ M). As avidin is small (15 kDa), it can be easily added as an affinity tag to a recombinant target protein. Lipid monolayers with biotinylated lipids could thus potentially be used to produce Affinity Grids with higher specificity than Ni-NTA lipid-based Affinity Grids.

Monolayer purification and Affinity Grids use lipids to create specific binding sites for proteins. Although these techniques have proved effective for producing EM specimens, a simpler design could be envisioned that no longer relies on functionalized lipids but instead on functionalized carbon or other support films. However, as long as such functionalized support films are not available, the lipid monolayer-based Affinity Grid is an easy, convenient, and versatile tool for the preparation of specimens for structure determination of macromolecular complexes by single-particle EM.

ACKNOWLEDGMENTS

Work on monolayer purification and Affinity Grids in the Walz laboratory is supported by NIH grant P01 GM062580 (to Stephen C. Harrison). TW is an investigator in the Howard Hughes Medical Institute.

REFERENCES

Abraham, J., Corbett, K. D., Farzan, M., Choe, H., and Harrison, S. C. (2010). Structural basis for receptor recognition by New World hemorrhagic fever arenaviruses. *Nat. Struct. Mol. Biol.* **4,** 438–444.

Al-Kurdi, R., Gulino-Debrac, D., Martel, L., Legrand, J. F., Renault, A., Hewat, E., and Venien-Bryan, C. (2004). A soluble VE-cadherin fragment forms 2D arrays of dimers upon binding to a lipid monolayer. *J. Mol. Biol.* **337,** 881–892.

Asturias, F. J., and Kornberg, R. D. (1995). A novel method for transfer of two-dimensional crystals from the air/water interface to specimen grids. EM sample preparation/lipid-layer crystallization. *J. Struct. Biol.* **114,** 60–66.

Asturias, F. J., Chang, W., Li, Y., and Kornberg, R. D. (1998). Electron crystallography of yeast RNA polymerase II preserved in vitreous ice. *Ultramicroscopy* **70,** 133–143.

Avila-Sakar, A. J., and Chiu, W. (1996). Visualization of β-sheets and side-chain clusters in two-dimensional periodic arrays of streptavidin on phospholipid monolayers by electron crystallography. *Biophys. J.* **70,** 57–68.

Azubel, M., Wolf, S. G., Sperling, J., and Sperling, R. (2004). Three-dimensional structure of the native spliceosome by cryo-electron microscopy. *Mol. Cell* **15,** 833–839.

Brisson, A., Mosser, G., and Huber, R. (1991). Structure of soluble and membrane-bound human annexin V. *J. Mol. Biol.* **220,** 199–203.

Carragher, B., Voss, N. R., Potter, C. S., and Smith, R. (2010). Software tools for molecular microscopy: An open-text Wikibook. *Methods Enzymol.* **482,** 381–392.

Darst, S. A., Kubalek, E. W., and Kornberg, R. D. (1989). Three-dimensional structure of *Escherichia coli* RNA polymerase holoenzyme determined by electron crystallography. *Nature* **340,** 730–732.

De Carlo, S., and Stark, H. (2010). Cryo-negative staining of macromolecular assemblies. *Methods Enzymol.* **481,** 127–145.

Ganser, B. K., Cheng, A., Sundquist, W. I., and Yeager, M. (2003). Three-dimensional structure of the M-MuLV CA protein on a lipid monolayer: A general model for retroviral capsid assembly. *EMBO J.* **22,** 2886–2892.

Golas, M. M., Sander, B., Will, C. L., Luhrmann, R., and Stark, H. (2003). Molecular architecture of the multiprotein splicing factor SF3b. *Science* **300,** 980–984.

Hussein, W. M., Ross, B. P., Landsberg, M. J., Levy, D., Hankamer, B., and McGeary, R. P. (2009). Synthesis of nickel-chelating fluorinated lipids for protein monolayer crystallizations. *J. Org. Chem.* **74,** 1473–1479.

Jurica, M. S., Sousa, D., Moore, M. J., and Grigorieff, N. (2004). Three-dimensional structure of C complex spliceosomes by electron microscopy. *Nat. Struct. Mol. Biol.* **11,** 265–269.

Kastner, B., Fischer, N., Golas, M. M., Sander, B., Dube, P., Boehringer, D., Hartmuth, K., Deckert, J., Hauer, F., Wolf, E., Uchtenhagen, H., Urlaub, H., *et al.* (2008). GraFix: Sample preparation for single-particle electron cryomicroscopy. *Nat. Methods* **5,** 53–55.

Kelly, D. F., Taylor, D. W., Bakolitsa, C., Bobkov, A. A., Bankston, L., Liddington, R. C., and Taylor, K. A. (2006). Structure of the α-actinin-vinculin head domain complex determined by cryo-electron microscopy. *J. Mol. Biol.* **357,** 562–573.

Kelly, D. F., Dukovski, D., and Walz, T. (2008a). Monolayer purification: A rapid method for isolating protein complexes for single-particle electron microscopy. *Proc. Natl. Acad. Sci. USA* **105,** 4703–4708.

Kelly, D. F., Abeyrathne, P. D., Dukovski, D., and Walz, T. (2008b). The Affinity Grid: A pre-fabricated EM grid for monolayer purification. *J. Mol. Biol.* **382,** 423–433.

Kelly, D. F., Lake, R. J., Middlekoop, T. J., Fan, H.-Y., Artavanis-Tsakonas, S., and Walz, T. (2010a). Molecular structure and dimeric organization of the Notch extracellular domain as revealed by electron microscopy. *PLoS ONE* **5,** e10532.

Kelly, D. F., Dukovski, D., and Walz, T. (2010b). Strategy for the use of affinity grids to prepare non-His-tagged macromolecular complexes for single-particle electron microscopy. *J. Mol. Biol.* **400,** 675–681.

Kettenberger, H., Armache, K. J., and Cramer, P. (2004). Complete RNA polymerase II elongation complex structure and its interactions with NTP and TFIIS. *Mol. Cell* **16,** 955–965.

Klaholz, B. P., Pape, T., Zavialov, A. V., Myasnikov, A. G., Orlova, E. V., Vestergaard, B., Ehrenberg, M., and van Heel, M. (2003). Structure of the *Escherichia coli* ribosomal termination complex with release factor 2. *Nature* **421,** 90–94.

Kubalek, E. W., Le Grice, S. F., and Brown, P. O. (1994). Two-dimensional crystallization of histidine-tagged, HIV-1 reverse transcriptase promoted by a novel nickel–chelating lipid. *J. Struct. Biol.* **113,** 117–123.

Lebeau, L., Lach, F., Venien-Bryan, C., Renault, A., Dietrich, J., Jahn, T., Palmgren, M. G., Kuhlbrandt, W., and Mioskowski, C. (2001). Two-dimensional crystallization of a membrane protein on a detergent-resistant lipid monolayer. *J. Mol. Biol.* **308,** 639–647.

Levy, D., Chami, M., and Rigaud, J. L. (2001). Two-dimensional crystallization of membrane proteins: The lipid layer strategy. *FEBS Lett.* **504,** 187–193.

McDowall, A. W., Dobro, M. J., Melanson, L. A., and Jensen, G. J. (2010). Plunge freezing for electron cryomicroscopy. *Methods Enzymol.* **481,** 63–82.

Medalia, O., Typke, D., Hegerl, R., Angenitzki, M., Sperling, J., and Sperling, R. (2002). Cryoelectron microscopy and cryoelectron tomography of the nuclear pre-mRNA processing machine. *J. Struct. Biol.* **138,** 74–84.

Muller, S. A., and Engel, A. (2001). Structure and mass analysis by scanning transmission electron microscopy. *Micron* **32,** 21–31.

Norville, J. E., Kelly, D. F., Knight, T. F., Jr., Belcher, A. M., and Walz, T. (2007). 7Å projection map of the S-layer protein sbpA obtained with trehalose-embedded monolayer crystals. *J. Struct. Biol.* **160,** 313–323.

Ohi, M., Li, Y., Cheng, Y., and Walz, T. (2004). Negative staining and image classification—Powerful tools in modern electron microscopy. *Biol. Proced. Online* **6,** 23–34.

Olofsson, A., Kaveus, U., Hacksell, I., Thelestam, M., and Hebert, H. (1990). Crystalline layers and three-dimensional structure of *Staphylococcus aureus* α-toxin. *J. Mol. Biol.* **214,** 299–306.

Qin, H., Liu, Z., and Sui, S. F. (1995). Two-dimensional crystallization of avidin on biotinylated lipid monolayers. *Biophys. J.* **68,** 2493–2496.

Radermacher, M., Wagenknecht, T., Verschoor, A., and Frank, J. (1987). Three-dimensional reconstruction from a single-exposure, random conical tilt series applied to the 50S ribosomal subunit of *Escherichia coli. J. Microsc.* **146,** 113–136.

Sjoquist, J., and Stalenheim, G. (1969). Protein A from *Staphylococcus aureus.* IX. Complement-fixing activity of protein A–IgG complexes. *J. Immunol.* **103,** 467–473.

Stagg, S. M., Lander, G. C., Pulokas, J., Fellmann, D., Cheng, A., Quispe, J. D., Mallick, S. P., Avila, R. M., Carragher, B., and Potter, C. S. (2006). Automated cryoEM data acquisition and analysis of 284742 particles of GroEL. *J. Struct. Biol.* **155,** 470–481.

Stroupe, M. E., Xu, C., Goode, B. L., and Grigorieff, N. (2009). Actin filament labels for localizing protein components in large complexes viewed by electron microscopy. *RNA* **15,** 244–248.

Tang, J., Taylor, D. W., and Taylor, K. A. (2001). The three-dimensional structure of α-actinin obtained by cryoelectron microscopy suggests a model for Ca^{2+}-dependent actin binding. *J. Mol. Biol.* **310,** 845–858.

Taylor, K. A., and Glaeser, R. M. (2008). Retrospective on the early development of cryoelectron microscopy of macromolecules and a prospective on opportunities for the future. *J. Struct. Biol.* **163,** 214–223.

Taylor, K. A., and Taylor, D. W. (1993). Projection image of smooth muscle α-actinin from two-dimensional crystals formed on positively charged lipid layers. *J. Mol. Biol.* **230,** 196–205.

Thess, A., Hutschenreiter, S., Hofmann, M., Tampe, R., Baumeister, W., and Guckenberger, R. (2002). Specific orientation and two-dimensional crystallization of the proteasome at metal-chelating lipid interfaces. *J. Biol. Chem.* **277,** 36321–36328.

Thomas, D., Schultz, P., Steven, A. C., and Wall, J. S. (1994). Mass analysis of biological macromolecular complexes by STEM. *Biol. Cell* **80,** 181–192.

Uzgiris, E. E., and Kornberg, R. D. (1983). Two-dimensional crystallization technique for imaging macromolecules, with application to antigen–antibody–complement complexes. *Nature* **301,** 125–129.

Venien-Bryan, C., Balavoine, F., Toussaint, B., Mioskowski, C., Hewat, E. A., Helme, B., and Vignais, P. M. (1997). Structural study of the response regulator HupR from *Rhodobacter capsulatus.* Electron microscopy of two-dimensional crystals on a nickel-chelating lipid. *J. Mol. Biol.* **274,** 687–692.

Ward, R. J., and Leonard, K. (1992). The *Staphylococcus aureus* α-toxin channel complex and the effect of Ca^{2+} ions on its interaction with lipid layers. *J. Struct. Biol.* **109,** 129–141.

CHAPTER FIVE

GraFix: Stabilization of Fragile Macromolecular Complexes for Single Particle Cryo-EM

Holger Stark[*,†]

Contents

Abstract

Here, we review the GraFix (Gradient Fixation) method to purify and stabilize macromolecular complexes for single particle cryo-electron microscopy (cryo-EM). During GraFix, macromolecules undergo a weak, intramolecular chemical cross-linking while being purified by density gradient ultracentrifugation. GraFix-stabilized particles can be used directly for negative-stain cryo-EM or, after a brief buffer-exchange step, for unstained cryo-EM. This highly reproducible method has proved to dramatically reduce problems in heterogeneity due

[*] MPI for Biophysical Chemistry, Göttingen, Germany
[†] Göttingen Center of Molecular Biology, University of Göttingen, Göttingen, Germany

Methods in Enzymology, Volume 481
ISSN 0076-6879, DOI: 10.1016/S0076-6879(10)81005-5

to particle dissociation during EM grid preparation. Additionally, there is often an appreciable increase in particles binding to the carbon support film. This and the fact that binding times can be drastically increased, with no apparent disruption of the native structures of the macromolecules, makes GraFix a method of choice when preparing low-abundance complexes for cryo-EM. The higher sample quality following GraFix purification is evident when examining raw images, which usually present a low background of fragmented particles, good particle dispersion, and high-contrast, well-defined particles. Setting up the GraFix method is straightforward, and the resulting improvement in sample homogeneity has been beneficial in successfully obtaining the 3D structures of numerous macromolecular complexes by cryo-EM in the past few years.

1. INTRODUCTION

The study of biological macromolecules at a molecular level has become a reality in the past decades. An important tool for this is high-resolution cryo-EM of single particles, which can be used to determine the three-dimensional (3D) structures of macromolecular complexes in their native forms. To prepare cryo-EM samples, purified complexes in solution are rapidly frozen, and the resulting vitrified complexes remain virtually artifact-free within their native buffer environments (Adrian *et al.*, 1990; Chapter 3, this volume). However, since the density of proteins is only slightly greater than that of vitrified ice, the particles have low contrast against the background. The use of negative-stained cryo-EM can increase the particle contrast, yet staining also introduces a number of limitations. Thus, while negative-stain cryo-EM can be beneficial in certain circumstances, unstained cryo-EM is the method of choice. Sample preparation and image analysis of vitrified samples is technically challenging, and much effort has been made recently to develop new and improved methods for cryo-EM.

It is now routine to obtain resolutions of higher than 10 Å for particles that are well-suited for the technique, that is, particles that are mostly symmetrical, rigid, and conformationally homogenous (Schuette *et al.*, 2009; Wolf *et al.*, 2010; Zhang *et al.*, 2008, 2010a,b). Viral particles are ideal for this method, and 3D reconstructions of several have been determined at near–atomic resolution (3.7–4 Å) (Chapter 15, Vol. 482). A recent breakthrough was made for a cryo-EM structure of the infectious subvirion particle (ISVP) of aquareovirus, which was resolved to 3.3 Å (Zhang *et al.*, 2010a,b). This resolution allowed *ab initio* model building for the final structure, in which a detailed protein structure could be distinguished. Reaching atomic-level resolution would allow cryo-EM to deliver as much structural information as X-ray crystallography yet with the advantage that one could work with complexes that are available at only very low concentrations and/or are too large or difficult to crystallize.

However, there are still a large number of technical issues to be dealt with prior to obtaining subnanometer or better resolution levels for the majority of macromolecular complexes that do not fit into this description. One major issue is dealing with the conformational flexibility within complexes, which is often linked to the ability of a complex to take on various functional states. Thus, even though a sample may be chemically pure, it can contain numerous different conformations. Structure determination of such samples requires the computational separation of the data into subpopulations of images that represent all possible conformational states that are present in the given sample. In recent years, new image processing strategies have been developed that indeed allow such a computational "purification" of images (Sander *et al.*, 2006; Scheres *et al.*, 2005, 2007) **(Chapter 13, Vol. 482; Chapter 10, Vol. 483)**. While such attempts are currently used in the intermediate resolution regime, they will also become a prerequisite for high-resolution structure determination of dynamic macromolecules by single particle cryo-EM in the future.

In the light of this, heterogeneity that arises from sample degradation during purification is an additional complication that should be avoided to the greatest degree possible. However, the ability to undergo conformational changes also often infers a flexibility to the particles that makes them more labile when being processed. Additionally, *in vitro* buffer conditions can lead to destabilization of the complexes. Difficulties in distinguishing between the conformational isoforms, the rotational degree of freedom of a single complex isoform, and partially degraded particles, in the presence of high levels of background noise, can drastically limit the probability of obtaining a high-resolution 3D reconstruction.

To avoid the problems arising from sample heterogeneity when purifying macromolecular complexes, we have recently introduced the GraFix (from *Gra*dient *Fix*ation) protocol (Kastner *et al.*, 2008). In this protocol, macromolecular complexes are exposed to a low concentration of a chemical cross-linker during sedimentation by ultracentrifugation through a density gradient. We have determined that, following the mild fixation and purification with GraFix, samples display drastically reduced to no degradation, as well as an improved quality of both the individual particle characteristics and the particle dispersion in the raw images. We and others have now used GraFix to determine 3D structures of numerous macromolecule particles (Golas *et al.*, 2009; Herzog *et al.*, 2009). For particles that are present only at an extremely low abundance in cells, using GraFix has proved to be not only helpful for improving sample quality but also essential for obtaining enough particle images for 3D reconstruction (Golas *et al.*, 2009). I discuss the particulars of this method and how we have implemented it in our laboratory, in the following review.

2. OVERVIEW OF THE GRAFIX PROCEDURE

The GraFix procedure combines purification by zonal ultracentrifugation with cross-linking through increasing exposure to a cross-linking reagent (see Fig. 5.1 for a schematic representation). Isolated particles can be directly used for negative-stain EM or for unstained cryo–EM following a buffer-exchange step (Fig. 5.1B). The GraFix procedure is highly

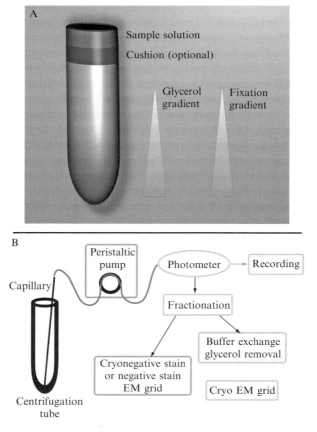

Figure 5.1 The GraFix method. (A) The centrifugation tube contains a gradient of both density (provided by glycerol) and cross-linking reagent, with an optional buffering cushion over the gradient. (B) Schematics of the GraFix method. Following ultracentrifugation, gradients are fractionated from bottom to top. Fractions can then be negatively stained or, following a one-step buffer exchange to remove the glycerol, plunge-frozen across EM grids.

reproducible, allowing for purification of a complex to proceed routinely after the initial setup. While we focus here mainly on using this procedure for the preparation of cryo-EM, it is important to point out that this is a universal procedure that could also benefit other structural analysis techniques that require the purification of high-quality, unbound complexes in their native state.

2.1. Chemical fixation of complexes during GraFix

2.1.1. Promotion of intramolecular cross-linking

Stabilization of large, fragile complexes by chemical cross-linking is a way to avoid disruption of complexes during the preparatory steps of the samples for cryo-EM. During cross-linking, covalent bonds are formed between the functional groups of the cross-linking reagent and those of the macromolecule, increasing the rigidity of the complex. However, the direct addition of chemical cross-linkers to the purified complexes is not usually a viable option for several reasons. Most purified macromolecular complexes are purified under buffer conditions that often promote weak aggregation of the complexes, so that the direct cross-linking of the complexes can result in intermolecular fixation, thereby increasing sample heterogeneity. Additionally, intermolecular cross-linked complexes are more likely to aggregate and precipitate out of the solution, leading to loss of sample material. The GraFix procedure avoids this problem through the increased pressure acting on the macromolecules as a result of the centrifugal force. This force is usually sufficient to disrupt weak aggregations, so that macromolecules are exposed to the chemical cross-linking reagent as individual complexes. Thus, the vast majority of the complexes will only undergo intramolecular, but not intermolecular, cross-linking. Obviously, if the concentration of the sample is too high, intermolecular cross-linking will occur. We have been able to avoid intermolecular cross-linking completely by applying less than 180 pmol of sample on a single GraFix tube (as tested for the ribosome). When using quantities higher than this, dimer cross-linking was observed. However, in general, amounts of sample higher than 180 pmol are not necessary when purifying samples for cryo-EM, since highly concentrated samples have to be subsequently diluted prior to preparing the EM grids. In any case, high sample concentrations are normally not obtainable for most macromolecules following commonly used biochemical purification procedures. For sample concentrations that are well-suited for EM grid preparation (e.g., that do not need to be diluted after GraFix), there is little danger that intermolecular cross-linking will occur.

2.1.2. Cross-linking in a compatible buffer

A further advantage of cross-linking during GraFix is that the buffer in which the cross-linking occurs be selected for compatibility with the cross-linker. Reactivity of buffer components with the cross-linking reagent is an important problem when cross-linking samples directly. For example, primary amino groups, such as those found in the TRIS buffer, are reactive with the commonly used aldehyde cross-linking agents. Reactive agents may also have been introduced during a previous purification step, such as by purifying a complex from an immunoaffinity column with peptides, which adds a significant amount of primary amine containing amino acids to the final buffer. This source of contamination is especially relevant due to the recent advances in purifying complexes by affinity-selection. The problem with the presence of cross-reactive agents in the sample buffer is by-passed by the GraFix method, by adding the cross-linking reagent to a density gradient rather than to the sample directly. In this way, the buffering environment of the macromolecular complex is completely exchanged prior to contact with the cross-linking reagent, without loss of sample due to an extra buffer-exchange step. An important point here is that there is a gradient not only of density but also of the cross-linking reagent, since the cross-linker is added only to the heavier, bottom solution and not to the lighter, top solution. The concentration of the cross-linking reagent at the top of the gradient is extremely low, and this is usually sufficient to prevent sample buffer artifacts during cross-linking. An additional (but not essential) precaution can be taken by replacing the top-most layer of the gradient with a cushion without a cross-linker reagent prior to applying the sample (Fig. 5.1A); in this manner, the sample buffer will never come in direct contact with the cross-linking reagent in the presence of the macromolecular complex, completely avoiding any chances of artifacts due to buffer reactivity with the cross-linking reagent. Thus, the composition of the original sample buffer has no impact on the cross-linking reaction.

2.1.3. Weak cross-linking with no apparent structural artifacts

Chemical cross-linking has always had a somewhat bad reputation of generating artifacts, mainly in more traditional cell biological EM applications (Hayat, 1986). Importantly, we were able to demonstrate that cross-linking with glutaraldehyde did not lead to any visible artifacts using the GraFix method up to the ~ 12 Å resolution level. This may be due in part to the fact that the cross-linking conditions that we have established using glutaraldehyde lead to a weak cross-linking, in which not all lysines within each particle are chemically modified. Nonetheless, as previous reports have determined that glutaraldehyde may lead to artifacts during cross-linking, it is important to note that other cross-linkers, such as formaldehyde and acroleine, can also be used. It is still unclear whether, and to which extent, cross-linking by GraFix will limit the obtainable resolution due to potential perturbations in the native structure that could only be visualized at higher resolution. However, we have established that GraFix does not interfere with sample integrity at intermediate resolutions of up to 12 Å, a resolution which can be extremely useful for determining reliable initial structures.

2.2. Purification over density gradients

In addition to providing a compatible environment for chemical cross-linking of macromolecular complexes, GraFix can also introduce a convenient purification step that may make it possible to eliminate previous steps of the purification. Density gradient centrifugation is a powerful technique for separating complexes based on their molecular masses, and the range of separation can be determined by selecting the density range. It should be noted that glycerol can be substituted for other sugars to create a density in the gradient, if required. To date, we have successfully tested sucrose, trehalose, and arabinose.

2.2.1. Glycerol removal prior to unstained cryo-EM

Direct processing of GraFix-purified complexes is possible when the samples are used for negative-stain or cryonegative stain EM (Chapter 6, this volume). However, high concentrations of glycerol (such as the 15–25% used in GraFix) interfere with the high–contrast image formation of macromolecules embedded in vitrified ice. Thus, prior to using the GraFix-purified particles for unstained cryo-EM, the glycerol in the buffer has to be removed. This can be done in a simple, one-step procedure using a buffer-exchange column. It is important to note that buffer-exchange columns cannot normally (e.g., in the absence of particle fixation) be included in the purification scheme, since unfixed complexes are easily damaged during this procedure. However, GraFix-stabilized complexes are usually not affected. This step allows a rapid removal of almost all the glycerol in the buffer, and the samples are then suitable for direct analysis by unstained cryo-EM.

2.3. Reduction of sample heterogeneity

The reduction of sample heterogeneity due to particle disintegration can be dramatically improved when handling GraFix-stabilized complexes during grid preparation. This is exemplified by images of the spliceosomal B complex, a macromolecular complex formed by three snRNAs and more than 100 proteins that sediments at 40S. The spliceosomal B complex is too labile to be subjected to any type of column purification but can be successfully purified over a glycerol gradient. When complexes were purified in the absence of chemical cross-linking and used directly for negative-staining cryo-EM, the resulting raw images were of poor quality, displaying heterogeneous particles, a high background of particle fragments, and poor particle distribution (Fig. 5.2A, top). The low quality of the raw image is reflected in the class averages following statistical analysis: only ~10% of the images could be ordered into class averages that display well-defined structures (Fig. 5.2A, top row of the bottom panel). In - contrast, purification with the GraFix method stabilized the spliceosomal B complexes sufficiently to almost completely prevent their disruption during grid preparation. The raw images of the GraFix particles show clearer outlines and improved fine structural features, and the number of smaller, broken particles was significantly reduced. This improvement in image quality can drastically increase the number of good quality class averages that can be obtained. For instance, following image alignment, multivariate statistical analysis, and classification, there was an approximately fivefold increase in the number of high-quality class averages for the spliceosomal B complex prepared by GraFix as compared to that prepared by gradients alone. Since the individual particles displayed a higher degree of structural homogeneity, these class averages also had a higher signal.

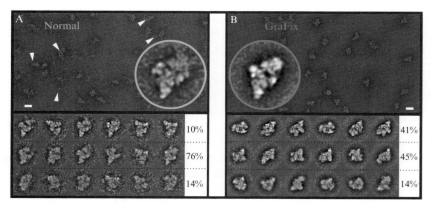

Figure 5.2 Negative-stain cryo-EM images of spliceosomal B complexes with or without GraFix. (A, B) Electron microscopic raw image of uranyl formate-stained spliceosomes prepared by a conventional glycerol gradient (A) or GraFix (B). Scale bars, 40 nm. Arrowheads, smaller broken parts and flexible elements. Insets, similarly oriented spliceosomal class average. Class averages obtained from a set of 5000 raw images of non-GraFix-prepared (A) or GraFix-prepared (B) samples are shown in the bottom panels. The average number of class members is 15 images. Class averages were sorted with respect to contrast and structural definition. GraFix treatment (B) generates computed class averages with much improved contrast (top and middle; 86% of images), as compared to samples prepared by the conventional method (A), where only 10% of class averages (top) show relatively well-defined structural features.

2.3.1. Structural heterogeneity

The question arises as to whether macromolecular complexes that have been chemically stabilized have been locked into a specific conformation, thereby reducing the conformational heterogeneity. This could be advantageous in simplifying the complexity of sorting the particles presented in the raw images. Our observations so far suggest that, while this may occur, it is a minor effect, which is perhaps not a bad thing. Aside from the technical difficulties that arise, dealing with various conformations of a macromolecular complex within the same conditions can be extremely informative. Often, the assumed conformations parallel the functional conformations, so that comparison of their structures can be key in understanding how the complex functions at a molecular level. We and others have recently observed this for the ribosome (Fischer *et al.*, 2010) working with a highly dynamic preparation of ribosomes in various stages of translocation (that had not been treated with GraFix). While the initial resolution was limited to ~ 20 Å, computational sorting allowed us to obtain a subnanometer resolution for certain subpopulations of the data set. The level of computational sorting was also sufficient to observe the molecular dynamics of the ribosome (i.e., the movement of tRNAs through the ribosome and the correlated motions within the ribosome itself). While dealing with structural heterogeneity currently represents one of the major resolution-limiting factors for 3D structure determination, it will make cryo-EM a powerful tool for analyzing molecular functions of macromolecular complexes once methods become routine to separate mixed population of images by computational image processing.

2.4. Advantages of using GraFix particles during carbon film binding

2.4.1. Increased particle binding

We have observed that GraFix purification of macromolecular complexes can often increase their binding to the carbon support film. Importantly, because of their chemical stabilization, GraFix-purified complexes can also be allowed to adsorb onto carbon film for extended periods of time without sacrificing the quality of the complexes. The combination of an improved binding over a longer time can lead to a significant increase in the number of bound particles. This may be able to compensate for any particle loss that may have occurred during the gradient purification. Importantly, this also increases the chances of being able to acquire enough images from particles that are present only in low-copy numbers in cells and cannot be easily purified at higher concentrations. As an extreme example, images of the *Trypanosoma brucei* kinetoplastid RNA editing complex following purification with GraFix are shown (Golas *et al.*, 2009). After a normal adsorption

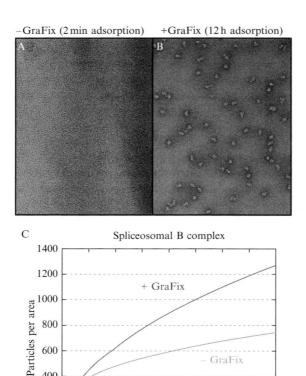

Figure 5.3 Enhanced GraFix particle binding to carbon support film. (A, B) Negative-stain cryo-EM image of the *T. brucei* kinetoplastid RNA editing complex after GraFix purification. After 2 min of adsorption time, no particles were detected on the grids (A). Extending the adsorption time up to 12 h led to good quality images with an acceptable particle distribution that allows further single particle analysis (B). (C) Binding rates of spliceosomal B complexes either treated with GraFix (red) or purified over a glycerol gradient in the absence of cross-linking (pink). No difference in particle concentration was observed after 2 min, but there was an increase of approximately twofold after ∼6 h of adsorption. There were no visible signs of particle disintegration even at the longest time points for the GraFix-stabilized particles. (For interpretation of the references to color in this figure legend, the reader is referred to the web version of this chapter.)

time of 2 min, no particles were visible in the raw image (Fig. 5.3, top left panel). In contrast, when the GraFix particles were allowed to adsorb for 12 h, a large number of good quality particles were distributed throughout the image (Fig. 5.3, top right panel). Additionally, as shown by comparing the particle binding of the spliceosomal B complex, there was a higher binding density over time of particles treated with GraFix as compared to nontreated particles (Fig. 5.3, bottom panel). The ability to increase particle binding during grid preparation is important as it could play a critical role in determining 3D structures of low-abundance complexes that can only be purified in extremely low concentrations.

2.4.2. Increased isotropic particle orientation on film

Many macromolecular complexes tend to bind to the carbon support film with only a few preferential orientations. This binding bias reduces the angles from which the particle is viewed, leading to nonisotropic sampling of information in 3D space. Following chemical fixation, the charge distribution of the complex changes. We have observed that this can sometimes lead to a more isotropic particle orientation when binding on carbon film, making it possible to obtain more particle views. This is beneficial not only for the initial structure determination phase, where a large number of different views make the initial model more reliable, but also for higher resolution structure determination with isotropic resolution.

2.5. Coanalysis of proteins within the cross-linked particles

Once cross-linking of the particles has occurred, it is no longer possible to directly analyze the composition of the complex by gel electrophoresis, since the cross-linking is irreversible and the proteins no longer dissociate following heat or detergent treatment. However, it is still possible to analyze the protein composition by running a parallel gradient without a cross-linking reagent which may then be used for SDS gel electrophoresis. Ideally, of course, one would like to analyze the GraFix-stabilized fraction directly. This could be done by using a reversible cross-linker, such as paraformaldehyde, in which case the cross-linking could be reverted prior to running the sample on an SDS gel. Another option that we are currently testing is to apply the chemically stabilized complexes to mass spectrometry analysis. Initial studies on several complexes indicate that it is indeed possible to study the protein composition of glutaraldehyde-cross-linked complexes by a combination of trypsin digestion and mass spectrometry (Richter *et al.*, 2010). Since trypsin requires accessible lysines (that have not been cross-linked) to enzymatically cleave the protein, it is important to note that this is possible following GraFix only because the cross-linking is weak and not all lysines have been modified.

3. Methods

3.1. Guidelines for determining centrifugation parameters

Determining how to set up a centrifugation gradient depends on the sedimentation value of a complex. In most cases, however, the S-value is not known, and often the exact size of the entire complex is unknown. To overcome this, we use a simulation program that predicts the centrifugation of a complex based on a rough estimate of its S-value. Based on this, we have compiled guidelines for determining the gradient conditions (Table 5.1). Use of these guidelines has proved very effective, which can avoid trial-and-error when setting up gradient conditions for the first time, reducing loss of sample and time.

3.2. Preparing a continuous density gradient

Gradients are created by mixing two solutions with a low and a high density and are obtained by adding glycerol, sucrose, or another carbohydrate in appropriate concentrations. Table 5.1 shows the density values for the gradient based on the molecular mass. Thus, for a macromolecular complex of 850 kDa, the top solution should contain 10% glycerol, and the bottom one, 30%. A successful gradient centrifugation will allow the complex-to-be-purified to move about two-thirds of the way down the gradient. This assures that the complex has been completely removed from smaller contaminants (such as found in the original buffer) yet is not too close to the bottom, where it could be contaminated with sediments.

Table 5.1 Ultracentrifugation guidelines for GraFix, based on a selection of various complexes

Molecular mass (kDa)	Gradient	RPM	Time
125	5–20%	40,000	18
450	10–30%	50,000	16
700	10–30%	33,000	16
850	10–30%	33,000	18
1500	10–40%	37,000	14
3600	15–45%	22,500	14

A rough estimate of the centrifugation conditions, based on the approximate molecular mass of the complex, is given. Gradient: the percentage of glycerol (or other sugar) to use in the top and bottom gradient solutions; RPM: the speed of the ultracentrifugation; and time: hours of centrifugation.

3.2.1. Solution preparation

Two buffer solutions should be prepared with the appropriate densities for creating the gradient (see Table 5.1). The cross-linker, such as glutaraldehyde, should be added only to the denser solution, at a concentration of 0.05–0.2% (v/v). Other cross-linkers can also be used. The buffers should not contain any primary amino groups (such as in TRIS). For instance, to prepare a typical 10–30% glycerol gradient, the top buffer would contain HEPES 50 mM, pH 7.5, ≤ 100 mM salt (as appropriate for the complex), and 10% (v/v) glycerol, while the bottom buffer would contain HEPES 50 mM, pH 7.5, ≤ 100 mM salt, 30% (v/v) glycerol, and 0.15% glutaraldehyde. Buffers should be filtered through a 0.3-μm filter prior to use.

3.2.2. Gradient formation

To form the gradient, the different-density solutions are layered in a 4.4-ml centrifuge tube (such as polyclear tubes, #S7010, Science Services), by first adding 2.1 ml of the less dense (top) solution to the tube. Next, 2.1 ml of the heavier (bottom) solution is drawn into a syringe with a blunt-end stainless steel needle (such as a Hamilton syringe). The end of the needle is placed at the bottom of the tube at a slight angle, and the solution is slowly expelled, so that the lighter solution is displaced upward. This must be done carefully to avoid disturbing the interface as much as possible; the interface should form a sharp line when finished. Tubes are then closed with a BioComp cap.

To form a continuous density gradient, tubes are placed into a specialized gradient mixer (such as the Gradient Master 107, BioComp Instruments) and rotated briefly, following the manufacturer's recommendations for determining the parameters (time[s]/angle/speed). Gradients should be prepared in advance and allowed to settle for an hour at 4 °C prior to centrifugation.

3.2.3. Adding a buffering cushion

As mentioned earlier, a buffering cushion can be used on the top of the gradient to completely avoid contact between the original sample buffer and the cross-linking reagent (Fig. 5.1). This prevents any reactivity of original buffer components (such as TRIS, or peptides) with the glutaraldehyde-cross-linking reagent. This is not usually necessary due to the extremely low concentration of the cross-linking reagent at the top of the gradient (we usually do not use this cushion for our gradients).

The cushion should contain the same buffer solution as the top solution, but with a slightly lower density; thus, if the top solution contains 10% glycerol, the cushion should contain 7% glycerol. To add a 200-μl buffering cushion, that amount plus any amount of the sample over 200 μl should be removed from the top of the gradient (e.g., if the sample will be 500 μl, remove the 200 μl for the cushion plus 100 μl for the sample; see in the following paragraphs). Carefully add the cushion to the top of the gradient and let the gradient settle at 4 °C as normal.

3.3. Sample concentrations

Since most macromolecular complexes can only be purified in small quantities, it is important to maintain a small gradient volume to keep the sample concentration within adequate ranges in the final gradient fractions. We usually use an ~4-ml gradient. The sample to be loaded should ideally be between 10 and 80 pmol of complex in a maximum volume of 400 μl (we routinely load 200 μl or less). The amount of the macromolecular complex that can be loaded onto the gradient can be increased, but we recommend loading less than 180 pmol onto a 4-ml gradient, as we observed artifacts due to intermolecular cross-linking when loading amounts higher than this. While it is preferable to have a minimum of 10 pmol of complex, this is not always feasible when handling low-abundance particles. As mentioned earlier, the apparent concentration of extremely low-concentration samples can be increased during the grid preparation, by increasing the binding times to the carbon support film. This can help overcome the problem of starting with low sample quantities in some cases. For example, only about 1 pmol of the RNA editing complex was loaded onto a GraFix gradient, yet we were able to obtain good quality raw images (Fig. 5.3).

When setting up the centrifugation, it might be useful to have an internal control of the sample under the same conditions except without cross-linker, centrifuged in parallel. This control reveals whether the fixation has changed the sedimentation of the macromolecular complex. An advantage of performing this control is that it provides aliquots for which the protein composition of the complex can be analyzed by SDS-PAGE, since the complex has not been cross-linked. However, this control is not essential (and we do not perform it once the centrifugal conditions have been determined).

3.3.1. Loading the sample

After removing the cap, there will be enough space at the top of the gradient to load up to 200 μl of sample. Since it is not important to completely fill the tube, sample volumes less than 200 μl can also be loaded. However, if the sample is larger than 200 μl, an amount of the gradient that is equivalent to the extra volume should be removed (e.g., if the sample is 300 μl, remove 100 μl prior to loading). After loading, tubes within the buckets need to be balanced prior to carefully placing them into the rotor that has been precooled to 4 °C.

3.4. Centrifugation

Ultracentrifugation is carried out at 4 °C in swing-out rotors, such as SW60 rotors (Beckmann) or TH-660 rotors (Kendor Laboratory). Note that, although both of these rotors use the same tubes, there are minor differences between centrifugation in the two rotors that can result in a shift of 1–2 fractions. Depending on the mass of the complex, centrifugation times are between 14 and 18 h with speeds of 22,500–40,000 rpm (Table 5.1).

3.5. Fractionation

Following centrifugation, the gradients are fractionated (at 4 °C) from the bottom, in order to minimize contamination with material from the top of the gradient. We fractionate using a capillary to pump the gradient out from bottom to top, taking fractions of five drops (corresponding to ~ 175 μl). Fractionation can also be performed by removing the fractions through the bottom of the tube, such as by using the Brandel Isco tube piercer (Isco, Inc., Lincoln, USA). The optical density of each fraction is measured with a photometer during fractionation (see Fig. 5.1). If required, the glutaraldehyde within the fraction can be neutralized by adding glycine to a final concentration of 80 mM (note that this has not been necessary for us).

3.6. Buffer exchange prior to unstained cryo-EM

If the gradient fraction sample is used for unstained cryo-EM, the glycerol must be removed prior to grid preparation. This is performed in a single step using a buffer-exchange column, such as the PD MINITRAP G-25 (GE Healthcare, following the manufacturer's protocol). This is a simple and quick procedure that is carried out by loading the fraction onto a prewashed microcolumn at 4 °C and, after a brief washing, eluting the complexes in a compatible buffer. Importantly, for many macromolecular complexes, it is only possible to use exchange columns if the complexes have been purified with GraFix due to the stability of the complexes afforded by cross-linking; most nonstabilized complexes will not survive this method intact. Note also that we prefer gravity-flow columns to spin columns, as they are less harsh on the complexes.

3.7. Sample preparation for EM grids

3.7.1. Placing carbon film onto samples

Following purification either from the gradient (for negative staining) or from the buffer-exchange columns (for cryo), the samples can be processed for grid preparation as normal. In our laboratory, we use thin (~ 10 nm)

Figure 5.4 Placement of carbon film over a sample. A homemade plastic block with holes drilled into it, each with a holding capacity of about 25 μl, is precooled on ice, and 30 μl of a sample is loaded into one hole. A thin carbon film is floated off a mica support by holding the mica with forceps and placing it into the sample at a slight angle. Once the carbon film is released, the mica is removed, and macromolecular complexes are allowed to adsorb for the appropriate time.

carbon films that have been coated onto a piece of mica and dried using a carbon vacuum evaporator system (Boc Edwards GmbH). To adsorb the complexes onto the carbon film, we first place the sample ($\sim$30 μl) into a hole drilled into a homemade black plastic block (made from polyoxymethylene) that had been previously placed on ice to cool (Fig. 5.4). Next, we cut a piece of the carbon-coated mica (about 2 mm $\times$ 2 mm) and carefully place this at an angle into the sample, with the mica side facing down. Contact with the sample releases the carbon from the mica. This method has the advantage that the carbon film side that is bound by the macromolecular complexes has never been exposed to the air. Using this method of introducing the carbon film to the sample has shown to be highly reproducible in our hands to produce a film with low background noise.

3.7.2. Adsorption times

While we still use a short adsorption time of a 1–2-min for samples with an acceptable concentration of macromolecular complexes, the adsorption time can be greatly extended for lower concentration samples. We have found that adsorption times of up to 24 h still give high-quality particle images (see, e.g., the 12-h adsorption for the RNA editing complex; Fig. 5.3). When we are working with very low-concentration samples, we usually allow the binding to occur overnight ($\sim$12 h). When using longer adsorption times, we cover the plastic block (containing the sample and the carbon film) with a lid sealed with an O-ring, to reduce sample evaporation. This is placed for the necessary time at 4 $^{\circ}$C.

3.7.3. Grid preparation

After the allotted adsorption time, the carbon film is picked up using a holey carbon grid. Negative-stain sample preparation differs only by the additional incubation with the heavy metal salt solution used for staining. For cryogrid preparation, an additional 3–4 μl of sample is placed over the carbon film prior to vitrification, to facilitate blotting of excess solution, and the films are plunged into liquid ethane. Grids can be stored in liquid nitrogen until they are used for cryo-EM.

4. CONCLUSIONS

Currently, the field of cryo-EM is experiencing an exciting expansion, following advances in methodology and instrumentation that has allowed it to deliver near-atomic resolution 3D structures of several well-behaved macromolecular complexes. The GraFix protocol expands the type of specimens that can be analyzed to include molecules that are more difficult to handle due to internal asymmetry, flexibility, and low concentrations of the complex following purification. It is important to note, however, that GraFix cannot be used to "repair" broken macromolecular complexes that were damaged already during sample purification. Rather, GraFix can only be helpful in stabilizing intact molecules. Optimization of biochemical tools and methods used for complex purification is still of utmost importance in single particle cryo-EM, and special care needs to be taken to determine the optimal buffer conditions in order to maintain the complex integrity to the greatest possible extent during purification. When starting with intact macromolecular complexes, it is reasonable to expect resolutions of 10 Å and better at the present for complexes purified at the end stage with GraFix. We also expect near-atomic resolution structure determination to be a viable goal in the foreseeable future for GraFix-treated macromolecular complexes.

REFERENCES

Adrian, M., et al. (1990). Direct visualization of supercoiled DNA molecules in solution. *EMBO J.* **9**(13), 4551–4554.

Fischer, N., Konevega, A. L., Wintermeyer, W., Rodnina, M. V., and Stark, H. (2010). Ribosome dynamics and tRNA movement by time-resolved electron cryomicroscopy. *Nature* **466,** 329–333.

Golas, M. M., et al. (2009). Snapshots of the RNA editing machine in trypanosomes captured at different assembly stages in vivo. *EMBO J.* **28**(6), 766–778.

Hayat, M. A. (1986). Glutaraldehyde: Role in electron microscopy. *Micron and Microsc. Acta* **17**(2), 115–135.

Herzog, F., et al. (2009). Structure of the anaphase-promoting complex/cyclosome interacting with a mitotic checkpoint complex. *Science* **323**(5920), 1477–1481.

Kastner, B., *et al.* (2008). GraFix: Sample preparation for single-particle electron cryomicroscopy. *Nat. Methods* **5**(1), 53–55.

Richter, F. M., Sander, B., Golas, M. M., Stark, H., and Urlaub, H. (2010). Merging molecular electron microscopy and mass spectrometry by carbon-film-assisted endoproteinase digestion. *Mol. Cell Proteomics.* Jun 8. [Epub ahead of print].

Sander, B., *et al.* (2006). Organization of core spliceosomal components U5 snRNA loop I and U4/U6 Di-snRNP within U4/U6.U5 Tri-snRNP as revealed by electron cryomicroscopy. *Mol. Cell* **24**(2), 267–278.

Scheres, S. H., *et al.* (2005). Maximum-likelihood multi-reference refinement for electron microscopy images. *J. Mol. Biol.* **348**(1), 139–149.

Scheres, S. H., *et al.* (2007). Disentangling conformational states of macromolecules in 3D-EM through likelihood optimization. *Nat. Methods* **4**(1), 27–29.

Schuette, J. C., *et al.* (2009). GTPase activation of elongation factor EF-Tu by the ribosome during decoding. *EMBO J.* **28**(6), 755–765.

Wolf, M., Garcea, R. L., Grigorieff, N., and Harrison, S. C. (2010). Subunit interactions in bovine papillomavirus. *Proc. Natl. Acad. Sci. USA* **107**(14), 6298–6303.

Zhang, X., *et al.* (2008). Near-atomic resolution using electron cryomicroscopy and single-particle reconstruction. *Proc. Natl. Acad. Sci. USA* **105**(6), 1867–1872.

Zhang, J., *et al.* (2010a). Mechanism of folding chamber closure in a group II chaperonin. *Nature* **463**(7279), 379–383.

Zhang, X., Jin, L., Fang, Q., Hui, W. H., and Zhou, Z. H. (2010b). 3.3 A cryo-EM structure of a nonenveloped virus reveals a priming mechanism for cell entry. *Cell* **141**(3), 472–482.

CHAPTER SIX

CRYONEGATIVE STAINING OF MACROMOLECULAR ASSEMBLIES

Sacha De Carlo* *and* Holger Stark[†]

Contents

Abstract

Cryoelectron microscopy (cryo-EM) combined with single-particle reconstruction methods is a powerful technique to study the structure of biological assemblies at molecular resolution (i.e., 3–10 Å). Since electron micrographs of frozen-hydrated biological particles are usually very noisy, improvement of the signal-to-noise ratio (SNR) is necessary and is usually achieved by image processing. We propose an alternative method to *improve the contrast at the specimen preparation stage: cryonegative staining.*

Cryonegative staining aims to increase the SNR while preserving the biological samples in the frozen-hydrated state. Here, we present two alternative procedures to efficiently perform cryonegative staining on macromolecular assemblies. The first is very similar to conventional cryo-EM, the main difference being that the samples are observed in the presence of an additional contrasting agent, ammonium molybdate. The second is based on a carbon-sandwich method and is typically used with uranyl formate or acetate. Compared to air-dried negative staining at room

* Department of Chemistry, Institute for Macromolecular Assemblies, City University of New York, City College Campus, New York, USA
† Max-Planck-Institut für Biophysikalische Chemie, Göttingen, Germany

Methods in Enzymology, Volume 481
ISSN 0076-6879, DOI: 10.1016/S0076-6879(10)81006-7

temperature, the advantage of both cryonegative-staining procedures presented here is that the sample is kept hydrated at all steps and observed at liquid nitrogen temperature in the electron microscope. The advantage over conventional cryo-EM is that the SNR is improved by at least a factor of three.

For each of these approaches, a few examples of attainable data are given. We cover the technical background to cryonegative staining of macromolecular assemblies, and then expand upon the different possibilities and limitations.

1. INTRODUCTION

At the end of the 1950s, the concept of *negative staining* as a light microscopical procedure that made an essentially transparent object visible by surrounding it with a colored solution was transferred to the growing field of electron microscopy. It is widely accepted that Bob Horne was the first electron microscopist to develop the technique. He presented a clear data showing that biological samples could be surrounded by a thin amorphous layer of *air-dried* stain (such as uranyl acetate), which considerably increased the amplitude contrast that lacks in the absence of stain (Brenner and Horne, 1959).

Perhaps the best that can be expected from negative staining is that it should reveal the solvent-excluded surface and shape of a biological molecule or other particle. In theory, intramolecular information such as α-helices or β-sheets is unlikely to be revealed by negative staining, which relies upon the relatively large mass–thickness difference between the biological material and the surrounding stain, rather than upon the more subtle difference of varying mass–thickness of biological material and the surrounding water, as is the case for cryoelectron microscopy of unstained vitrified specimens (Adrian *et al.*, 1984; Dubochet *et al.*, 1988). The addition of a negative stain is still not sufficient to image biological macromolecules in great details. Defocus-induced phase contrast, which is important for unstained biological specimens, is also thought to contribute to electron imaging by negative stain (Massover, 2008). The use of trehalose alone as a negative stain generates a thin supportive film of mass–thickness marginally greater than that of a layer of vitreous water, within which viruses and large protein molecules and polymers can be revealed (Harris and Scheffler, 2002), whereas contrast matching tends to occur for smaller molecules (i.e., they are no longer visible since the background contrast is just as high). These intermediate mass density conditions have yet to be fully exploited, although glucose and trehalose have been widely used as preservation and contrast-inducing materials for electron imaging of 2D protein crystals (Hirai *et al.*, 1999) or as a protective agent for cryo-EM of viruses (De Carlo *et al.*, 1999).

One point that is often neglected is that following air-drying of a negatively stained specimen, a considerable quantity of water remains bound to the biological material and within the seemingly amorphous surrounding stain. Once inserted into the electron microscope and subjected to high vacuum, this bound water is rapidly removed. However, if a negatively stained specimen is cooled with liquid nitrogen in a cryotransfer holder, transferred to and maintained within the electron microscope under low temperature conditions, the bound water will not be removed, as indeed is the case throughout the cryonegative-staining procedure.

In reality, the concept of cryonegative staining is not new. Already at the beginning of the 1980s, Lepault *et al.* (1983) tried to add solutes to the solution prior to vitrification. The fact that solutes are not segregated in vitreous ice makes it possible to increase the density of the medium, thus changing the contrast of embedded particles. The problem is that when a compound like metrizamide (a tri-iodinated benzamido derivative of glucose) is added to the buffer prior to vitrification, a situation is reached where the contrast is dramatically reduced (contrast matching). When the surrounding solute concentration is increased, contrast inversion is obtained (with ∼30% metrizamide), and the particles in the EM image appear to be negatively stained, surrounded by the higher density metrizamide. However, in this situation, the fine structural internal details of metrizamide-embedded biological particles are not revealed, making this embedding technique essentially useless for high–resolution 3D structure determination.

Other attempts were carried out by Christoph Böttcher at the Fritz-Haber-Institute of the Max-Planck-Society in Berlin but published after his move to Imperial College. Böttcher *et al.* (1999) tried to mix conventional negative stains with the sample in the solution, immediately before vitrification, with the stain concentration kept below 2%. The actual stain concentration on the grid is completely unknown because the freezing does not follow immediately after the final blotting. There is thus always some time for the stain to become more concentrated and it depends on how much of the stain solution is being blotted before freezing. Due to this low stain concentration within the very thin vitrified specimen, the contrast obtained by Böttcher and coworkers did not appear to be significantly different from negative-stain preparations, maybe a little less because of the water still present (it was of course because it was used to visualize a 180–kDa protein, which is considered at the "limit" of visibility for unstained cryo-EM).

Cryonegative staining combines the advantages of frozen–hydrated preparations with the electron scattering power of a heavy–metal salt solution such as the ones typically used for conventional negative staining. Two distinct alternative procedures are presented: we shall refer to them as the *Adrian* and the *Stark* methods, after their "inventors" Marc Adrian and Holger Stark, respectively.

Cryonegative staining following Adrian's protocol is a fairly recent technique that allows the samples to be observed in the vitrified state at low temperature and at physiological pH, but in the presence of ammonium molybdate at saturating conditions (Adrian et al., 1998, 2002; De Carlo et al., 2002, 2003, 2008; El-Bez et al., 2005; Jawhari et al., 2006; Kostek et al., 2006; Tsitrin et al., 2002).

The alternative Stark method is introduced in Section 3. It differs slightly from the ammonium molybdate staining technique introduced by Adrian in that it requires the samples to be trapped between two carbon films, thus the "carbon-sandwich" name (Tischendorf et al., 1974). The sample is then frozen in liquid nitrogen as opposed to liquid ethane or propane, the commonly used cryogens for conventional cryo-EM of unstained, frozen-hydrated samples. Stark and Luhrmann (2006) introduced it, although in the literature, the cryonegative sandwich technique is also frequently called cryonegative staining (Golas et al., 2003, 2005).

The cryonegative-staining methodology using ammonium molybdate developed from a productive collaboration between J. Robin Harris and the Lausanne-based cryoelectron microscopists Marc Adrian and Jacques Dubochet (see Adrian et al., 1998), in an attempt to produce superior image contrast while maintaining samples in their hydrated state. This topic was reviewed briefly by Harris and Adrian (1999) and Harris et al. (2006). Here, we present a more complete and up-to-date survey of cryonegative staining, together with a detailed protocol.

Adrian's current version of the cryonegative-staining technique is—from a methodological point of view—very similar to the thin film vitrification technique, with the exception of an additional step: a short time of contact between the biological sample and the negative stain (Fig. 6.1). As a result, the biological sample is entrapped in a vitrified ammonium molybdate salt solution, used essentially at saturation concentration (~ 0.8 M). However, despite the analogy with the conventional cryomethodology, the results obtained

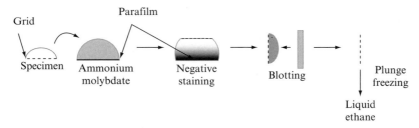

Figure 6.1 A schematic outline of the cryonegative-staining procedure (Adrian et al., 1998). A droplet of the ammonium molybdate solution is deposited on a small piece of parafilm. The grid is floated (sample facing down) onto the staining droplet for 10–60 s. The sample remains hydrated at all steps in the procedure.

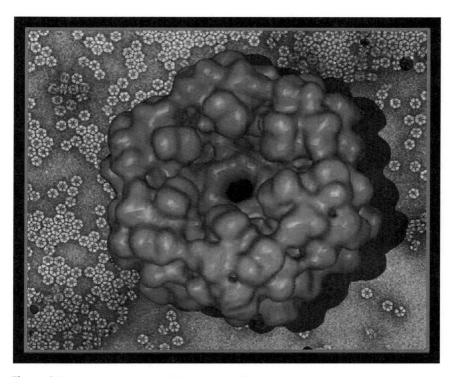

Figure 6.2 *Nereis virens* hemoglobin observed by cryonegative staining. A raw micrograph is shown overlaid with a 3D reconstruction of the molecule calculated from the data (De Carlo and Boisset, unpublished data).

with this cryonegative-staining technique are completely different, and in our opinion, exceptional. The SNR is higher, making it possible to visualize biological samples with an extraordinary level of detail and clarity (see Fig. 6.2). In the first published work (Adrian *et al.*, 1998), a large number of different samples were presented as successful examples of the technique: icosahedral viruses, isolated proteins, and even thin catalase crystals. The gain in SNR has been demonstrated qualitatively by the straightforward comparison of the same samples, imaged in both unstained vitrified and cryonegatively stained conditions. A quantitative analysis was performed by comparing the spectral signal-to-noise ratio (SSNR) obtained with unstained versus cryonegatively stained GroEL (De Carlo *et al.*, 2002). The gain in SNR was measured and was found to be a 10-fold increase for GroEL particles embedded in vitrified ammonium molybdate versus unstained ones. We repeated the measurement with tobacco mosaic virus (TMV) by comparing the signal strength of the 23-Å layer line in several diffraction images obtained in the presence of ammonium molybdate (Fig. 6.3C) versus unstained TMV.

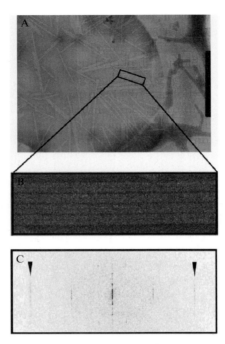

Figure 6.3 (A) TMV in cryonegative staining with saturated ammonium molybdate. Scale bar = 0.5 μm. (B) Inset: selected TMV rods that were used to calculate a power spectrum (shown in reversed contrast in (C)). The 11.5-Å layer lines are faint but still visible (arrowheads). Defocus = 700 nm (De Carlo, unpublished data).

The results were more realistic; the SNR of cryonegatively stained TMV was increased by a factor of three (De Carlo and Wang, unpublished data). However, the major interest is the increased contrast of cryonegatively stained particles; they can be imaged close to focus, thus better preserving the high-resolution information (e.g., 9–10 Å resolution). For instance, the DNA packaging was clearly shown in bacteriophage T2 capsids (Adrian *et al.*, 1998). Nevertheless, this advantage was not demonstrated quantitatively: the 3D reconstruction of the TBSV virus 2D crystal did not show an improved result, perhaps because of the small number of virus particles present and the 2D crystal disorder. An improved resolution was claimed, but this first encouraging pioneering work left unanswered many important biological questions. The most important is whether the biological structure is preserved down to atomic resolution for noncrystalline objects. This question can be asked using other words: does the negative stain penetrate deeply within the protein, and if yes, to what extent (what kind of internal details can be revealed)? We present data to address this difficult question (see Section 2.1).

2. Cryonegative Staining — The "Adrian" Method: *Frozen-Hydrated Specimens in the Presence of a Saturated Ammonium Molybdate Solution at Neutral pH*

A very detailed protocol that includes materials needed as well as the step-by-step procedure is already available in the literature (De Carlo, 2008b). However, the procedure can be easily summarized as follows:

1. Add 1.2–1.3 g of ammonium molybdate tetrahydrate (Fluka, Sigma) in a 2.0-ml tube.
2. Add 0.875 ml of water.
3. Shake.
4. Add 0.125 ml 10 M NaOH.
5. Shake (it is a saturated solution, some powder remains and forms a slurry).
6. The EM-grid can be previously glow-discharged if needed.
7. The carbon film can be continuous or holey depending on your needs.
8. Avoid plastic (e.g., Formvar) on grids. If still present, dissolve plastic with ethyl-acetate/chloroform prior to experiment.
9. Typically, use 3–5 μl of sample on the EM grid, let the sample adsorb for 10–30 s if you are using a grid coated with continuous carbon film.
10. Prepare a petri dish (plastic or glass) with cover.
11. Put a square 2 by 2 in. piece of parafilm paper in the petri dish.
12. Deposit a 80–100-μl droplet of the staining solution on parafilm paper.
13. Place the grid on the staining-solution droplet (sample face down, toward the stain).
14. Remove the grid after 10–60 s, depending on the sample sensitivity to highly concentrated salts. Longer times of contact allow better stain distribution.
15. Mount the specimen on the plunge-freezing apparatus as fast as you can to avoid further water evaporation, or use commercially available units that allow environment-controlled vitrification (e.g., Vitrobot, FEI Company or Cp3 from Gatan).
16. Apply filter paper to remove excess liquid.
17. Remove filter paper and wait for 1–3 s before plunging the sample in the cryogen.
18. Continue as you would for a conventional cryo-EM grid.

The negative-staining solution consists of ammonium molybdate at saturated concentration. Briefly, we add 1.2–1.3 g of ammonium molybdate to 0.875 ml water, neutralized to pH 7.2–7.4 with the addition of 0.125 ml of 10 M NaOH at room temperature. This forms a saturated slurry that can be stored as such at room temperature for 2–3 h. Immediately before use, we quickly shake the slurry, let it sediment for a few seconds and take the supernatant as the negative staining solution. At this point, we have measured a solution density of 1.45 ± 0.05 g/cm^3. Finally, 80–100 μl aliquots of this staining solution are deposited on parafilm and subsequently used for the staining procedure (De Carlo et al., 2002, 2008). A short evaporation period (1–3 s) after blotting and before plunge freezing is essential for production of satisfactory thin films of vitrified ammonium molybdate, but care must be taken not to blot the grid dry (Fig. 6.4, bottom). Under these conditions, the vitrified specimens are still hydrated (in terms of protein-bound water), as the vitrified medium contains a saturated heavy-metal salt with $\sim$30% water by volume (Adrian et al., 1998).

We usually follow the cryonegative-staining technique of Adrian et al. (1998), using lab-made holey carbon films on 200-mesh copper grids. Commercially available grids like Quantifoil (Ted Pella, Inc.) or C–Flat (Protochips, Inc.) can also be used, but care should be taken in choosing the right hole size. Holes smaller than 3 μm tend to give thick stain layers which

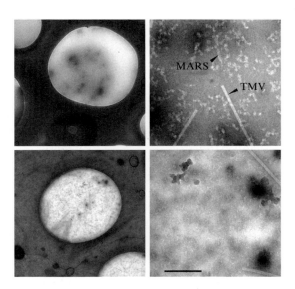

Figure 6.4 A field of view of cryonegatively stained MARS (multiaminoacyl-tRNA-synthetase) complexes in the presence of TMV (rod-like viruses). Scale bar = 1 μm (left panels) and 0.1 μm (right panels) (De Carlo and Boisset, unpublished data). *Top*: Low-magnification view of a hole in the carbon support, showing a typical cryonegatively stained area suitable for imaging, as illustrated in the top right panel at 40,000× magnification. *Bottom*: Low-magnification view of an area that is not suitable for high-quality imaging, as shown on the bottom right panel, the particles are suspended in a dry layer of stain. The decrease in SNR is striking when comparing the bottom right to top right panels.

are not suitable for imaging biological particulates at high resolution. The best results we have had so far were obtained after evaporating a thin layer of Au/Pd on the side of the holey carbon grid to assist sample spreading and stabilizing the carbon, a useful trick introduced to one of us (S. D. C.) by Marc Adrian (De Carlo, 2002).

Cryonegative staining can also be done with continuous carbon-coated grids. Controlling the thickness of the hydrated stain mix is difficult. However, regions of the grid where molecules are well embedded can easily be found and used for imaging. We successfully used continuous carbon grids with samples that are purified at low yield, such as human transcription factor TFIIE (Jawhari et al., 2006) and human RNA polymerase II (Kostek et al., 2006). The fact that samples were adsorbed on a continuous carbon film did not affect the structural analysis since we did not detect any specimen flattening, a common artifact due in part to adsorption of biological material on a continuous surface.

2.1. Results obtained with the Adrian method (saturated ammonium molybdate)

Results obtained through the past decade and more recently have clearly demonstrated that although not in their fully hydrated state, samples prepared with the Adrian cryonegative-staining technique are well preserved in the vitreous/frozen-hydrated state. Examples where a direct side-by-side comparison can be made are now available in the literature; among these are the TMV (Fig. 6.3; see also Adrian et al., 1998); the bacterial chaperonin GroEL (Fig. 6.5; see De Carlo et al., 2002, 2008), and several eukaryotic RNA polymerases (see De Carlo et al., 2003; Kostek et al., 2006). An example is presented in Fig. 6.6 with yeast RNA polymerase I. In other cases, cryonegative staining has revealed molecular details (i.e., between 10 and 15 Å) for proteins considered very small for cryo-EM, such as the human transcription factor TFIIE, which is a dimeric protein complex of $\sim$110 kDa (Jawhari et al., 2006).

The improved resolution resulting from cryonegative staining as compared to conventional room temperature negative staining seems to be due to several factors: these may include hydration of the biological particles, suspension in a thin layer of vitreous ice instead of adsorption to a continuous support, and the documented increase in SNR ratio. Partial specimen hydration is an obvious advantage, since it allows the embedded protein to maintain its 3D shape, thus avoiding fine structural collapse that is likely to occur in the total absence of water. The stain does not dry, thus preventing the formation of a dried cast onto and around the protein, the molybdate anions remain free in the vitreous solution and may rapidly diffuse into the protein accessible areas, allowing the high-resolution fine structure of the protein to be imaged in reversed contrast instead of imaging only the surface stain-excluded protein volume, as is the case for air-dry negative staining at room temperature. The key is in

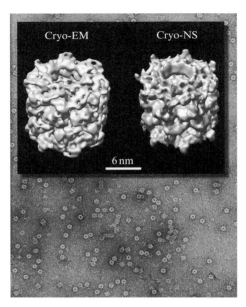

Figure 6.5 A field of view of GroEL prepared with ammonium molybdate according to the "Adrian" cryonegative-staining method (De Carlo *et al.*, 2002). Inset: side-by-side comparison of GroEL 3D reconstructions obtained by unstained cryo-EM (EMD-1081; Ludtke *et al.*, 2004) and cryonegative staining (De Carlo *et al.*, 2008).

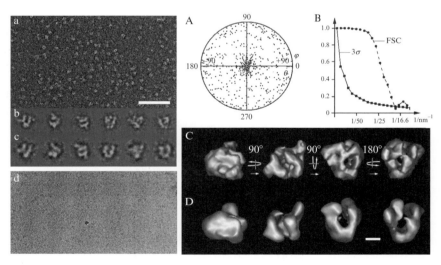

Figure 6.6 (a) Yeast RNA polymerase I cryonegatively stained with ammonium molybdate (from De Carlo *et al.*, 2003). (b) Representative class-averages obtained with the data set in (a). Compare with an unstained cryo-EM data set of the same preparation: (c) Class-averages representing the same views as in (b), note the much weaker SNR for this small particle in unstained cryo-EM in (d). (A) Angular distribution of the cryo-NS data set. (B) FSC plot of the cryo-NS data set and 3-σ noise plot. (C) 3D reconstruction of this data set in cryo-NS. (D) 3D reconstruction of the same data set in unstained cryo-EM.

the combination of adding solutes to the solution that preserves the hydration of protein, as has been proved for trehalose, with the cooling of samples (and their hydrated surfaces) to liquid N_2 temperatures. Cooling allows bound water to be preserved and it helps in reducing beam-irradiation damage (De Carlo et al., 1999).

A more detailed analysis of the possibility of reaching subnanometer resolution with cryonegative staining was presented by De Carlo et al. (2008), for the comparison between frozen-hydrated GroEL as compared to cryonegatively stained samples in the presence of a vitrified saturated ammonium molybdate solution. Imaging only 14,000 particles in the vitrified stain allowed the authors to reach about 10 Å resolution, as opposed to the hundreds of thousands of unstained GroEL particles necessary to reach subnanometer resolution, as demonstrated by several research groups (Ludtke et al., 2004; Ranson et al., 2006; Stagg et al., 2006). It should be noted, however, that the use of a very large number of particles is needed to achieve 4–6 Å resolution, as was the case for the aforementioned studies.

Studying biological enzyme activity has proved to be possible in the presence of the vitrified stain on a continuous carbon film. We showed that human RNA polymerase II could be imaged at ~ 20 Å resolution, fully preserved in the thin vitreous stain layer supported by a continuous carbon film (Kostek et al., 2006). Moreover, using statistical analysis methods such as the 3D variance bootstrap procedure (Penczek et al., 2006a,b), we could distinguish between two distinct conformational states of the enzyme.

Conventional air-drying negative staining is achieved with a heavy-metal salt, usually used in the 0.5–3% (w/v) concentration range. There are several heavy-metal salts listed in the literature that are very effective contrasting agents (Harris, 1997). However, for cryonegative staining according to Adrian, we need the staining solution to be at saturated concentration. Besides ammonium molybdate, all the common salts used for conventional air-drying negative-staining precipitate if their concentration is raised.

Another advantage of using anionic molybdate over the commonly used cationic uranyl salts is that even at saturated concentrations it can be buffered to any desired physiological pH by the addition of a very small volume of strong base, although this comment could also apply to the phosphotungstate/silico-tungstate negative-staining salts. Cryonegative staining can readily be performed in a semiautomated fashion inside a blotting and vitrification device that is fully computer controlled, such as the Vitrobot (FEI) or the Cp3 (Gatan). It should be noted, however, that there can be a difference between specimens prepared by single versus double-sided blotting, particularly if it is desirable to induce 2D crystal formation at one untouched fluid–air interface prior to plunge freezing. Special precautions may need to be taken if the blotting chamber is used at 100% relative humidity. In our experience, it is better to keep the sample environment relatively dry immediately before blotting. If the sample buffer concentrations need to be maintained at defined humidity values, the blotting parameters on the automated vitrification device need to be changed accordingly.

2.2. Advantages and limitations

Does the technique always work? Quoting Yogi Berra "In theory, there is no difference between theory and practice. In practice, there is." Provided the protein complex under investigation remains native in the concentrated ammonium molybdate (this may be cross-checked with conventional negative staining and cryo-EM and can be judged from the fitting parameters of individual particles to each other), cryonegative staining has been found to have several advantages. Our main findings obtained so far from cryonegative staining on biological particulates imaged in a field-emission gun (FEG) electron microscope under strict cryoconditions can be summarized as follows:

- The negative-staining solution produces a higher SNR than conventional unstained cryo-EM. This has several positive effects on 3D structure determination of macromolecular complexes, as further outlined below.
- For reaching a certain resolution on a given structure, fewer particles have to be collected than with conventional cryo-EM.
- The minimal size of particles may be lower because they are more easily visualized and classified due to the higher contrast.
- The defocus values during image recording can be lowered, extending the first zero-node of the contrast transfer function (CTF) further out. Unstained GroEL is hardly detectable on electron micrographs recorded with ~800 nm defocus at 200 kV, while they are clearly visible in the presence of ammonium molybdate.
- Samples are less beam-sensitive in the presence of stain as shown by the absence of damage despite the threefold increase in the exposure dose (see Fig. 1 in De Carlo et al., 2002). However, a precise quantification remains to be done.
- The vitrified and stained particles deliver a stronger Thon ring pattern, facilitating image quality control and CTF correction.

However, we have also encountered a few limitations of cryonegative staining. Most importantly, the structure under study must withstand the exposure to the concentrated ammonium molybdate. Over the last decade, a few biological assemblies showed sensitivity to the saturated stain and tend to partially break apart into their smaller subunits (e.g., KLH). These rare, yet important, unsuccessful cases include microtubules and some multimeric ATPases, such as NtrC (De Carlo et al., 2006), which dissociate completely into their monomeric components (data not shown). If the sample allows it, the presence of high amounts of glycerol should be avoided. The same applies to detergents for isolated membrane proteins.

In the presence of high amounts of glycerol (or other sugars) or detergent, we observed that ammonium molybdate does not mix with the sample and large areas of the grid are simply unstained (De Carlo, unpublished data).

The other important finding is that, while the macromolecules that are well preserved in vitrified saturated ammonium molybdate, and their structure is modulated with a good accuracy to about 10 Å resolution, internal fine structures are sometimes lost, that is, the accuracy may not be isotropic in the 3D map. This may be due to insufficient stain penetration at these locations, resulting in structural details being imaged at reverse contrast (positive contrast as in conventional cryo-EM instead of negative contrast as from cryonegative staining). Alternatively, cryonegative staining may overemphasize the contrast of low-resolution features. Hence, the range in contrast from low- to high-resolution structures seems much larger than in conventional cryo-EM. Also, contrast enhancement may not be the same for all spatial frequencies, and it is difficult to estimate the real contribution of stain to the amplitude contrast in the CTF, especially at low spatial frequencies. The resolution achieved so far is about 9–10 Å for the Adrian method (De Carlo *et al.*, 2008) as well as for the Stark method introduced in Section 3 (Golas *et al.*, 2003).

3. Cryonegative Staining with the Sandwich Method: The "Holger Stark" Alternative

An alternative method to the one described earlier is to basically use normal negative-stain grid preparation protocols. The only difference of a cryonegative-stain grid according to the Stark method as opposed to a normal negative-stain grid is that the cryonegative-stain grid is not left for drying at room temperature prior to transfer into the microscope. Instead, after binding the molecules to a carbon support film, incubation with usually 2% negative-stain solution and final blotting of excess staining solution, grids are left at room temperature for 1–2 min before they are frozen in liquid nitrogen (Golas *et al.*, 2003). The procedure is outlined here and illustrated in Fig. 6.7.

1. A slide coated with a supporting carbon film (1) is inserted in the sample-containing solution (step A).
2. The sample is now adsorbed onto the carbon surface and is floated upside-down into a well containing uranyl formate (step B).
3. A grid is used to pick up the sample adsorbed onto the carbon film (1).
4. A second, thin carbon film (2) is floated onto a well containing uranyl formate (step C).
5. The grid with the sample is inserted in the well with the samples facing up, underneath carbon film "2," such that the samples are "sandwiched" between the supporting carbon (1) and the thin carbon (2).
6. The grid rests for 1–2 min and then frozen in liquid nitrogen (step D).

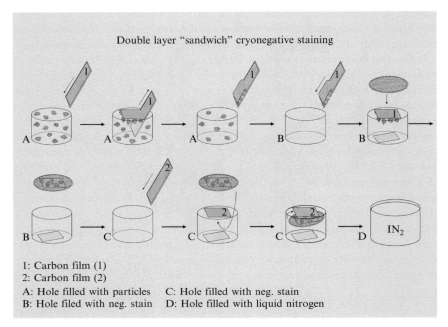

Double layer "sandwich" cryonegative staining

1: Carbon film (1)
2: Carbon film (2)
A: Hole filled with particles   C: Hole filled with neg. stain
B: Hole filed with neg. stain   D: Hole filled with liquid nitrogen

Figure 6.7 A schematic outline of the "Stark" alternative for cryonegative staining (Golas *et al.*, 2003). The "sandwich" term comes from the use of two carbon films "trapping" the particle in between (Tischendorf *et al.*, 1974). See text for details.

The main advantage of this method compared to conventional air-drying negative staining is that the biological sample is not dried and that it is transferred into the microscope in a fully hydrated state. Since sample transfer into the microscope as well as imaging is carried out at liquid nitrogen temperature, the molecules remain fully hydrated very similar to unstained cryopreparations. This can easily be tested by evaporating water under the electron beam in cryonegatively stained samples (Herzog *et al.*, 2009), similar to the freeze-drying procedure performed by Adrian *et al.* (1998). The image contrast and the overall appearance of the molecules are almost identical to conventional negative-stain grids. Depending on the water content, the contrast might be only slightly reduced when compared to completely dried negative-stain grids. The cryonegative-stain grid preparation alternative proposed by Stark was successfully applied to a number of ribonucleoprotein complexes (RNPs) and protein complexes in the past 10 years (Boehringer *et al.*, 2004; Chari *et al.*, 2008; Deckert *et al.*, 2006; Golas *et al.*, 2003, 2005, 2009; Herzog *et al.*, 2009; Sander *et al.*, 2006; Fig. 6.8). The main reasons for using this technique were the following:

1. snRNPs and spliceosomes like many other complexes are not stable during sample purification and grid preparation without glycerol being present in

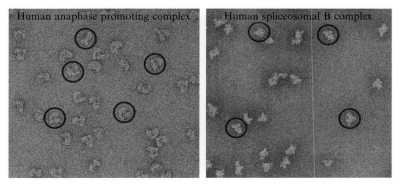

Figure 6.8 Cryonegative-stain raw images of the human anaphase-promoting-complex (Herzog *et al.*, 2009) and the human spliceosomal B complex (Boehringer *et al.*, 2004).

solution. While the presence of glycerol is acceptable for cryonegative-stain grid preparation, it is incompatible with unstained cryogrid preparation.

2. The sample concentration that is required for this cryonegative-stain grid preparation is lower than for unstained cryo-EM and also lower compared to the "Adrian" version of cryonegative-staining technique.

The "Stark" technique was tested for a number of complexes by comparing with either known X-ray structure or unstained cryo-EM 3D structures. For the human U4/U6.U5 tri-snRNP, the major subcomplex of the spliceosome, the unstained cryo-EM and the cryonegative-stain 3D structures are virtually identical down to 20 Å resolution (Sander *et al.*, 2006). The same holds true for the human anaphase-promoting-complex which is involved in cell-cycling regulation (Herzog *et al.*, 2009). Another example of a cryonegative-stain 3D structure obtained with the "Stark" alternative is the influenza A virus protein hemagglutinin, which at ∼14 Å resolution, reveals a striking similarity to the high-resolution X-ray structure of this protein (Böttcher *et al.* (1999; Fig. 6.9). Even 9 Å resolution structure determination is possible with cryonegative staining as shown for the structure of the human spliceosomal subcomplex SF3b (Golas *et al.*, 2003).

4. CONCLUSION

Cryonegative staining represents a complementary method to conventional cryo-EM of unstained vitrified samples and provides a valuable alternative, in particular, for samples too small for unstained cryo-EM, or where a preliminary 3D model is required to initiate a cryo-EM image averaging

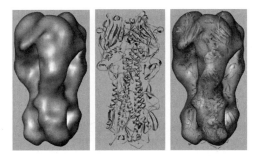

Figure 6.9 Comparison of the cryonegative-stain 3D reconstruction of influenza A hemagglutinin with the structure determined by X-ray crystallography. The EM reconstruction was determined at 14 Å resolution and shows excellent agreement with the X-ray structure (Böttcher *et al.*, 1999).

approach. Cryonegative staining preserves the structure of biological macromolecules to fine detail (i.e., 9–10 Å). The vitrified ammonium molybdate solution surrounding the sample can access and reveal internal densities of a protein, such as α-helices. This finding is of great importance for structural biology studies of macromolecules that are too small for single-particle reconstruction (less than 200 kDa), and especially with cases where subunits or even smaller functional domains of a protein complex cannot be detected in cryo-EM density maps due to a lack of contrast.

While it is now obvious that the cryonegative-staining technique can be helpful for small macromolecular complexes, there are a number of limitations. With the "Adrian" method, the protein complex under study must withstand the concentrated ammonium molybdate.

With the "Stark" alternative, while the molecules remain hydrated, the well-known flattening effect that can be observed in conventional air-drying negative-stain grid preparation cannot be avoided completely. A combination of cryonegative-stain with the carbon-sandwich technique can even lead to ~30% flattening of the sample. This amount of flattening was observed for TBSV, while it is usually considerably less in case of smaller macromolecular complexes in the size range of a few hundred kDa. Single-carbon film cryonegative-stain grids show much less flattening effects, indicating that in the carbon-sandwich technique, the pressure between the two carbon films is very high even when the sample is kept in a fully hydrated state. A summary of pro and cons for each technique is given in Table 6.1.

The maximum achievable resolution by cryonegative staining still remains unknown. It may well depend on the cluster size of the heavy-metal salt in solution which would make resolutions of up to 5–6 Å possible to be achieved in the future.

Table 6.1 *Pro and cons* of each cryonegative-staining method

Which cryonegative-staining procedure is right for you?		
Specimen conditions, expected outcomes,...	*A*	*S*
Produces outstanding contrast enhancement (helpful for small particles)	✓	✓
Has demonstrated capabilities of achieving at least 10 Å resolution	✓	✓
Requires liquid Nitrogen temperature for data collection	✓	✓
Is guaranteed a 100% "artifact-free"	✗	✗
Maintains the samples at neutral (physiological) pH	✓	✗
Does not use radioactive substances	✓	✗
Can be used in the presence of 10–30% glycerol	✗	✓
Can be used in the presence of detergents	✗	✓
Can be used with a single-carbon film	✓	✗
Can in principle be achieved with any negative stain (i.e., uranyl formate/actetate, etc.)	✗	✓
Can be performed with small amounts of sample	✓[a]	✓
May cause specimen flattening artifacts	✗	✓
May damage the specimen due to highly concentrated stain	✓	✗

A, "Adrian" method; S, "Stark" method.
[a] Provided that a continuous carbon film support is used.

ACKNOWLEDGMENTS

S. De Carlo would like to thank the Research Centers in Minority Institutions for support through an NIH/NCRR/RCMI Grant G12-RR03060 to CCNY.

REFERENCES

Adrian, M., Dubochet, J., Lepault, J., and McDowall, A. W. (1984). Cryo-electron microscopy of viruses. *Nature* **308**, 32–36.

Adrian, M., Dubochet, J., Fuller, S. D., and Harris, J. R. (1998). Cryo-negative staining. *Micron* **29**, 145–160.

Adrian, M., Cover, T. L., Dubochet, J., and Heuser, J. E. (2002). Multiple oligomeric states of the *Helicobacter pylori* vacuolating toxin demonstrated by cryoelectron microscopy. *J. Mol. Biol.* **318**, 121–133.

Boehringer, D., Makarov, E. M., Sander, B., Makarova, O. V., Kastner, B., Luhrmann, R., and Stark, H. (2004). Three-dimensional structure of a pre-catalytic human spliceosomal complex B. *Nat. Struct. Mol. Biol.* **11**, 463–468.

Böttcher, C., Ludwig, K., Herrmann, A., van Heel, M., and Stark, H. (1999). Structure of influenza haemagglutinin at neutral and at fusogenic pH by electron cryo-microscopy. *FEBS Lett.* **463**, 255–259.

Brenner, S., and Horne, R. W. (1959). A negative staining method for high resolution electron microscopy of viruses. *Biochim. Biophys. Acta* **34**, 60–71.

Chari, A., Golas, M. M., Klingenhager, M., Neuenkirchen, N., Sander, B., Englbrecht, C., Sickmann, A., Stark, H., and Fischer, U. (2008). An assembly chaperone collaborates with the SMN complex to generate spliceosomal SnRNPs. *Cell* **135,** 497–509.

De Carlo, S. (2002). Cryo-negative staining: Advantages and applications for three-dimensional electron microscopy of biological macromolecules. PhD Thesis, Université de Lausanne, Switzerland. Available online at, http://www.planetesacha.com/PDF/DeCarlo_2002_PhD.pdf.

De Carlo, S. (2008). Cryo-negative staining. *In* "Handbook of Cryopreparation Methods for Electron Microscopy," (Annie Cavalier, Danielle Spehner, and Bruno M. Humbel, eds.), 9780849372278. CRC Press, Boca Raton, FL.

De Carlo, S., Adrian, M., Kälin, P., Mayer, J. M., and Dubochet, J. (1999). Unexpected property of trehalose as observed by cryo-electron microscopy. *J. Microsc.* **196,** 40–45.

De Carlo, S., El-Bez, C., Alvarez-Rúa, C., Borge, J., and Dubochet, J. (2002). Cryo-negative staining reduces electron-beam sensitivity of vitrified biological particles. *J. Struct. Biol.* **138,** 216–226.

De Carlo, S., Carles, C., Riva, M., and Schultz, P. (2003). Cryo-negative staining reveals conformational flexibility within yeast RNA polymerase I. *J. Mol. Biol.* **329,** 891–902.

De Carlo, S., Chen, B., Hoover, T. R., Kondrashkina, E., Nogales, E., and Nixon, B. T. (2006). The structural basis of regulated assembly and function of the transcriptional activator NtrC. *Genes Dev.* **20,** 1485–1495.

De Carlo, S., Boisset, N., and Hoenger, A. (2008). High-resolution single-particles 3D analysis on GroEL prepared by cryo-negative staining. *Micron* **39,** 934–943.

Deckert, J., Hartmuth, K., Boehringer, D., Behzadnia, N., Will, C. L., Kastner, B., Stark, H., Urlaub, H., and Luhrmann, R. (2006). Protein composition and electron microscopy structure of affinity-purified human spliceosomal B complexes isolated under physiological conditions. *Mol. Cell. Biol.* **26,** 5528–5543.

Dubochet, J., Adrian, M., Chang, J. J., Homo, J. C., Lepault, J., McDowall, A. W., and Schultz, P. (1988). Cryo-electron microscopy of vitrified specimens. *Q. Rev. Biophys.* **21**(2), 129–228.

El-Bez, C., Adrian, M., Dubochet, J., and Cover, T. L. (2005). High resolution structural analysis of *Helicobacter pylori* VacA toxin oligomers by cryonegative staining electron microscopy. *J. Struct. Biol.* **151,** 215–228.

Golas, M. M., Sander, B., Will, C. L., Lührmann, R., and Stark, H. (2003). Molecular architecture of the multiprotein splicing factor SF3b. *Science* **300,** 980–984.

Golas, M. M., Sander, B., Will, C. L., Luhrmann, R., and Stark, H. (2005). Major conformational change in the complex SF3b upon integration into the spliceosomal U11/U12 di-snRNP as revealed by electron cryomicroscopy. *Mol. Cell* **17,** 869–883.

Golas, M. M., Bohm, C., Sander, B., Effenberger, K., Brecht, M., Stark, H., and Goringer, H. U. (2009). Snapshots of the RNA editing machine in trypanosomes captured at different assembly stages in vivo. *EMBO J.* **28,** 766–778.

Harris, J. R. (1997). Negative staining and cryo-electron microscopy: The thin film techniques. RMS Microscopy Handbook No. 35. BIOS Scientific, Oxford.

Harris, J. R., and Adrian, M. (1999). Preparation of thin-film frozen-hydrated/vitrified biological specimens for cryoelectron microscopy. *In* "Electron Microscopy Methods and Protocols, Vol. 117," (M. A. Nasser Hajibagheri, ed.), pp. 31–48. Humana Press, New Jersey.

Harris, J. R., and Scheffler, D. (2002). Routine preparation of air-dried negatively stained and unstained specimens on holey carbon support films: A review of applications. *Micron* **33,** 461–480.

Harris, J. R., Bhella, D., and Adrian, M. (2006). Recent developments in negative staining for transmission electron microscopy. *Microsc. Anal.* **20,** 17–21.

Herzog, F. I., Primorac, I., Dube, P., Lenart, P., Sander, B., Mechtler, K., Stark, H., and Peters, J. M. (2009). Structure of the anaphase-promoting complex/cyclosome interacting with a mitotic checkpoint complex. *Science* **323,** 1477–1481.

Hirai, T., Murata, K., Mitsuoka, K., Kimura, Y., and Fujiyoshi, Y. (1999). Trehalose embedding technique for high-resolution electron crystallography: Application to structural study on bacteriorhodopsin. *J. Electron Microsc.* **48,** 653–658.

Jawhari, A., Uhring, M., De Carlo, S., Crucifix, C., Tocchini-Valentini, G., Moras, D., Schultz, P., and Poterszman, A. (2006). Structure and oligomeric state of human transcription factor TFIIE. *EMBO Rep.* **7,** 500–505.

Kostek, S., Grob, P., De Carlo, S., Lipscomb, S., Garczarek, F., and Nogales, E. (2006). Molecular architecture and conformational flexibility of human RNA polymerase II. *Structure* **14,** 1691–1700.

Lepault, J., Booy, F. P., and Dubochet, J. (1983). Electron microscopy of frozen biological suspensions. *J. Microsc.* **129,** 89–102.

Ludtke, S., Chen, D., Song, J., Chuang, D., and Chiu, W. (2004). Seeing GroEL at 6 Å resolution by single particle electron cryomicroscopy. *Structure* **12,** 1129–1136.

Massover, W. H. (2008). On the experimental use of light metal salts for negative staining. *Microsc. Microanal.* **14,** 126–137.

Penczek, P. A., Yang, C., Frank, J., and Spahn, C. M. (2006a). Estimation of variance in single-particle reconstruction using the bootstrap technique. *J. Struct. Biol.* **154,** 168–183.

Penczek, P. A., Frank, J., and Spahn, C. M. (2006b). A method of focused classification, based on the bootstrap 3D variance analysis, and its application to EF-G-dependent translocation. *J. Struct. Biol.* **154,** 184–194.

Ranson, N. A., Clare, D. K., Farr, G. W., Houldershaw, D., Horwich, A. L., and Saibil, H. R. (2006). Allosteric signaling of ATP hydrolysis in GroEL–GroES complexes. *Nat. Struct. Mol. Biol.* **13,** 147–152.

Sander, B., Golas, M. M., Makarov, E. M., Brahms, H., Kastner, B., Luhrmann, R., and Stark, H. (2006). Organization of core spliceosomal components U5 snRNA loop I and U4/U6 Di-snRNP within U4/U6.U5 Tri-snRNP as revealed by electron cryomicroscopy. *Mol. Cell* **24,** 267–278.

Stagg, S. M., Lander, G. C., Pulokas, J., Fellmann, D., Cheng, A., Quispe, J. D., Mallick, S. P., Avila, R. M., Carragher, B., and Potter, C. S. (2006). Automated cryoEM data acquisition and analysis of 284742 particles of GroEL. *J. Struct. Biol.* **155,** 470–481.

Stark, H., and Luhrmann, R. (2006). Cryo-electron microscopy of spliceosomal components. *Annu. Rev. Biophys. Biomol. Struct.* **35,** 435–457.

Tischendorf, G. W., Zeichhardt, H., and Stoffler, G. (1974). Determination of the location of proteins L14, L17, L18, L19, L22, L23 on the surface of the 50S ribosomal subunit of *Escherichia coli* by immune electron microscopy. *Mol. Gen. Genet.* **134,** 187–208.

Tsitrin, Y., Morton, C. J., El-Bez, C., Paumard, P., Velluz, M.-C., Adrian, M., Dubochet, J., Parker, M. W., Lanzavecchia, S., and van der Goot, F. G. (2002). Conversion of a transmembrane to a water-soluble protein complex by a single point mutation. *Nat. Struct. Biol.* **9,** 729–733.

CHAPTER SEVEN

LIPOSOMES ON A STREPTAVIDIN CRYSTAL: A SYSTEM TO STUDY MEMBRANE PROTEINS BY CRYO-EM

Liguo Wang *and* Fred J. Sigworth

Contents

Abstract

In this chapter, we describe the preparation of cryo-EM specimens for random spherically constrained (RSC) single-particle reconstruction of membrane proteins. The specimen consists of liposomes into which the purified membrane protein is reconstituted at low density. The substrate is a 2D streptavidin crystal, which serves as an affinity surface that tethers the liposomes, which are doped with biotinylated lipids; the crystal can also serve as an image-quality

Department of Cellular and Molecular Physiology, Yale University, New Haven, Connecticut, USA

Methods in Enzymology, Volume 481
ISSN 0076-6879, DOI: 10.1016/S0076-6879(10)81007-9

and image-calibration reference. After subtraction of the crystal and lipid membrane contributions to the image, the remaining particle images can be used for 3D reconstruction.

1. INTRODUCTION

A long-sought goal in structural biology has been the imaging of membrane proteins in their lipid environments. This goal has been achieved in several cases where the proteins can form 2D crystalline arrays in planar membranes (Chapter 4) or in helical membrane tubes (Chapter 2 of this volume and Chapter 7 of vol. 483). As an alternative approach, we have sought to extend the generality and power of the single-particle reconstruction technique to membrane proteins. Tilley *et al.* (2005) were the first to obtain single-particle reconstructions of membrane proteins in lipid membranes from hundreds of "side views" (within 20° of the equator) of the proteins in liposomes of variable size. We have recently described a more general method, in which membrane protein particles are imaged in spherical liposomes. The spherical shape allows the lipid membrane contribution to each image to be computationally removed, regardless of the liposome size, and also aids the determination of the orientation of each protein particle. This method has been successfully employed to study the structure of the BK "big K^+" calcium-activated potassium channel (Wang and Sigworth, 2009).

In this chapter, we describe the steps in assembling the particular type of cryo-EM specimen that we have used for this method (Fig. 7.1). Membrane proteins are reconstituted at low density into liposomes, and the resulting proteoliposomes are applied to a nanostructured substrate. The substrate is a 2D streptavidin crystal, which spans the holes in a perforated carbon film. The streptavidin serves as an affinity surface, tethering liposomes at a uniform density. The ordered lattice of this crystal serves as a built-in image-calibration and evaluation reference. The periodic lattice also allows efficient subtraction of the signal from the streptavidin layer, yielding less background noise than traditional ultrathin carbon films.

A typical image of this specimen is shown in Fig. 7.1C. In order to determine the structure of the membrane protein, the contributions to the image of the 2D crystal and the lipid membrane are first removed, and then the protein particles are picked from the resulting images (Fig. 7.1D). From the position of each imaged particle with respect to the center of the liposome, two of the particle's three Euler angles can be estimated. This greatly reduces the computational cost of determining the orientation of each particle and increases the reliability of the orientation assignment.

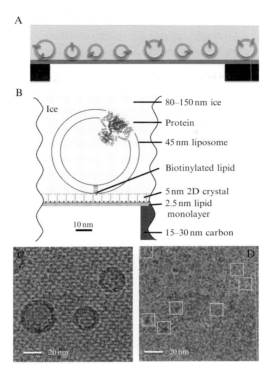

Figure 7.1   Liposomes on a streptavidin crystal. (A) Scheme of the tethering of liposomes in a vitreous ice layer (light gray) on a 2D crystal substrate (gray) spanning a hole in the carbon support film (black). (B) Scale drawing of the tethering system. The 2D crystal layer (5.0 nm thick) is bound to a lipid monolayer (2.5 nm) which spans a hole in the perforated carbon film (15–30 nm in thickness). (C) Portion of a cryo-EM micrograph showing proteoliposomes and the streptavidin-crystal background. (D) The same image, after the contribution of lipid membrane and 2D crystals, is removed. Manually picked particles are marked by white boxes.

Given orientations of the particles, the 3D structure of the protein can then be determined using conventional 3D reconstruction methods.

 ## 2. METHODS

2.1. Streptavidin crystals

Streptavidin is a 15-kDa protein of 159 residues (Sano *et al.*, 1995). It is proteolyzed naturally at both ends, and the most stable and well-studied form of this protein (called core-streptavidin) contains 125–127 residues.

The atomic structure has been determined by X-ray diffraction (Hendrickson *et al.*, 1989; Weber *et al.*, 1989). Other structures (full length, biotin-bound complex, and mutants) have also been determined (Freitag *et al.*, 1997, 1999; Izrailev *et al.*, 1997). Streptavidin forms a 2D crystal on a monolayer of biotinylated lipid; this assembly has been studied by electron crystallography (Avila-Sakar and Chiu, 1996; Darst *et al.*, 1991; Kubalek *et al.*, 1991; Le Trong *et al.*, 2006; Wang *et al.*, 1999). The crystal structures show that the protein is a homotetramer of identical subunits. Each subunit contains a β-barrel, with the biotin-binding site located at one end (Fig. 7.2A).

Two-dimensional crystals grown on a lipid monolayer also show a tetrameric arrangement of subunits (Fig. 7.2A). The 2D crystal has a unit cell with $a = b = 8.23$ nm, $\gamma = 90°$ (Avila-Sakar and Chiu, 1996). Due to the mirror symmetry in the projection map, the reflection spots with indices $h + k = 2n + 1$ are absent; thus in projection, the crystal appears to have a square lattice with $a = b = 5.82$ nm and $\gamma = 90°$. We use these parameters in analyzing our cryo-EM images.

The 2D streptavidin crystal has been shown to be readily picked up by an EM grid coated with a perforated carbon film (Avila-Sakar and Chiu, 1996; Crucifix *et al.*, 2004; Kubalek *et al.*, 1991), and crystal sizes up to a few square micrometers have been observed. Crystals fast-frozen in vitreous ice show electron diffraction to atomic resolution (Avila-Sakar and Chiu, 1996), and the 2D crystals have been shown to be useful as tethering surfaces, for example, tethering biotinylated DNA molecules (Crucifix *et al.*, 2004).

2.2. Growth of the streptavidin crystal

The 2D streptavidin crystal is grown on a lipid monolayer containing biotinylated lipid (Darst *et al.*, 1991), as shown in Fig. 7.2B. Streptavidin solution is loaded into a well, onto whose surface a lipid monolayer containing biotinylated lipids has been formed. As a first step in the formation of the 2D crystal, two of the four binding sites on the streptavidin tetramer bind sequentially to biotin lipid headgroups. Then, by diffusing in the plane of the monolayer, the tetramers interact to form 2D arrays. In the resulting crystal, each streptavidin tetramer has two vacant biotin-binding sites that are available to bind to other molecules.

Streptavidin was purchased from Sigma-Aldrich (St. Louis, MO). Biotinylated dipalmitoyl-phosphatidylethanolamine (biotin-DPPE), 1,2-dipalmitoyl-sn-glycero-3-phosphoethanolamine-N-(cap biotinyl) (biotin-cap-DPPE), and dioleyl-phosphatidylcholine (DOPC; Avanti, Alabaster, AL) were used as received. Two-dimensional streptavidin crystals are grown at room temperature using a procedure similar to that described by Kubalek *et al.* (1991). Sixty-seven microliters of Tris buffer (150 m*M* NaCl, 50 m*M*

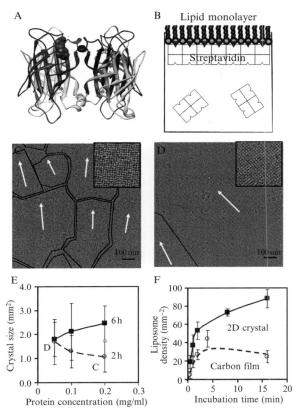

Figure 7.2 The streptavidin crystal as a tethering substrate. (A) Side view of the streptavidin tetramer. One bound biotin is shown as a space-filling model, and the other three binding sites are marked with asterisks. (B) Two-dimensional crystals are formed by attaching streptavidin molecules to a lipid monolayer via specific biotin–avidin binding. Biotinylated lipid molecules are indicated by triangles. (C, D) Negative-stain EM images show crystals grown from 0.2 to 0.05 mg/ml streptavidin solutions for 2 h at room temperature. The black lines outline the domain boundaries, the arrow shows the crystal orientation in each region, and the insets are 3× close-ups. (E) Dependence of the average domain size on the protein concentration and growth time. The crystal growth time was varied from 2 h (dashed line) to 6 h (solid line) and overnight (data not shown, but the crystal covered entire 2 μm-diameter holes for 0.2 mg/ml of protein concentration). Error bars are the standard deviation of crystal domain sizes with $n = 20$. (F) The dependence of the average liposome density on the incubation time. The lines represent the liposome densities on 2D crystal (solid line) and on a continuous carbon support film (dashed line), respectively.

Tris, pH 7.0) containing 0.05–0.40 mg/ml of streptavidin is deposited in a microwell formed by the polypropylene cap of a 0.2-ml PCR tube. Then 0.5–1.5 μl of lipid solution is spread on the surface; the lipid solution is chloroform/hexane (1:1) containing 0.1 mg/ml biotin-DPPE and 0.4 mg/ml DOPC. The PCR caps are placed in a humid chamber and kept at room temperature for 2–20 h.

Several crystal growth conditions were investigated. We found that when the growth time was short (e.g., 2 h), the average crystal domain size decreased as the streptavidin concentration in the subphase under the biotinylated lipid monolayer was increased from 0.05 to 0.2 mg/ml (Fig. 7.2C and D). However, the average domain size increased with the protein concentration when the growth time was 6 h, as shown in Fig. 7.2E. With an overnight growth at a protein concentration of 0.2 mg/ml, individual crystal domains were usually large enough to cover the entire 2-μm-diameter holes. Furthermore, the 2D crystal layer covered more than 90% of the holes in an EM grid, and the crystal over holes remained intact even in negative stain.

2.3. Crystal transfer and liposome tethering

The streptavidin crystal is picked up by a perforated carbon film on an EM grid. We prefer to use "aged" perforated carbon film, at least 4–7 days old, since freshly deposited carbon film is hydrophilic and cannot be used to pick up the hydrophobic lipid monolayer where the 2D crystal is grown. We have used standard 400-mesh EM grids coated with a perforated carbon film made by the stamping method (Chester et al., 2007), or alternatively, C-flat^{TM} Holey Carbon Grids (Electron Microscopy Sciences, Hatfield, PA). A grid is washed with hexane and dried in air for an hour, and then is placed for 1 min on the surface of the well where the crystal has been grown. Then, the grid is lifted vertically away from the surface by tweezers having a bent tip and placed onto a drop (25 μl on a sheet of Parafilm) of Tris buffer to wash away free streptavidin molecules, as shown in Fig. 7.3A. In an early version of this procedure, we continued to use tweezers to transfer the grid from one drop to another, but we have found that the large surface-tension effects of lifting the grid from a drop (Fig. 7.3B) break the crystal into small domains, smaller than 0.5 μm^2 (see also Chapter 4).

To minimize the disruption, we now use a "loop-fishing" method for transferring the grid from one solution to another. The 2D crystal is removed from the crystallization well as described earlier but is then transferred onto a well of the wash buffer to remove free streptavidin molecules. This well is the cap of a 1.5-ml Eppendorf tube (Fig. 7.3C). Starting from this point, the handling of the grid is by a metal loop instead of a pair of tweezers. The loop is about 3.5 mm in diameter, slightly larger than the EM grid; it can be made by wrapping a piece of metal wire around a metal rod. The important point is that the loop should be as circular and planar as possible. Such a loop may also be purchased from Ernest F. Fullam, Inc.

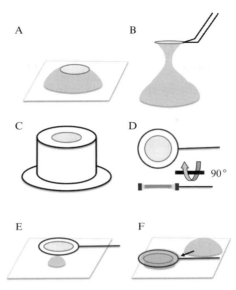

Figure 7.3 Crystal washing and transfer techniques. (A) An EM grid with the 2D crystal is placed onto a droplet of wash buffer. (B) When the EM grid is picked up by tweezers, strong surface-tension effects deform the crystal. (C) An EM grid with the 2D crystal is placed onto a well filled with wash buffer. (D) The grid shown in (C) is picked up by a wire loop. The side view shows the grid floating on the water layer trapped in the loop. (E) The grid is placed onto a droplet of proteoliposome suspension. (F) A droplet of dilution buffer is placed near the grid and then dragged to the grid with a pipette tip.

(Latham, NY). The loop is inserted into the buffer at an angle perpendicular to the surface, rotated to be parallel to the grid and moved under the grid, and then withdrawn straight up to pick up the grid. In this process, the loop receives the surface-tension forces during the removal of the grid from the well; the grid itself floats on the water layer trapped in the loop, as shown in Fig. 7.3D. The grid is transferred to two additional wash-wells in the same manner. Note that the water surface in the wells should be flat. A concave or convex surface causes the grid to jump during its landing from the loop to the surface.

After the free streptavidin is washed away, the grid, floating on the water layer in the loop, is placed onto a drop of the sample solution—here, the proteoliposome suspension—on a sheet of Parafilm (Fig. 7.3E). The sample volume is 6–10 μl, chosen small to allow a gentle merging of the loop buffer with the drop. The side with the streptavidin crystal faces down and makes contact with the liposome suspension in the 3× diluted running buffer (composition is given below). The liposomes are doped with biotin-cap-DPPE at 1:3600 molar ratio with 1-palmitoyl-2-oleoyl-*sn*-glycero-3-phosphocholine (POPC) to yield about three copies per 30 nm liposome. The low density of biotinylated lipid is chosen for two reasons. First, we seek to avoid the flattening of liposomes which can occur when multiple tethers attach to the streptavidin crystal. We do not know to what extent the liposome is distorted when a few tethers are present, but we do not observe any distortion effects in our projection images and assume the distortion to be small. Second, we find that at higher biotinylated lipid concentrations, considerable direct cross-linking of liposomes occurs through residual free streptavidin in the solution.

The specific binding between biotin and streptavidin allows the desired density of liposomes on the EM grid to be achieved despite the relatively low concentration of liposomes in the sample solution. By controlling the tethering time, the density of tethered liposomes can be tuned, as shown in Fig. 7.2F. This is especially critical for membrane proteins that are hard to express. It turns out that the biotin–DPPE is inferior to the biotin–cap–DPPE for doping the liposomes, because it binds preferentially to disordered regions but poorly to ordered crystalline regions of the streptavidin surface. This presumably arises from the position of the binding pocket at one end of the β-barrel (Fig. 7.2A) and the fact that in biotin–cap–DPPE, the longer linker gives more mobility to the biotin group. It can be imagined that biotin binding might cause a local distortion of the crystal, but we have not investigated this possibility.

After the desired time of incubation, typically 5–20 min in a humid atmosphere, a 25-μl drop of dilution buffer having the same composition as the sample solution but 3× diluted with water, is put onto the Parafilm and dragged to the loop via a 200-μl pipette tip, as shown in Fig. 7.3F. This dilution represents the first of several osmotic shocks that are used to swell the liposomes, as described later. The dragging process should be slow to prevent abrupt merging of the droplet with the loop, again to protect the integrity of the 2D crystal. The grid is then picked up by tweezers and either stained or fast-frozen for negative-stain or cryoimaging, respectively. In preparing cryosamples, 6 μl of 5× diluted buffer is added right before freezing to ensure that the external osmotic pressure is lower than the internal osmotic pressure even if water evaporates during the freezing process.

Plunge-freezing is carried out by standard techniques (Chapter 3). We blot grids manually from one side. The frozen ice layer is flat, because the crystal presents its hydrophilic surface on the side opposite the carbon support film. We seek to have an ice thickness of 100–150 nm, considerably greater than the liposome diameter, to avoid flattening of the liposomes.

2.4. Analysis and removal of crystal information from micrographs

A simple method was employed by Crucifix *et al.* (2004) to remove the periodic signal from images. Basically, the image is Fourier transformed, the data at the positions of the reciprocal-lattice peaks are set to zero, and then the inverse Fourier transform is performed. With this simple method, the authors were able to obtain, without artifacts from the crystal, a 3D reconstruction of RNA polymerase to a resolution of 3.0 nm. From simulations of a liposome on the 2D crystal, we found that a slightly better strategy is to scale the pixels in the lattice peaks to the locally averaged amplitude, instead of zero. This improved method was used to remove the crystal information from our cryo-EM images (Wang *et al.*, 2008). Firstly, a cryo-EM micrograph (0.51 μm × 0.51 μm of specimen area) is Fourier transformed. Then, a mask having the same reciprocal lattice as the 2D crystal, with a 3-pixel-radius disc at each lattice point, is applied to the Fourier transform. An annulus (inner and outer radii 3 and 4 pixels) defines the region over which the local background value is computed. From theory, one expects that optimum removal of the crystal information would occur when the mean of the complex values in the annulus is applied to each of the masked points. We found, however, that slightly better results are obtained by scaling the masked Fourier pixel values so that their magnitude is equal to the rms magnitude in the annulus, while the phase of each masked pixels is left unchanged. There is no visible crystal residue left in the resultant liposome images, as illustrated in Fig. 7.1D.

The signal from the 2D crystal can be used as a built-in reference to calibrate the image formation process, as described previously (Wang *et al.*, 2008). We found that Fourier components up to fourth order (index 4,1) are readily visible in the power spectrum of a micrograph. There is often distortion in the crystal lattice, undoubtedly due to mechanical strain of the unsupported crystal during transfer and blotting. If, however, we perform image-warping to unbend the lattice (Henderson *et al.*, 1986) components up to fifth order are significant, providing information to about 1 nm resolution from a single micrograph. For quantitative evaluation of the crystal image, we constructed a model image of a hydrated streptavidin tetramer, which we take to represent a unit cell of the crystal. By comparing the data with the model in Fourier space, the ratios of the intensities at different spatial frequencies can be used to define the envelope function (Wang *et al.*, 2008). The magnitudes and phases of the Fourier components can also be used to determine the contrast transfer function (CTF) parameters. In addition, the signal-to-noise ratio can be estimated by comparing the intensities of the reflection spots to the surrounding background.

2.5. Proteoliposomes

As most membrane proteins are isolated and purified in detergent solution, the most successful and frequently used strategy for reconstitution of membrane proteins is detergent removal. The detergent-solubilized membrane proteins are mixed with detergent-solubilized lipid; then the detergent is removed and liposomes with incorporated proteins (proteoliposomes) form spontaneously. The molecular mechanism for the formation of proteoliposomes has not yet been well understood. The simplest theory is that the proteins participate in the membrane-formation process (Rigaud *et al.*, 1995). Presumably, the membrane-formation process is the reverse of the liposome solubilization process (Fig. 7.4), in which case there are three stages of liposome formation: micelles corresponding to stage III in the solubilization scheme, transformation of micelles into detergent-saturated liposomes corresponding to stage II, and transformation of detergent-saturated liposomes to detergent-free liposomes corresponding to stage I.

The main approaches to remove detergents are dialysis, gel chromatography, and adsorption on hydrophobic resins. The rate of detergent removal depends largely on the critical micelle concentration (CMC), which is defined as the concentration at which the detergent monomers begin to form micellar aggregates. The higher the CMC, the more easily can the detergent be removed. Detergents with a low CMC are not readily removed by gel chromatography, and even less by dialysis, but can be efficiently removed through adsorption on hydrophobic resins such as the polystyrene Bio-Beads (Bio-Rad Laboratories).

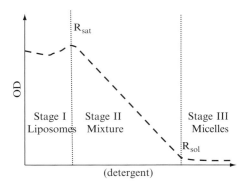

Figure 7.4 Schematic representation of the liposome solubilization process. Optical density (as a reporter of light scattering due to particle size) is plotted as a function of the detergent concentration. The main lipid phase is the lipid bilayer (liposomes) in stage I. Bilayers and micelles coexist in stage II, and stage III represents complete solubilization of lipid in mixed micelles. R_{sat} and R_{sol} are the critical detergent-to-lipid ratios where the lipid phase changes from stages I to II, and II to III.

Membrane proteins are extracted from cellular membranes using a detergent that is selected to provide high extraction efficiency and to stabilize the protein in solution. We use dodecylmaltoside (DDM) to extract the BK potassium channel (Wang and Sigworth, 2009), but DDM has an extremely low CMC (0.2 mM). Although it can be removed by Bio–Beads, we prefer to use gel chromatography. The biggest concern is the amount of detergent to be removed. Since only a few tens of micrograms of the BK channel is used for every reconstitution, only about 1 μmol of DDM needs to be removed, which requires only one or two Bio–Beads. It is impractical to control the rate of detergent removal in this case because of the large size variation of the Bio–Beads, and because the removal of detergent is highly localized around the hydrophobic beads. Therefore, we prefer to replace the DDM with decylmaltoside (DM), which has a much higher CMC ($\sim$2 mM) and can be removed easily by dialysis or gel chromatography.

2.6. Purification and reconstitution of BK channels

The BK channel α-subunit is stably expressed in an HEK293 cell line as the full-length KCNMA1 (gi:507922) gene product, carrying an N-terminal FLAG tag. Cells are disrupted and membranes are collected and solubilized in DDM. BK protein is subsequently bound to FLAG-affinity beads (Sigma–Aldrich) in the presence of 8 mM DDM in a buffer containing 200 mM KCl, 50 mM Tris, pH 7.4. The beads are collected into a column, which is first washed with 10–20 volumes of the same buffer containing 8 mM DDM followed by 10 volumes of buffer containing 4 mM DM. Finally, the BK protein is eluted with five sequential applications, at intervals of 20 min, of 0.4 ml DM buffer containing 0.1–0.2 mg/ml FLAG peptide. The eluate is concentrated to 0.2–0.3 ml using a concentrator with a cutoff of 100 kDa (Sartorius North America Inc., Bohemia, NY). A detergent–lipid mixture is then added to the detergent-solubilized membrane protein solution. For the BK channel, POPC was used since the thickness of the POPC bilayer matches the transmembrane region of another potassium channel, KcsA (Doyle et al., 1998). The detergent-to-lipid ratio is chosen to be larger than R_{sol}, which is usually about 1:1. To ensure a well-solubilized lipid solution, the molar ratio was chosen to be DM:POPC = 3:1.

If there are many copies of the membrane protein in a proteoliposome, the protein particles will overlap in the cryo-EM projection images, rendering the images uninterpretable. Thus, it is important to control the protein-to-lipid ratio for reconstitution. A final protein-to-lipid molar ratio of 1:5000 is used in the reconstitution of the BK channel, which yields about two copies of the BK channel in a 30-nm liposome. To ensure complete mixing between the detergent-solubilized membrane protein and the lipid–detergent micelles, the solution is shaken gently for an hour or overnight at 4 °C before detergent removal.

The detergent is removed using gel chromatography at room temperature. A column (1 cm^2 × 25 cm) is packed with well-soaked Sephadex® G-50 resin (Sigma-Aldrich) and equilibrated with running buffer (135 mM KCl, 5 mM NaCl, 1 mM EDTA, 10 mM HEPES, pH 7.4). About 250 μl of the protein–lipid–detergent (~0.2 nM protein, 1 mM POPC, ~3 mM DM, 135 mM KCl, 5 mM NaCl, 1 mM EDTA, 10 mM HEPES, pH 7.4) is loaded onto the top of the column to initiate slow detergent removal by diffusion. After an hour, running buffer is added to the column at 0.2 ml/min, and the eluate is fractionated. The fractions are analyzed using dynamic light scattering. Proteoliposomes elute at 30–45% of the packed volume, and the detergent elutes at about 100% of the packed volume. During the detergent-removal process, the flow rate is controlled by the height difference between the buffer container and the end of the eluate tubing. If the flow rate is faster than 0.5 ml/min, the detergent is not removed completely. Slower detergent removal is expected to yield larger liposomes. Although our target size was a diameter of 40–50 nm, even with the attempts to reduce the detergent removal rate, most of the resulting liposomes were smaller than 40 nm.

After detergent removal, the proteoliposomes, empty liposomes, and free protein aggregates coexist in the solution. To separate them, the sample is incorporated into the 20% layer of a discontinuous Nycodenz gradient (Fig. 7.5A). All the layers have the same osmotic pressure (~300 mOsm) to ensure no swelling or shrinkage of liposomes. After ultracentrifugation (238,000×g for 18 h), the protein-free liposomes float up to the 3% Nycodenz band, while proteoliposomes appear at the 5–15% boundary

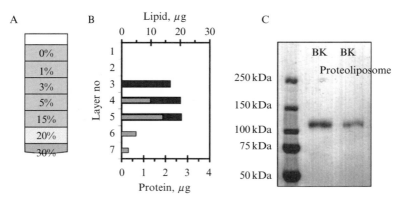

Figure 7.5 Separation of BK proteoliposomes. (A) Density gradient flotation of proteoliposomes; the weight percent Nycodenz is given for each layer. (B) Distribution of protein (gray bars) and phospholipid (black bars) in the gradient layers after flotation (18 h at 238,000×g). Empty liposomes are seen to float to layer 3. (C) A coomassie-stained SDS-PAGE gel shows protein extracted from the BK proteoliposomes as compared to the purified, solubilized BK protein before reconstitution.

(Fig. 7.5B and C). Lipid concentrations are determined by measuring phosphate using a colorimeter; protein is determined by the BCA (bicinchoninic acid) assay (Pierce), and the relative fraction of BK in each layer is determined by the densitometry of a Western blot.

In the density gradient, the position of the liposomes depends on the liposome size (smaller liposomes are more dense than larger ones) as well as on the presence of incorporated protein. The clear segregation of empty liposomes and proteoliposomes in the 3% and the 5–15% boundary, respectively, is consistent with an effective separation due to the added density from protein. The reconstitution efficiency (reconstituted slo/total slo protein) is about 70%, and about 80% of liposomes are proteoliposomes (fraction of lipid in proteoliposomes/total lipid). The layers containing both liposomes and proteins are collected and the Nycodenz is removed by the concentration–dilution method using a Vivaspin ultrafiltration spin column (Sartorius North America Inc.) with a cutoff of 100 kDa. The solution was concentrated at around $1000 \times g$ and diluted with running buffer for 3–4 cycles to reach 10,000 times dilution in total.

In the random spherically constrained (RSC) method, it is essential that the liposomes have a spherical shape. To make them highly spherical, we swell them by repeated osmotic shocks, adding water to the liposome suspension (11%, 14%, 18%, 24%, and 33% of the original volume) at 1-h intervals at 4°C. We found that when we used a single swelling step, many of the liposomes were not spherical. We speculate that the reason is the following: to change from an ellipsoid to a sphere, some of the lipid molecules in the inner leaflet of the liposome membrane have to move to the outer leaflet. The spontaneous flip-flop rate is very slow, but with an osmotic shock, the transient formation of a membrane defect allows flip-flop to occur. The multiple osmotic shocks allow equilibration of the two leaflets to be achieved to form highly spherical liposomes.

2.7. Flux assay to test the function of reconstituted proteins

The function of reconstituted BK channels is assayed using the cationic voltage-sensitive dye, 5,5′,6,6′-tetrachloro-1,1′,3,3′-tetraethylbenzimadazolyl-carbocyanine iodide (JC-1, Invitrogen), to monitor K^+-induced changes in membrane potential (Chanda and Mathew, 1999; Reers, 1991). BK proteoliposomes loaded with 135 mM KCl are incubated with channel blockers if desired and diluted into 5 mM KCl, 135 mM NaCl solution containing 1.6 μM of JC-1, giving a total lipid concentration of $\sim$6 μM. The fluorescence signal of the J-aggregates ($\lambda_{ex} = 480$ nm, $\lambda_{em} = 590$ nm) is monitored as the external K^+ concentration is increased by the addition of small volumes of 2 M KCl solution. The formation of aggregates appears to be a nonequilibrium process, so for reproducibility, all steps following the addition of JC-1 are held to a fixed schedule, with the KCl additions made at intervals of 20–40 min.

The large permeability of BK channels ($\sim 10^8$ ions/s) and the small liposome size mean that the membrane potential of a liposome will be established very quickly once a channel opens, as the movement of only ~ 100 ions is required. The membrane potential in turn drives an influx of the cationic JC-1 dye, so that an exchange of an unknown number of JC-1 molecules for K^+ ions occurs. We perform the flux assays in nanomolar free Ca^{2+}, and the potassium-dependent membrane potentials (~ 80 mV maximum) are such that the inside-out channels are expected to have an open probability up to 10^{-2}, while right-side-out channels have an open probability of 10^{-6}–10^{-5}. Even so, a functional right-side-out channel is highly likely to open during a 100-s interval, so that a single channel may provide a maximal fluorescence signal in a given liposome. The time scale of charging is so much faster than the time scale of redistribution of JC-1 across the membrane (tens of seconds) that the block of BK channels must be very complete if a reduction in the fluorescence response to K^+ gradients is to be observed. Thus, we applied channel blockers at about $1000 \times K_d$ in order to reduce the net permeability to control levels.

2.8. Computational removal of the lipid membrane contribution

In the RSC method, entire proteoliposomes are imaged. It is, however, not practical to perform a 3D reconstruction of entire proteoliposomes because of the large variation in their diameters. Instead, we subtract the membrane contribution from the proteoliposome images. To do this, we need an accurate, scalable model of a liposome image, which in turn requires an accurate profile of electron scattering by the phospholipid bilayer. Although a model for the electron scattering can be derived from molecular dynamics simulations and computations of electrostatic potentials, we prefer to obtain the membrane profile experimentally (Wang et al., 2006).

The model is constructed in the following way. Having identified 10–20 liposomes in a representative micrograph, we apply an inverse filter to compensate for the low-frequency variations in the CTF and compute the circularly averaged intensity of each liposome image. The Fourier transform of this 1D function then yields experimental structure factors. From these, the scattering profile of a slice through the spherical vesicle is obtained (Wang et al., 2006) by the Hankel transform (Bracewell, 2000). Finally, the averaged membrane profile is obtained after shifting each profile to align the membrane center, as in Fig. 4A of Wang et al. (2006). This profile is used in computing the image of a liposome of arbitrary radius.

For each crystal-subtracted micrograph, the CTF parameters are first determined, and then the CTF-modified liposome model is fitted, by nonlinear least-squares, to each liposome in the micrograph. The center and radius of each fitted liposome model is recorded, and the model is subtracted from the micrograph. The subtraction is imperfect, because a "pure" liposome model is fitted to images of liposomes containing protein particles, but the protein contribution is so weak that the fits are little affected by the extra density. In the case of the BK channel where no initial 3D model was available, we picked the particles manually from the liposome-subtracted micrographs. Picking was done with reference to the corresponding unsubtracted micrographs so that particles in liposomes could be distinguished from artifacts in the images.

2.9. Using the liposome image to aid the orientation determination of each particle

The most difficult and time-consuming computational step in the single-particle reconstruction process is the determination of the orientation of each protein particle that is imaged (Chapter 9 of vol. 482). Usually, this is done by comparing each particle image with reprojections of the particle model in all possible orientations. One has to search over the three Euler angles ϕ, θ, and ψ that define the orientation of the protein. With the use of liposomes, two of the three Euler angles (θ and ψ) can be determined from the location of the particle in the liposome image, relative to the liposome center (Fig. 7.6B; Jiang et al., 2001; Wang and Sigworth, 2009). The angles can be estimated up to a fourfold ambiguity: the protein can be either on the top or on the bottom half of the liposome and can be inserted either inside-out or right-side-out. This fourfold ambiguity can be resolved while

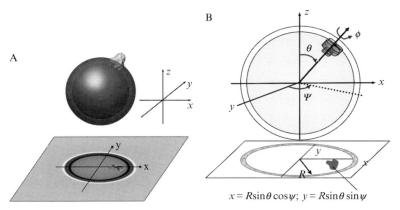

$$x = R\sin\theta\cos\psi; \quad y = R\sin\theta\sin\psi$$

Figure 7.6 The spherical geometry of a liposome allows estimation of the two Euler angles θ and ψ. (A) A 3D model of a proteoliposome and its projection image. (B) The geometry used to determine the two Euler angles.

searching over the Euler angle ϕ by considering the two angles θ and $\pi - \theta$ that are consistent with the particle location, and also considering inside-out and right-side-out orientations. Nevertheless, the use of the spherical geometry as a constraint greatly eases the determination of particle orientations. In the case of the BK channel, we attempted to make a 3D reconstruction from images of the detergent-solubilized protein, but this reconstruction was of poor quality, presumably because of the difficulty of determining all three Euler angles from particles of a roughly spherical shape in the absence of lipid membranes, a difficulty also encountered by other researchers (Alabi *et al.*, 2007; Jiang *et al.*, 2003).

The determination of the two angles allows a cylindrically symmetric initial model to be built directly. For the BK channel, we constructed such a model directly from the estimated θ and ψ values by applying a 32-fold symmetry around the self-rotation axis. We then forced a reduction to the assumed C4 symmetry by subtracting density in four wedges normal to the rotation axis. This initial model was refined to the final density map using the Frealign algorithm (Grigorieff, 2007), with the search over ϕ varying from 0 to $\pi/2$, but the search over the other variables (θ, ψ, x, y) being tightly constrained. All of the image processing and reconstruction steps were carried out using homemade computer programs in the Matlab programming environment.

The resulting map has an artificially low density in the transmembrane region, because a continuous membrane density was subtracted from each liposome image. The interpretation of this map can be approached in two ways. First, the density that is subtracted shows little variation near the center of the membrane plane (say, within ± 1 nm of the center) so that the interpretation of transmembrane density variations in that region is straightforward. Second, it is possible to computationally restore the membrane density. Given the estimated position and θ and ψ angles of the protein particle, the modeled image of a circular membrane patch surrounding the particle can be restored to each particle image and the final 3D reconstruction step rerun. The result is a map of the protein with the membrane density restored, which can be interpreted in the same way as a cryo-EM map from a 2D crystal of a membrane protein embedded in a membrane.

3. CONCLUSION

The use of a streptavidin crystal as a substrate allows the tethering of liposomes at a uniform density over the entire EM grid. The streptavidin crystal also serves as a built-in reference for image-quality control and the calibration of the image formation process. Meanwhile, the reconstitution of membrane proteins into liposomes has the considerable advantage that the integrity and activity of the proteins are preserved, while functional assays as well as structural studies can be carried out on the same sample. The spherical geometry of the liposome also eases the task of determining the orientation of each particle image and increases the accuracy of particle alignment and thus the resolution of the determined structure. This new system has been successfully applied to study the structure of a potassium channel, the BK channel (Wang and Sigworth, 2009), and can be applied to study other membrane proteins and their complexes.

REFERENCES

Alabi, A. A., *et al.* (2007). Portability of paddle motif function and pharmacology in voltage sensors. *Nature* **450,** 370–376.

Avila-Sakar, A. J., and Chiu, W. (1996). Visualization of beta-sheets and side-chain clusters in two-dimensional periodic arrays of streptavidin on phospholipid monolayers by electron crystallography. *Biophys. J.* **70,** 57–68.

Bracewell, R. N. (2000). The Fourier transform and its applications. McGraw Hill, Boston, MA.

Chanda, B., and Mathew, M. K. (1999). Functional reconstitution of bacterially expressed human potassium channels in proteoliposomes: Membrane potential measurements with JC-1 to assay ion channel activity. *Biochim. Biophys. Acta: Biomembr.* **1416,** 92–100.

Chester, D. W., *et al.* (2007). Holey carbon micro-arrays for transmission electron microscopy: A microcontact printing approach. *Ultramicroscopy* **107,** 685–691.

Crucifix, C., *et al.* (2004). Immobilization of biotinylated DNA on 2-D streptavidin crystals. *J. Struct. Biol.* **146,** 441–451.

Darst, S. A., *et al.* (1991). Two-dimensional crystals of streptavidin on biotinylated lipid layers and their interactions with biotinylated macromolecules. *Biophys. J.* **59,** 387–396.

Doyle, D. A., *et al.* (1998). The structure of the potassium channel: Molecular basis of K+ conduction and selectivity. *Science* **280,** 69–77.

Freitag, S., *et al.* (1997). Structural studies of the streptavidin binding loop. *Protein Sci.* **6,** 1157–1166.

Freitag, S., *et al.* (1999). A structural snapshot of an intermediate on the streptavidin–biotin dissociation pathway. *Proc. Natl. Acad. Sci. USA* **96,** 8384–8389.

Grigorieff, N. (2007). FREALIGN: High-resolution refinement of single particle structures. *J. Struct. Biol.* **157,** 117–125.

Henderson, R., *et al.* (1986). Structure of purple membrane from halobacterium halobium: Recording, measurement and evaluation of electron micrographs at 3.5 Å resolution. *Ultramicroscopy* **19,** 147–178.

Hendrickson, W. A., *et al.* (1989). Crystal-structure of core streptavidin determined from multiwavelength anomalous diffraction of synchrotron radiation. *Proc. Natl. Acad. Sci. USA* **86,** 2190–2194.

Izrailev, S., *et al.* (1997). Molecular dynamics study of unbinding of the avidin–biotin complex. *Biophys. J.* **72,** 1568–1581.

Jiang, Q. X., *et al.* (2001). Spherical reconstruction: A method for structure determination of membrane proteins from cryo-EM images. *J. Struct. Biol.* **133,** 119–131.

Jiang, Y. X., *et al.* (2003). X-ray structure of a voltage-dependent K+ channel. *Nature* **423,** 33–41.

Kubalek, E. W., *et al.* (1991). Improved transfer of 2-dimensional crystals from the air–water-interface to specimen support grids for high-resolution analysis by electron-microscopy. *Ultramicroscopy* **35,** 295–304.

Le Trong, L., *et al.* (2006). Crystallographic analysis of a full-length streptavidin with its C-terminal polypeptide bound in the biotin binding site. *J. Mol. Biol.* **356,** 738–745.

Reers, M. (1991). J-aggregate formation of a carbocyanine as a quantitative fluorescent indicator of membrane potential. *Biochemistry* **30,** 4480–4486.

Rigaud, J. L., *et al.* (1995). Reconstitution of membrane-proteins into liposomes—Application to energy-transducing membrane-proteins. *Biochim. Biophys. Acta Bioenerg.* **1231,** 223–246.

Sano, T., *et al.* (1995). Recombinant core streptavidins. A minimum-sized core streptavidin has enhanced structural stability and higher accessibility to biotinylated macromolecules. *J. Biol. Chem.* **270,** 28204–28209.

Tilley, S. J., *et al.* (2005). Structural basis of pore formation by the bacterial toxin pneumolysin. *Cell* **121,** 247–256.

Wang, L., and Sigworth, F. J. (2009). Structure of the BK potassium channel in a lipid membrane from electron cryomicroscopy. *Nature* **461,** 292–295.

Wang, S. W., *et al.* (1999). Two-dimensional crystallization of streptavidin mutants. *J. Phys. Chem. B* **103,** 7751–7761.

Wang, L., *et al.* (2006). Using cryo-EM to measure the dipole potential of a lipid membrane. *Proc. Natl. Acad. Sci. USA* **103,** 18528–18533.

Wang, L., *et al.* (2008). Streptavidin crystals as nanostructured supports and image-calibration references for cryo-EM data collection. *J. Struct. Biol.* **164,** 190–198.

Weber, P. C., *et al.* (1989). Structural origins of high-affinity biotin binding to streptavidin. *Science* **243,** 85–88.

CHAPTER EIGHT

Micromanipulator-Assisted Vitreous Cryosectioning and Sample Preparation by High-Pressure Freezing

Mark S. Ladinsky

Contents

Abstract

Cryo-electron microscopy (cryo-EM) of unfixed, unstained, frozen-hydrated samples conveys the most reliable view of the "live" state of cells and tissues. Advances in sample preparation methods and electron microscope technology

Division of Biology, California Institute of Technology, Pasadena, California, USA

Methods in Enzymology, Volume 481
ISSN 0076-6879, DOI: 10.1016/S0076-6879(10)81008-0

over the last decade have made this approach more routine and available to a broader segment of the structural biology community. Many cryo-EM samples are thin enough to be imaged as wholemounts, but most cells and tissues are too thick and must be cryosectioned to obtain samples that are sufficiently thin for successful imaging. Cryosectioning of vitreous material is a challenging and time-consuming task. A number of laboratories have worked hard to develop approaches and technologies for cryosectioning and to carefully characterize the nature of vitreous sections and the artifacts associated with them. Several different cryosectioning methods are in use in cryo-EM laboratories, each of which is effective and routinely yields high-quality structural data. Here, we describe a particular method that utilizes a micromanipulator to aid the cryomicrotomist in controlling vitreous sections as they are being cut and to facilitate transfer of the sections to an EM grid. Each step in the process, from preparing samples by high-pressure freezing to affixing vitreous sections to a grid, is covered in detail, including discussions of cryosectioning hardware, environmental conditions, and sectioning artifacts.

1. INTRODUCTION

Imaging vitreously frozen biological samples is now at the forefront of modern structural cell biology. Because such samples are free of the artifacts caused by chemical fixation and heavy-metal stains, cryo-electron microscopy (cryo-EM) of frozen, hydrated cells and tissues allows the most reliable view of the "live" state that can be achieved with the transmission electron microscope. Cryo-EM studies have been undertaken since the 1980s (Adrian et al., 1984; see also the chapter "Historical Perspective"), but it is only within the last decade or so that cryo-EM has become a powerful tool for the structural cell biologist. This is due to the refinement of methods for efficiently and repeatably vitrifying specimens, improved technology of cryoelectron microscopes and associated detectors, and the application of high-resolution tomography and 3D reconstruction.

Cryo-EM is not without its limitations. The absence of fixation and polymer embedment makes cryo-EM samples very sensitive to damage from the electron beam (see Chapter 15). Also, the absence of electron-opaque fixatives (i.e., osmium tetroxide) and heavy-metal stains means that samples are very low in contrast. Most importantly, samples must be very thin in order to acquire interpretable images from electron microscopes operating at standard or intermediate accelerating voltages (i.e., 100–300 KeV). Some samples, such as small bacteria, are naturally thin enough to be imaged successfully as whole mounts. Most samples, however, such as large bacteria and most eukaryotic cells, are simply too thick to image whole. Even electron tomography does not solve this problem, most notably because when a sample is tilted, it becomes progressively thicker relative to the electron beam. Just as with traditional, plastic-embedded specimens, the solution to cryoimaging of thick vitreous samples is to section them, with a cryo-ultramicrotome, to a suitable thickness and view the resulting "cryosections" in the cryo-EM.

Cryo-ultramicrotomy is a difficult technique that is not free of its own inherent artifacts. A number of laboratories around the world have worked hard to improve the technique and develop methods to make cryosectioning more straightforward and applicable to a wider range of samples (Al-Amoudi et al., 2004a,b; Bouchet-Marquis et al., 2005; Hsieh et al., 2005; Ladinsky et al., 2006; Matias et al., 2003; McEwen et al., 2002; Pierson et al., 2010). These studies have resulted in many excellent approaches and new tools that simplify the method and improve the quality of the resulting sections. Nonetheless, successful cryo-ultramicrotomy requires specialized tools and a patient, well-practiced operator.

2. WHY IS VITREOUS CRYOSECTIONING SO DIFFICULT?

Most of the difficulties associated with cryosectioning lie in the fact that cryosections must be cut "dry." With room-temperature microtomy of plastic-embedded samples, sections are cut with a diamond knife attached to a concave "boat" that is filled with water. As sections are cut, they float on the surface of the water, where they can be collected onto an EM grid or with a wire loop. Furthermore, sections floating on water, which are slightly compressed during the cutting action, can be easily reexpanded to their original size by exposing them to the vapor of a solvent, such as chloroform.

Obviously, water cannot be used to collect vitreous cryosections, which are typically cut at temperatures between -140 and $-170\,^{\circ}C$. Experiments have been made to use solvents that remain liquid at temperatures suitable for cryosectioning, but for the most part, cryosections must be cut without the aid of a liquid medium.

Dry sections cut at low temperatures are very sensitive to static charges in the cryomicrotome's chamber, making them difficult to control; they can easily be blown away from the knife, or they can stick to the surface of the knife and be effectively crushed by the next section that is cut from the block. Furthermore, the compression that occurs when a section is cut by the diamond knife is compounded for dry, low-temperature sectioning (Al-Amoudi et al., 2005), and there is no effective means to reexpand a compressed section after it is cut. Tools and techniques must therefore be employed to reduce (or redirect) static charging in the microtome chamber, control sections as they are being cut, prevent sections from sticking to the knife, and minimize section compression as much as possible, all at the same time! Finally, once sections are cut, they must be transferred and affixed to an EM grid for subsequent cryoimaging, which in itself is a challenging, multistep task.

Every experienced cryo-ultramicrotomist has developed their own specific methods for preparing vitreous cryosections, each of which is arguably effective and yields high-quality cryo-EM data. This chapter will focus on one particular method that utilizes a micromanipulator to hold and control cryosections as they are being cut and to subsequently transfer them to an EM grid (Ladinsky *et al.*, 2006). The micromanipulator facilitates slower cutting speeds, better control of a forming ribbon of cryosections, and more precise transfer of ribbons to a support grid. The chapter will cover preparation of cryosectioning samples by high-pressure freezing and suitable cryoprotectants, mounting the vitreous sample block in the cryo-ultramicrotome, trimming the block, the sectioning step itself, and attachment of ribbons of cryosections to an EM grid. The tools and hardware involved in each step will be discussed in detail. Lastly, there will be a brief discussion of environmental factors that affect cryosectioning and also the artifacts associated with cryosections and how to minimize them. Specimen support grids used for cryosections and imaging of cryosections in the electron microscope are similar to that of other vitrified specimens, which are covered in other chapters of this volume (Chapters 10 and 12).

3. High-Pressure Freezing for Vitreous Cryosectioning

There are two key factors to preparing samples for vitreous cryosectioning: (1) Vitrifying the sample effectively, uniformly, and in a manner that is conducive to good sectioning; and (2) Holding the sample in such a way that it is easily accessible to the knife once it is placed in the cryomicrotome. Since samples that are to be cryosectioned are inherently too thick to be imaged as whole mounts, they are also usually too thick to be effectively vitrified by plunge freezing (see Chapter 3). Instead, they are vitrified by high-pressure freezing. There are essentially three commercially available high-pressure freezing machines and an array of methods for using them to prepare samples for either cryosectioning or freeze-substitution fixation (reviewed, McDonald and Auer, 2006; McDonald *et al.*, 2007). Two freezing methods, both using the BalTec HPM-010 high-pressure freezer, will be discussed here.

When freezing samples for vitreous sectioning, it is necessary to contain the sample in a carrier that not only fits specifically in the high-pressure freezer, but can also be held securely in the cryo-ultramicrotome. There are two devices that fit this bill; two-piece metal planchettes, or hats, that enclose the sample; or thin metal tubes (Fig. 8.1).

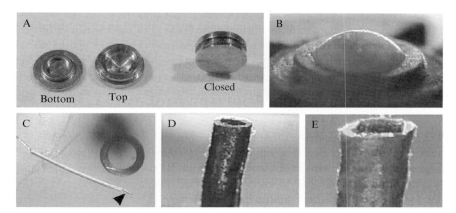

Figure 8.1 High-pressure freezer sample carriers used for vitreous cryosectioning. (A) A two-part brass planchette carrier commonly used for high-pressure freezing. Sample is placed in the flat bottom part of the planchette, while the inner surface of the dome-shaped top is lubricated to prevent sticking. The planchette is closed and frozen. When opened, the sample remains in the lower part, forming a "dome" that projects from the metal base (B). Very efficient freezing is obtained with samples frozen inside of copper or gold tubes. (C) A gold tube is filled with sample by using a thin metal wire as a syringe-like plunger. The ends of the tube are crimped shut and the tube is frozen. One of the crimped ends is removed to reveal the vitrified sample within (D). The tube is then placed in the cryomicrotome where the metal walls are trimmed away to expose the sample for sectioning (E).

The best planchettes for cryosectioning are made of brass and are composed of a bottom part that has a flat base with an inner raised ring, and a top part with a dome-shaped inner surface (Fig. 8.1A). The inside of the top part is lubricated with either lecithin or carbon to prevent the sample from sticking to it during freezing. A pellet of cells is loaded into both parts of the planchette, either by pipetting or transferring with a toothpick. The two parts of the planchette are closed together, placed in the freezer's sample holder, frozen, and then transferred to a container of liquid nitrogen. The planchette is then separated, using either a specific tool for this task or simply a pair of forceps. The sample should remain entirely in the bottom part of the planchette, forming a dome that stands away from the metal base (Fig. 8.1B). This is the most advantageous shape and size for a cryosectioning sample, as it gives a great deal of latitude in trimming various sized blockfaces and also does not require removal of any metal prior to trimming and sectioning. However, the large volume of the dome sometimes results in progressively poorer vitrification as one cuts further into the block. Often only the upper ~25% of the dome's volume is frozen well enough to yield high-quality cryosections.

Use of thin metal tubes as sample carriers for cryosectioning has become popular with the introduction of the Leica EM-PACT series of high-pressure freezers, which feature sample holders designed for 250 μm diameter copper tubes (see McDonald *et al.*, 2007). Metal tubes have been adapted for use in the BalTec freezer as well. Tubes have the advantage of yielding very good freezing quality because they are considerably thinner than planchettes and more of their surface (relative to their volume) is directly exposed to the jets of cryogen in the high-pressure freezer. The disadvantages of using tubes are that the metal walls of the tube must be trimmed away prior to cutting a blockface, and also that the size of a blockface that can be trimmed is limited by the inner diameter of the tube. Often this is not of great concern because typically small blockfaces yield more consistent sections. However, as will be discussed later, some types of samples and the characteristics of some cryo-diamond knives necessitate the use of larger blockfaces.

Both copper and gold tubes are available. Gold tubes (Fig. 8.1C), which are smaller in diameter than copper tubes, are cut to a length of approximately 1 cm and then loaded with sample either by capillary action or by sucking sample into the tube with a plunger fashioned from a thin metal wire that matches the inner diameter of the tube (Fig. 8.1C, arrowhead). Once filled, an approximately 3 mm segment of tube is cut and placed in the sample holder of the high-pressure freezer along with a washer (shown in Fig. 8.1C) for support. When the sample holder is closed, the ends of the tube are crimped shut and the sample is frozen. Prior to insertion into the cryomicrotome, one of the tube's crimped ends is cut off with a cooled razor blade or scalpel to expose the vitrified sample (Fig. 8.1D). Once mounted in the microtome, the walls of the tube are trimmed away and a blockface is formed from the protruding sample (Fig. 8.1E). Tubes typically give more complete vitrification, meaning that most, if not all, of the tube's contents will yield high-quality cryosections. On the other hand, tubes are more difficult to handle when mounting in the microtome and far more difficult to properly trim before sections can be cut.

4. EXTRACELLULAR CRYOPROTECTANTS FOR VITREOUS CRYOSECTIONING

The goal of rapid-freezing is to encase the sample in vitreous or crystal-free ice. When ice crystals form within or around cells, their growth will alter or destroy the native biological structure. In high-pressure freezing, the high pressure itself acts as a cryoprotectant in that it retards the formation of ice crystals, but alone it is often not enough to allow samples to be fully vitrified (Dubochet, 1995). Since the advent of rapid-freezing as a method for preserving biological structure, various compounds, primarily large, branched sugars, and other polymers have been mixed with the samples to prevent, or at least substantially slow down, the nucleation of ice crystals (covered in detail by Echlin *et al.*, 1977; Franks *et al.*, 1977; Skae *et al.*, 1977).

In addition to providing protection from ice crystal damage, certain cryoprotectants can be used to facilitate good cryosectioning of samples that are otherwise difficult or impossible to cut (Matias *et al.*, 2003). Pellets of cultured mammalian cells and mammalian tissues (with the possible exception of skin; Norlen, 2008) are particularly difficult to vitrify and cryosection without adding cryoprotectants. The most common and routinely successful cryoprotectant is Dextran, used at a concentration of 20% or more (Al-Amoudi *et al.*, 2004a; Bouchet-Marquis *et al.*, 2005; Zhang *et al.*, 2004) and mixed with the sample just prior to high-pressure freezing. Sucrose, of course, has proved to be the best sugar for facilitating cryosections and has been used as such for decades (Tokuyasu, 1971, 1986). However, sucrose will quickly change the osmotic balance of cells, causing often severe osmotic damage to organelles and membranes. As such, it is not recommended as a cryoprotectant for the cryosectioning of unfixed, frozen-hydrated samples.

Some organisms are naturally conducive to good-quality cryosectioning and do not require addition of a polymer cryoprotectant. This includes some bacteria and, most importantly, the budding yeast *Saccharomyces cerevisiae* (Ladinsky *et al.*, 2006). In fact, mixing difficult-to-section samples with *S. cerevisiae* will often confer sectioning characteristics that are as good or better than sugar cryoprotectants. For example, we found that mammalian tissue culture cells, such as PtK or NRK cells, which ordinarily section very poorly, could be successfully cryosectioned when mixed with yeast cells, pelleted and high-pressure frozen. When imaging the resulting cryosections, the distinct shape, size, and cell walls of the yeast cells allowed the mammalian cells to be easily distinguished.

5. MOUNTING A VITRIFIED SAMPLE IN THE CRYO-ULTRAMICROTOME

It is very important that the specimen be mounted rigidly in the cryomicrotome. A sample that is slightly loose in its holder will vibrate as it contacts the knife, resulting in artifactual "chatter" that will be glaringly visible when the section is imaged. A very loose sample will, at best, give extreme variations in section thickness and at worst, may shatter when it contacts the knife and even damage the knife itself. Unfortunately, the standard sample holder on most cryomicrotomes is a collet designed to hold a pin that is 2 mm in diameter (Fig. 8.2A). This is too small to effectively clamp a planchette specimen and too big to hold tube specimens.

For planchette specimens, I have designed a holder that is effectively a "secondary collet." This device consists of a brass pin with a 2-mm diameter

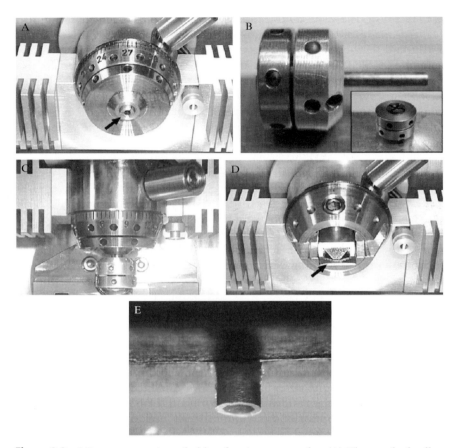

Figure 8.2 Microtome specimen holders for vitreous samples. (A) The standard collet-type holder on the UC6/FC6 system. The opening in the collet (arrow) is designed for a 2-mm diameter pin: too small for a planchette-style sample and far too big to firmly hold a tube-type sample. (B) A homemade "secondary collet" designed to hold a planchette-style sample. The design consists of a central pin that fits into the microtome's collet holder, a countersunk base where the planchette is seated (inset), and threaded aluminum rings that tighten to lock the planchette in place. (C) Top view of the planchette holder in position for sectioning. (D) A clamp-type holder suitable for use with tube-style samples. The tube can be positioned within the jaws of the clamp (arrow) to align it with the knife. (E) A gold tube held firmly in the clamp, faced-off and ready for trimming.

tail that will fit securely in the microtome's holder (Fig. 8.2B). The head of the pin is cut and countersunk to accept the bottom part of a brass planchette containing a domed vitreous sample (Fig. 8.2B, inset). Threaded aluminum rings tighten around the pin head to hold the planchette firmly in place. A planchette sample is loaded under liquid nitrogen, after which the assembly is inserted into the microtome's sample collet and locked into place (Fig. 8.2C). This kind of holder device has several advantages; First, it will effectively hold any planchette, even if a particular planchette's diameter is slightly undersized. Second, it is easy to load under liquid nitrogen and, if a sample is to be stored and reused, it can be easily unloaded without damaging the planchette or the trimmed dome.

For tube specimens, a clamp-type microtome specimen holder is available (Fig. 8.1D). This clamp is again too small to accommodate planchettes, but it will securely hold the tubes. Positioning the tube in the clamp and making sure it is properly secured are important steps prior to trimming and sectioning. After a tube has been trimmed of its crimped end (see Fig. 8.1), it is transferred to the cryomicrotome's chamber and inserted into the clamp. The tube is held in the proper position with a pair of cooled forceps while the clamp is tightened around it. Once the tube is secured in the clamp (Fig. 8.2E), it is ready for trimming and sectioning. Using a flat clamp to hold the tube allows it to be oriented properly relative to the knife in the event that the tube becomes bent during freezing or subsequent steps (which is unfortunately common).

6. HARDWARE: CRYO-ULTRAMICROTOMES

Ultramicrotomes equipped with cryostages have been available since the late 1970s. The first truly practical system was the Reichert (now Leica Microsystems) UltraCut-E microtome equipped with an FC-E cryostage, introduced in the 1980s. Although this system has been eclipsed by more modern technologies, it is an extremely robust machine that remains in use in many laboratories today. Cryostage systems are usually sold as optional "accessories" that can be purchased as a package with a new ultramicrotome or added separately to an existing ultramicrotome unit. Typically, the cryostage system will cost almost as much as the microtome itself. The cryostage system consists of a chamber unit that fits onto the base of the ultramicrotome, a liquid nitrogen dewar and pump that are connected to the chamber via an insulated hose, a control unit that powers the pump, heaters, and lights, and an active static ionization device (discussed later in this chapter). The chamber contains a knife stage that usually holds two cryo-knives and a specimen holder that attaches to the microtome's cutting arm. The chamber is further fitted with independent heating elements that regulate the temperature of the chamber, the knife stage, and the specimen holder.

The micromanipulator-assisted cryosectioning method has been used on two cryo-ultramicrotome systems, the Leica UltraCut-UCT with FCS cryostage and the current Leica UC-6 with FC-6 cryostage. A newer, FC-7 cryostage is now available, which features a novel electrostatic

charging device that facilitates attaching vitreous sections to the EM grid
(Pierson *et al.*, 2010).

The mechanics of the cryosectioning method are the same for both
microtomes, but the design and operation of each system has some
notable differences. The older UltraCut-UCT/FCS system is operated via
three control boxes oriented for the microtomist's left hand (Fig. 8.3A).
These include the cryostage control box, which manages the liquid nitrogen
pump and the three chamber heaters, the microtome control box, which
manages all of the cutting parameters, and a separate rheostat for the active
static ionizer. The micromanipulator is placed on the right side of the unit

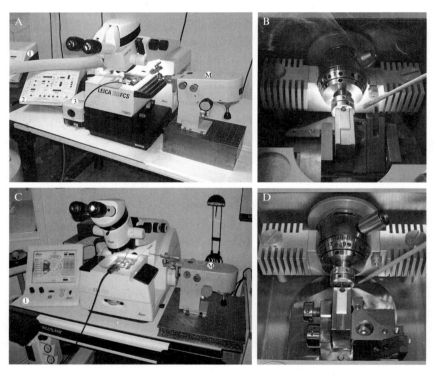

Figure 8.3 Two generations of Leica cryo-ultramicrotomes, equipped with micromanipulators. (A) The UltraCut-UCT/FCS system with its three control boxes for the
cryostage (1), microtome (2), and static ionizer (3) oriented for the microtomist's left
hand. The Leitz Micromanipulator-M (M) is positioned on the right side of the unit. (B)
Detail of the FCS cryochamber, showing the parallel arrangement of the two-knife
holder. (C) The EM-UC6/FC6 system positioned with its single integrated control box
(1) on the left of the unit and the Leitz Micromanipulator-M (M) on the right. (D)
Detail of the FC6 cryochamber showing the rotating knife holder with the two knives at
90° offset. Both microtome systems are placed on air-charged vibration isolation tables
that support both the microtome chassis and the micromanipulator.

and is controlled by the microtomist's right hand. The chamber of the FCS cryostage (Fig. 8.3B) features a knife holder that holds two cryodiamond knives in a parallel configuration. Most microtomists will place a trimming knife in one position and their favorite cryodiamond knife in the other. The holder is affixed to the base of the cryostage via a dovetail mount. This knife holder design has the advantage of allowing quick change from one knife to the other, by simply moving the base laterally via a knob on the outer right side of the cryostage. The multiple control boxes of the UCT/FCS is somewhat disadvantageous in that the microtomist must have mastery of two different control sets (the microtome control and the ionizer rheostat) in addition to the micromanipulator while sectioning is under way.

The newer UC6/FC6 system (Fig. 8.3C) has a single integrated, touch-screen control panel that manages all aspects of microtome control, cryostage features, and the static ionizer. This greatly simplifies the job for the micro-tomist, since during sectioning, the left hand can remain in a single position while the right hand controls the micromanipulator. The UC6/FC6 also features an improved lighting system, which improves accuracy in the trimming step and control of sections as they are cut. The chamber of the FC6 cryostage (Fig. 8.3D) is built to tighter tolerances than the older system and uses a rotating knife holder that positions two knives at 90° angle to each other. The holder is fixed to the stage by a central bolt. A knife is selected by loosening this bolt with an Allen wrench, rotating the holder to bring the knife into alignment with the sample, then retightening the bolt. Although this holder design is accurate with respect to knife alignment and is very stable, great care must be taken when rotating the holder to avoid damaging the knives or the specimen. The stage should be retracted to its most rearward position prior to loosening the holder, to insure that the knife does not slam into the specimen as it is rotated into alignment with the block. This is of greatest concern when using a collet-type planchette holder since there is virtually no gap between the knife and the specimen, and contact of a sensitive diamond knife with the metal edge of the collet could cause severe damage to the blade. It is even advisable to mount the knife a few millimeters back from its stop in the holder to insure a safe rotation.

The microtome should be placed on a high-quality, air-charged vibra-tion isolation table. The table should be large enough to accommodate the microtome, control box(es), and the micromanipulator (Fig. 8.3A and C). Leica sells tables that have a platform in the middle that specifically isolates the microtome chassis; devices on either side of the chassis are not isolated. These tables are excellent for room-temperature sectioning or cryosection-ing without a micromanipulator. However, they are not recommended for the micromanipulator cryosectioning method because the microtome can move independently from the micromanipulator, often causing gross move-ments that can result in damage to or loss of the cryosections.

7. THE MICROMANIPULATOR

Micromanipulators of various designs and operations are used for a variety of tasks in the cell biology laboratory, often in conjunction with light microscopes for cell microinjection studies. Any commercially available manipulator could probably be adapted for use in cryosectioning. The unit I use is a Leitz (now Leica Microsystems) Micromanipulator-M (shown to the right of the microtome in Fig. 8.3A and C). This design dates to the 1970s and is still sold by Leica. Due to its excellent manufacture and robustness, many used and refurbished units are available through microscope and laboratory supply companies. The Micromanipulator-M was selected for several reasons; first, it is fully manual and does not complicate the workspace with electrical or pneumatic connections. Second, its compact size makes it portable and easy to align with the microtome without interfering with the microtome's controls (i.e., the manual arm control wheel). Third, the manipulator's controls are ergonomically positioned, requiring minimal movement of the operator's hand. The controls consist of two knobs on the side of the unit for rough X and Y movement of the tool holder, a large knob for adjusting pitch, a double knob at the back of the unit for rough and fine vertical movement, and a downward-facing adjustable joystick for fine X and Y movement. For improved stability, the micromanipulator should be mounted to a heavy metal base that is of appropriate height to allow the unit to control a tool through its full range of motion. The manipulator is shown attached to a homemade base in Fig. 8.3A and to a commercially manufactured base in Fig. 8.3C. The tool holder is essentially a clamp, originally designed to hold micropipettes or small syringes, connected to the end of the manipulator's arm. Typically, the tool holder will have a ball-and-socket design that allows it to be adjusted in three dimensions relative to the manipulator arm.

The "tool" for cryosectioning is a thin wooden dowel, approximately 6–8-in. long, with a single fiber glued to its tip. The dowel is locked into the tool holder and the tip with the fiber attached is lowered into the cryomicrotome's chamber. The fiber, now controlled by the micromanipulator, is used to hold and move cryosections as they are being cut and to subsequently transfer them to an EM grid. Traditionally, the fiber is a human eyelash or hair from certain shorthair breeds of dog, such as the Dalmatian. Lately, we have been employing nylon fibers from a fine paintbrush. These are used because nylon fibers tend to be stronger than hair and can more firmly hold, or even "pull" a ribbon of cryosections without losing tension. Also, the end of a nylon fiber can be easily bent to a convenient angle for manipulating cryosections.

8. CryODIAMOND KNIVES

Cryosectioning knives have often been made of glass and produced from glass ingots by microtomists themselves in the laboratory. Glass knives are very economical and can be very sharp. However, variability in sharpness and blade angles is inevitable, and the softness of the glass means that they will quickly become dull and yield progressively poorer cryosections. Commercially produced cryoknives made of diamond have overcome these problems (Matzelle *et al.*, 2003; Michel *et al.*, 1992). Diamonds can be honed to an edge of less than 3 nm and, if properly cared for, will maintain their sharpness and cutting characteristics for hundreds of sectioning hours. Furthermore, diamond knives that do become dull or damaged can be resharpened to their original specifications. There are a few companies that manufacture diamond knives for ultramicrotomy. One of the leaders in the field, especially for knives specifically designed for cryo-ultramicrotomy, is Diatome, Ltd. of Switzerland (marketed in the United States as Diatome-US). Diatome produces several styles of cryodiamond knife, three of which will be covered here.

The most commonly used cryodiamond knife consists of a right triangle-shaped metal base with the diamond blade affixed at the tip of the hypotenuse (Fig. 8.4A, left). The diamond is embedded in the holder with low-temperature epoxy, the blade parallel with the edge of the holder (Fig. 8.4B, left). These knives are extremely stable and suitable for a wide range of cryosectioning applications. The epoxy embedment affords the blade a modicum of protection from damage during use, but over a period of time, it can become cracked or warped with repeated cooling and warming. Furthermore, some microtomists report that this design may decrease the efficiency of the active ionizer, possibly making it more difficult to control sections as they are cut. This design was improved upon by flattening the upper edge of the metal base and attaching the diamond to a freestanding post (Fig. 8.4A, middle). Viewed face-on (Fig. 8.4B, middle), the full facet of the diamond is exposed and stands clear of the metal base. This design is often used for cryosectioning methods where sections are collected on the face of the diamond and subsequently removed with a loop or other instrument. It is also useful for vitreous cryosectioning and does indeed yield better sectioning performance than its predecessor.

The triangular base design presents one particular challenge to the microtomist; there is no room to place an EM grid for transferring newly cut sections. After sections are cut, users of this type of knife must introduce an EM grid to the chamber, held by a separate clamp device. The grid is maneuvered close to the diamond knife and sections are transferred to it using a fiber tool (hair or nylon). The "loaded" grid must then be moved to a flat surface within the chamber for pressing/affixing the sections to the grid. Although this set of steps is not insurmountable and can be done in

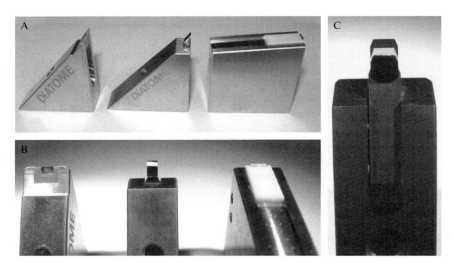

Figure 8.4 Diamond knives designed for vitreous cryosectioning. Side (A) and face (B) views of three current cryodiamond knife styles. Standard triangular bases are available with the blade mounted flush to (left), or standing clear (middle) from the metal base. Both designs have advantages and disadvantages, described in the text. The Cryo-Platform knife (right) features a ceramic platform onto which an EM grid can be placed to receive cryosections right after they have been cut. We have found this to be the best style for use in the micromanipulator-assisted cryosectioning method. A specifically designed diamond trimming knife with 20° angled sides (C) facilitates the accurate blockface trimming required for successful cryosectioning.

conjunction with a micromanipulator, it adds a level of complexity and increases the chances of the sections being damaged or lost.

This problem was effectively solved by introduction of the "Cryo-Platform" diamond knife (Fig. 8.4A, right). This design consists of a roughly square metal base, the top of which is angled slightly downward. The freestanding diamond blade is affixed to the uppermost edge and is abutted by a small ceramic platform that sits in groove on the top surface. Before sectioning begins, an EM grid is placed on the ceramic platform, against the base of the diamond (Fig. 8.4B, right; and shown with a grid in place in Fig. 8.3B and D). During cutting, a ribbon of sections can be "pulled out" over the grid, using the fiber controlled by the micromanipulator. Once the ribbon is long enough, the fiber is lowered to transfer the ribbon directly onto the grid. This greatly simplifies the process and makes the Cryo-Platform the best design for use with the micromanipulator cryosectioning method. When the method is described in detail, later in this chapter, it will be discussed with respect to a Cryo-Platform knife being used.

All Diatome cryodiamond knives can be purchased with included angles (the angle of the blade relative to the sample) ranging from 25° to 45°. Higher angle knives (35° and 45°) are more durable and are the best choice for routine work and for novice cryomicrotomists. Cryosections cut with these knives, however, often show more compression and alteration to fine structure. A 25° knife will usually produce the best-quality sections, that is, noticeably less section compression and better preservation of fine structure. The tradeoff for this, however, is a considerably more sensitive blade that can be easily damaged if not used and maintained with the utmost care. It is recommended that only very experienced cryomicrotomists use 25° knives, or if it is affordable, more than one such knife be purchased in the event that one is damaged. Theoretically, a knife with even less included angle will yield sections with even less compression. However, any such knives are so delicate that they will be immediately dulled or damaged just in the course of regular use. Thus, lower angle knives are not commercially available.

9. The Cryotrimming Knife

Regardless of the method being used, successful vitreous sectioning is critically dependent on having a properly trimmed block (detailed below). High-quality diamond knives are not used for trimming because a great deal of material, sometimes including metal, is often removed during this step. Glass, or even steel, knives have often been used for trimming, but a specifically designed diamond trimming knife (Fig. 8.4C) gives the best results. A diamond trimming knife typically consists of a freestanding diamond blade mounted to a lightweight aluminum triangular base with a 35° included angle. The robust blade is often larger than that of a bona fide cryosectioning knife with sides that are beveled to 20° or 45°. The bevel imparts greater stability to the blockface and reduces the chances of the sides of the blockface cracking or splintering during trimming. A diamond trimming knife is almost as sharp as a sectioning knife and maintains its edge just as long. This results in precise and repeatable blockface trimming.

10. Setting up the Cryo-Ultramicrotome for Sectioning

It takes approximately 15–20 min for the cryo-ultramicrotome to cool down and equilibrate before sectioning can begin. Knives should be loaded into the knife holder and placed in the cryostage chamber before the unit is cooled. All control boxes are switched on and the desired sectioning temperature is selected for all three of the chamber heaters (a typical starting temperature is −150 °C). The liquid nitrogen pump assembly, normally stored at room temperature, is placed into the filled dewar (*note*: leaving the pump assembly in the dewar when the microtome is not in use will result in ice accumulation in the pump that will block liquid nitrogen flow and possibly damage the pump's mechanism. Always remove the pump assembly from the dewar when cryo-sectioning is completed). The pump is connected to the chamber with the insulated nitrogen feed hose and the pump switched on. After the chamber has reached the selected temperature and all three heaters have equilibrated, the sample can be transferred to the chamber and fastened into the specimen holder.

11. TRIMMING A BLOCKFACE

As mentioned previously, a properly trimmed blockface is absolutely crucial to successful vitreous cryosectioning. Although it is recommended to make a blockface as small as the needs of the sample will allow, the shape of the face and the evenness of its sides are much more important. In room-temperature sectioning of plastic-embedded material, a blockface is typically trimmed into a trapezoid, with the parallel sides oriented parallel to the knife for cutting; the inclined sides reduce drag as the section passes off the knife. A frozen-hydrated sample should be trimmed to a square or rectangle, with all four sides being straight and even. If the sides are not parallel or if one side is pitted or uneven, the result will be sections that are unevenly compressed as they pass the knife-edge, which will induce greater artifacts and make proper ribboning and collection of the sections nearly impossible.

Depending on the cryo-ultramicrotome model, trimming can be done manually or automatically. Some microtomists will set the microtome to produce relatively thick (∼500 nm) sections, start the automatic cutting cycle, and allow the machine to trim into the block until a desired facet is produced. The knife is then moved and/or the sample rotated by 90°, and the process is repeated. The Leica UC6/FC6 system features an automatic trimming mode that can be used to trim each facet to a preset depth and then stop. Depending on the cutting speed and thickness selection, automatic trimming may take several hours; some microtomists may even spend a whole day on the trimming task. This requires considerable patience but usually results in the most precise blockfaces.

Manual block trimming involves using the rough and fine knife advance controls on the microtome's control panel to cut a face into the block while moving the sample across the knife using the manual flywheel on the side of the microtome's chassis. This approach requires a lot of practice to master the necessary cutting speed and knife advance needed to cut a proper facet without damaging the sample. However, this method can yield a properly trimmed block in as little as 15 min.

The goal of the trimming step is to produce a properly shaped blockface and remove all unnecessary surrounding material, so that only the sample of interest is presented to the knife for sectioning. If a domed, planchette-type sample is being used, it is mounted in the microtome's sample holder (discussed earlier) and the trimming knife is advanced to the surface of the dome (Fig. 8.5A). Initial cuts are made to produce a smooth face in the front of the block that is slightly larger than the perceived final blockface. It is at this stage where one can determine if the sample is properly vitrified for cryosectioning: A vitreous sample will yield smooth, even cuttings that increase in size as the block is trimmed. The resulting face should be smooth, mirror-like, and free of pitting. If the initial cuttings are broken or come off the knife as powdered ice, and the resulting face is pitted and not smooth, the ice in that region is not vitreous and should not be used. Sometimes a different area of the block can give better results, but more than likely the block will need to be replaced. After the initial face is trimmed, the trimming knife is moved laterally to position a corner of the blade with the edge of the face, where the first facet will be made. Trimming is commenced and the facet is cut to the desired depth. Cutting is

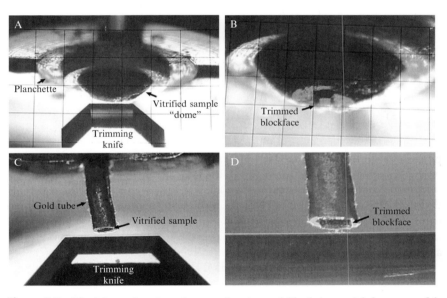

Figure 8.5 Blockface trimming. A properly trimmed block is crucial for successful cryosectioning. (A) A planchette-style vitrified sample in the cryomicrotome, ready to be trimmed. The long edge of the diamond trimming knife is used first, to trim a smooth face into the sample "dome." The knife is then moved laterally to use its corners to cut perpendicular facets on either side of the face. The specimen is then rotated 90° and the process repeated to create a square or rectangular blockface with even and parallel sides, shown in (B). Trimming of tube-style samples (C) is similar, except that the metal tube walls must be carefully removed to expose the vitrified sample. After a smooth face is trimmed with the long edge of the knife, the corners of the blade are used to first trim away a sufficient amount of metal, then trim the four side facets as aforementioned. Care must be taken to avoid removal of too much sample, as the size of the final blockface is limited by the inner diameter of the tube (D).

stopped, the knife retracted away from the sample and repositioned laterally for the other corner of the blade to make the parallel facet on the opposite side of the face. The knife is then retracted to its rearward–most position. Tools included with the cryomicrotome are used to loosen the sample holder and rotate the sample 90°. Care must be taken to insure that the first two facets are then exactly parallel with the edge of the trimming knife. The trimming process is then repeated to cut the orthogonal facets. The final blockface should be an even square or rectangle (Fig. 8.5B). If a facet is chipped or pitted during trimming, it must be retrimmed to remove the damaged area so that the facet matches its opposite. If too much of the block needs to be removed to repair the damage, it should be refaced and the trimming process started again.

If a tube-style sample is used, it is mounted into the microtome via the clamp-type specimen holder and the trimming knife advanced to its surface (Fig. 8.5C). In order to expose the sample in the center of the tube, the metal walls must be trimmed away. Since diamond is much harder than either copper or gold, trimming the metal is unlikely to damage the blade, provided that care is taken and only very thin cuts are made with each pass of the sample across the knife. During trimming, metal shards will accumulate in the chamber and can possibly blow around due to static charging. It is important to make sure that shards do not fall onto the sectioning knife, as they may damage the sensitive edge. As for planchette-type specimens, the first step is to trim a flat face that is perpendicular to the knife. Since the inner diameter of the tube is so small, the amount of sample available for trimming is very limited. When cutting the side facets, the corner of the blade must be carefully aligned to minimize removal of sample as the metal wall of the tube is trimmed away. Four parallel facets can then be trimmed, as per the aforementioned method, to give a proper block-face that is free of interfering metal (Fig. 8.5D). The sample is now ready for sectioning. The trimming knife is retracted and the knife holder is moved to bring the sectioning knife into alignment with the sample.

12. Preparing to Cryosection

Before sectioning begins, all components necessary for the process must be placed in the chamber. If a Cryo-Platform knife is being used, a grid will be placed on the ceramic platform, abutting the base of the diamond. The arm of the micromanipulator will be extended and the tool holder lowered to bring the wooden dowel into the chamber with the fiber at its tip brought close to the diamond, just above the grid. The micromanipulator's movements are checked to confirm that the fiber can be translated over a distance of at least 1 cm from the knife blade in all three dimensions. The static ionizer head is positioned in the chamber (attached with a screw for the UCT/FCS system and via a magnetic base for the UC6/FC6

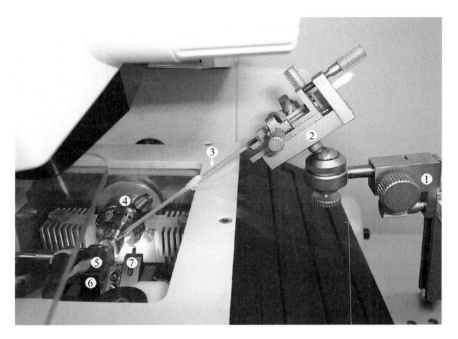

Figure 8.6 Detail of an UltraCut-UCT/FCS cryo–ultramicrotome as cryosectioning begins. The arm of the micromanipulator (1) supports a tool holder (2), which in turn holds a wooden dowel with a single fiber attached to its tip (3). The fiber is positioned near the blade of a Cryo-Platform diamond knife (6) on which an EM grid has been placed. The vitrified sample is held in a collet, which itself is locked into the microtome's specimen holder (4). The head of the static ionizer (5) is positioned ∼2 cm from the knife and aimed directly at the diamond blade. The FCS's knife holder supports both the diamond knife and a diamond trimming tool (7) in a parallel configuration.

system). The tip of the ionizer should be placed approximately 2–3 cm behind and above the knife and aimed directly at the edge of the blade. Lastly, the section thickness, cutting window, and cutting speed must be set. Initial settings of 30–60 nm for thickness and 1.0–1.4 mm/s for the cutting speed are recommended. These settings can be (and often are) adjusted to optimize cutting conditions. Figure 8.6 shows a detail of an UCT/FCS chamber with all components in place for sectioning to begin.

13. THE ACTIVE STATIC IONIZER

It is a consensus among cryomicrotomists that an active ionization system is essential for successful vitreous cryosectioning (Michel *et al.*, 1992). An ion beam directed at the diamond knife prevents sections from sticking and allows ribbons of sections to "flow" along the surface of the knife, which decreases section compression and allows the ribbons to be more easily manipulated. Often, the ionizer's power is adjusted continuously during sectioning, as too much power may cause a ribbon of cryosections to bounce or bend, while too little power will not prevent sections from sticking to the knife. Both cryomicrotome systems discussed in this chapter include ionizers, control systems, and necessary mounting hardware. Prior to each use, the ionizer should be tested at full power (using a handheld detection device, usually included with the system or sold separately by the ionizer manufacturer) to confirm that it is producing a 3–4 cm ion beam at room temperature. During long sectioning sessions, the tip of the ionizer may become contaminated with ice, decreasing its performance. The tip should be gently brushed periodically to remove the ice. During breaks from sectioning, it is recommended that the ionizer be removed from the chamber, warmed to room temperature, and dried of any accumulated moisture. It can then be placed back in the chamber for the next round of cutting.

14. CRYOSECTIONING

As sectioning begins, the static ionizer is brought to full power and the micromanipulator-controlled fiber is placed in contact with the diamond knife, just below the edge of the blade. The microtomist's left hand is positioned to control cutting parameters (mainly section thickness and the microtome's start/stop control) and the rheostat for the static ionizer. The right hand is positioned to control the micromanipulator. As successive sections are cut, they will stick together at their edges and move down the face of the knife to form a ribbon. The fiber should be positioned such that the ribbon will "flow" over it as it lengthens. After the first few sections have passed the fiber, it should be manipulated to gently lift the forming ribbon away from the knife. Ideally, the first few sections will wrap around the fiber, allowing the ribbon to be held fast (Fig. 8.7A). Once secured, the fiber is moved gently away from the knife to put the forming ribbon under slight tension. Now, as each new section is cut, the fiber is progressively advanced to maintain tension on the ribbon and to insure that only the most nascent section is in actual contact with the knife blade. In this manner, a ribbon of cryosections can be produced that is as long as, or longer, than the diameter of an EM grid (Fig. 8.7B).

The manipulator is used to keep the forming ribbon as straight as possible; if the ribbon should twist or bend (due to the ionizer beam or possibly a contaminant on the knife blade), the fiber can be adjusted laterally to recenter the ribbon and minimize any distortions. In addition to controlling the ribbon, the tension afforded by the micromanipulator stretches the sections slightly, which reverses some of the section compression induced by the knife (Ladinsky et al., 2006). Controlling a forming ribbon becomes more difficult as it grows longer.

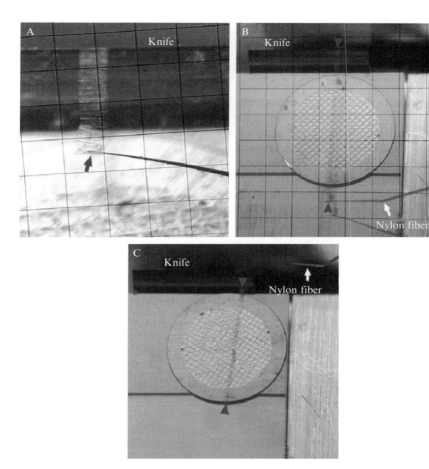

Figure 8.7 Steps in micromanipulator-assisted cryosectioning. (A) When sectioning begins, the leading end of the ribbon is picked up with the manipulator-controlled fiber. The first few sections should wrap around the fiber (arrow), allowing the ribbon to be lifted up and away from the knife. (B) As more sections are cut and the ribbon lengthens (arrowheads), the manipulator-controlled fiber is gently pulled back to hold the ribbon away from the knife, above the underlying EM grid, and under slight tension. Sectioning is stopped when the ribbon is slightly longer than the diameter of the grid ($\sim$3.5 mm). (C) In a three-step process, the micromanipulator is used to lower the ribbon (arrowheads) onto the grid, where it is subsequently detached from the knife and from the manipulator-controlled fiber. (See Color Insert.)

The micromanipulator will need to be adjusted both laterally and vertically to keep the ribbon under adequate tension and to prevent it from bowing downward and prematurely contacting the underlying EM grid. Also, as mentioned earlier, the power of the static ionizer may need continuous adjustments during the cutting process.

Under ideal circumstances, a ribbon will remain firmly attached to the fiber until it is of sufficient length for transfer to the grid. If the ribbon should prematurely disconnect from the fiber, the ribbon will often remain intact and curl back toward the knife. The micromanipulator can be used to "rescue" the ribbon; pick it back up, restore tension, and continue sectioning. If the ribbon breaks along its length (which is not usual unless the blockface is improperly trimmed, resulting in poor contact between adjacent sections), the two halves of the ribbon will most likely snap away and become tangled. These sections are usually not recoverable. They should be discarded and sectioning started again.

15. TRANSFERRING CRYOSECTIONS TO THE EM GRID

One of the most difficult and time-consuming steps in vitreous cryo-sectioning is transferring a ribbon of sections to a grid. It is at this step where the ribbon is released from its supports (the knife-edge at one end and the fiber at the other) and is easily damaged or lost due to static charges, air motion, or errant movements by the microtomist. The micromanipulator facilitates a three-step transfer process in which the ribbon of sections remains supported during each step. It is, nevertheless, a difficult procedure that requires a great deal of patience and practice.

After a ribbon of cryosections is of sufficient length (slightly longer than the diameter of an EM grid, or $\sim$3.5 mm; see Fig. 8.7B), sectioning is stopped and the static ionizer is switched off. In the first step, the manipulator is adjusted to move the fiber slightly back toward the knife to release the tension on the ribbon and align the leading edge of the ribbon (held by the fiber) with the edge of the underlying EM grid. Next, the manipulator is adjusted downward to bring the leading edge of the ribbon into contact with the grid. At this point, the ribbon itself will be slightly bowed. In the second step, the microtomist will use another fiber tool (identical to the one held by the micromanipulator) to gently push the ribbon down so that it contacts the grid along most of its length. Great care must be taken at this point to avoid disrupting the ribbon or damaging the grid's support film. The ribbon of cryosections will now be supported by the grid but still attached to the fiber and the knife blade at either end. In the third step, the microtomist uses the fiber tool to disconnect the ribbon. The knife end of the ribbon is released first, by using the fiber to break the ribbon near the base of the blade. This allows for a few sections to remain on the knife to serve as the starter of a new ribbon for a subsequent round of sectioning. The now-freed end of the ribbon should fall onto the grid with little overlap. Next, the leading end of the ribbon is gently teased off of the micromanipulator-held fiber. This is typically accomplished by using the handheld fiber to hold the edge of the ribbon in place; the manipulator is then moved laterally to slide its fiber away from the sections. The manipulator arm is moved clear of the sectioning area, leaving the ribbon fully supported by the grid (Fig. 8.7C).

16. Pressing the Sections onto the Grid

Following transfer of a ribbon to the grid, the sections will remain in position and are relatively unaffected by static, air motion, or gentle movements of the grid. However, the sections are not affixed strongly enough to survive subsequent cryotransfer steps, storage in liquid nitrogen, or insertion into the vacuum of the cryoelectron microscope. To assure the firmest possible attachment to the underlying substrate of the EM grid, the ribbon of sections is pressed down by physical contact with a hard, flat instrument. Pressing serves not only to affix the sections, but also to flatten them out against the substrate. Sections that are bent or bowed will result in uneven focus in the cryo–EM and, if imaged for electron tomography, will yield a distorted final reconstruction. Section pressing is challenging because it involves direct contact with the sections, which can easily result in damage to the ribbon or substrate or loss of the sections if they become stuck to the pressing device. It is important that the pressing device be properly precooled and free of any contaminants. Pressing must be done with the grid on a flat surface and pressure applied in the vertical direction only; any lateral motion of the pressing device while it is in contact with the grid will damage the substrate and cause loss of the sections.

Several devices for pressing sections have been designed and implemented over the years. The most common one is a pen–like tool (Fig. 8.8A) with flattened Teflon or aluminum tips at either end. This device is included with all cryomicrotome systems. The tip is precooled in liquid nitrogen and carefully held over the grid. The microtomist applies gentle vertical pressure for a few seconds to firmly affix the sections and then removes the tool from the cryochamber. Use of this device requires a steady hand and considerable practice to learn the amount of pressure and time needed to successfully attach the sections.

Another device, introduced with the Reichert UltraCut-E/FC-E system, is a piston–like tool consisting of a spring-loaded Fiberglass or Teflon block held above a polished aluminum wheel (Fig. 8.8B). The device was designed to fit in the back right corner of the FC-E cryochamber but also fits in the chamber of the UCT/FCS system. A grid containing cryosections is placed on the exposed portion of the aluminum wheel. The wheel is rotated to bring the grid underneath the block, which is subsequently lowered onto the grid by tightening a screw on the top of the unit. Pressure is applied to the grid as the screw is tightened down. After pressing for several seconds or minutes, the screw is loosened, the block raised, and the grid removed for storage. This device is considerably more difficult to use than the pen but yields more even and consistent pressure. The pressing block and the aluminum wheel must be kept clean to prevent sticking to either surface.

Figure 8.8 Devices for pressing cryosections onto the EM grid. (A) The most common pressing device, a pen-like instrument with Teflon (arrow) or aluminum tips at either end. The instrument is held by the microtomist to apply pressure onto the grid. (B) An earlier pressing device designed for the UltraCut-E/FC-E system. A grid with cryosections is placed on the slotted wheel and rotated under the spring-loaded block. The block is screwed down to apply pressure to the grid. (C) A simple, but very effective pressing device consists of a small chip of glass placed over the grid. A pair of precooled fine forceps is used to apply direct and even pressure along the ribbon of cryosections (arrowheads), firmly affixing it to the grid. (See Color Insert.)

Both of the devices described above have a distinct disadvantage in that the grid cannot be observed during the pressing step. Seeing the grid while pressing it facilitates even application of pressure and more readily insures that damaging lateral motions do not occur. Designs that utilize flat sapphire or other mineral crystals, held with forceps or attached to a base by means of a hinge, have proved effective in pressing sections with minimal associated damage. We have developed a very effective and inexpensive method that simply uses a small piece of glass cut from a standard microscope slide. A small (~5–6 mm) square chip of glass cut from the edge of a slide is thoroughly cleaned and rinsed with ethanol to remove any traces of oil or other contaminants. Prior to pressing, the grid containing cryosections is moved to the far end of the knife platform, away from the diamond blade. The glass chip is cooled in liquid nitrogen and quickly transferred to the cryochamber with a pair of fine forceps. The glass is briefly touched to the side of the knife base to discharge any static and then gently placed on top of the loaded grid. The ribbon of sections on the grid can be easily seen through the glass (Fig. 8.8C). The forceps are then used to apply pressure to the top of the glass, along the length of the ribbon. The glass is lifted away, revealing a finished grid that is ready to be imaged in the cryoelectron microscope.

A recent advance has done away with physical pressing by utilizing electrostatic discharge to efficiently attach sections to a grid (Pierson *et al.*, 2010). Contact between the sections and the substrate is as good or better than what is achieved by physical pressing, with much less risk of damage or section loss. In this method, a device similar to the active static ionizer applies an electrical charge to the grid, causing the sections on it to flatten out and become firmly affixed to the grid substrate. This device is now being sold as part of the Leica FC-7 cryosectioning system.

After the sections have been pressed, the finished grid is transferred to a cryostorage box and stored under liquid nitrogen until it is ready to be placed in the cryoelectron microscope. In a typical session, we will prepare four grids from a given sample, each holding 1–3 ribbons of vitreous cryosections (Fig. 8.9C). Grids will often be photographed prior to storage, to document their condition and the position of the sections, as an aid to the microscopist when the grids are imaged.

17. THE SECTIONING ENVIRONMENT AND RELATIVE HUMIDITY

If cryosectioning was not complicated enough, its success is acutely affected by several environmental conditions. These include vibration (discussed earlier with vibration isolation tables), air movement in the sectioning room and, most importantly, relative humidity. Ideally, the cryo–ultramicrotome should be installed in the smallest possible room that can accommodate the instrument and all its components, the operator, and a small table for tools and other accessories. The room should be free of ventilation ducts or have accessible ducts that can be closed off to minimize air flow (a note of caution; there must be at least some air flow to prevent dangerous accumulation of nitrogen gas. In fact, it is highly recommended that an oxygen sensor be placed in the room). Often a large closet, a small office, or a converted walk-in refrigerator will be selected as a cryosectioning room. Some laboratories have designed custom facilities for their cryomicrotomes, in which all environmental conditions can be controlled and maintained.

Figure 8.9 Vitreous cryosections prepared in a very humid environment ($>50\%$ relative humidity). Under such conditions, moisture in the air within the cryosectioning chamber will freeze and contaminate the sections and support grid. (A) A newly cut ribbon of cryosections held between the knife and the micromanipulator-controlled fiber (arrowheads). Ice particles are seen on the knife-edge, the ceramic platform, and on the ribbon itself, where they have caused the ribbon to twist (arrow). (B) An ice-contaminated ribbon of cryosections on an EM grid (arrowheads). The presence of ice makes proper placement and section pressing difficult. In the cryo-EM, ice particles appear as large spheres that limit the imagable area of the sections. (C) A grid of cryosections prepared under optimal conditions. No ice particles are present and ribbons are straight and evenly affixed to the support film. The micromanipulator method facilitates accurate placement of multiple ribbons onto a single grid.

In plunge freezing and other rapid-freezing methods, humidity is important in preventing samples from drying before they are vitrified (see Chapter 3). High relative humidity is, however, the bane of cryo–ultramicrotomy. When a cryosectioning environment is very humid ($>\sim50\%$), moisture in the air inside the cryomicrotome's chamber will freeze and fall like snow onto the knife, the grid, and the cryosections (Fig. 8.9A). Ice particles on the knife's edge will act as an abrasive on the sections, causing pits and scratches as the sections are cut. Accumulation of ice on a ribbon of cryosections may cause it to destabilize, twist, and even break as it is being pulled away from the knife. Ice particles on the grid or the sections cannot be easily removed and will appear as large, black spheres of contamination in the electron microscope, effectively reducing the imagable area of the sample. Furthermore, even if a ribbon can be successfully produced, ice contamination will prevent it from being sufficiently attached to the grid (Fig. 8.9B).

My colleagues and I have been blessed in that our cryosectioning work has been carried out in two parts of the United States (Eastern Colorado and inland Southern California) where relative humidity is naturally very low. As such, measures to control humidity are seldom necessary; cryosectioning could simply be postponed on the few days when relative humidity exceeded $\sim45\%$. In other places, where humidity is naturally much higher, the cryosectioning room must be dehumidified before work can begin. If the room is small enough, a portable dehumidifier, run for several hours before the sectioning session, may be sufficient for reducing humidity to a manageable level. A portable air-conditioning unit may also be used, but it must be positioned to avoid increased airflow over the cryomicrotome. Some laboratories in especially humid parts of the world have installed elaborate air-conditioning and filtration systems for isolating their sectioning rooms and maintaining them precisely at, or below, 30% relative humidity. Recently, an enclosure similar to a glove-box has been designed that surrounds and isolates the cryomicrotome's operating area, providing a localized humidity-controlled environment (Pierson *et al.*, 2010). This "cryosphere" is sold as an optional accessory for the Leica FC7 cryosectioning system.

18. CRYOSECTIONING ARTIFACTS

There are several distinct kinds of section damage, or artifacts, that are caused by cryosectioning. These artifacts have been studied extensively for over three decades (Al-Amoudi *et al.*, 2003, 2005; Dubochet *et al.*, 2007; McDowall *et al.*, 1983; Richter, 1994; Richter *et al.*, 1991), so the topic will only be discussed briefly here.

The most important cryosectioning artifacts are section compression and "crevassing." These artifacts are common to all cryosectioning methods and are directly related to sectioning with a dry knife (as apposed to a wet knife used for room-temperature sectioning of plastic-embedded samples, described in Section 1). Current technology and approaches are helping to minimize these artifacts.

Compression is caused by friction between the edge of the knife and the section being cut and manifests as a decrease in the length of the section, in the cutting direction, with a concomitant increase in section thickness (Al-Amoudi et al., 2003; McDowall et al., 1983). Simply put, the answer to compression is to "stretch" the sections back out after they have been cut. Of course, this is much easier said than done. Plastic sections floating on a pool of water can be fully restored to their original length by exposure to a solvent vapor, such as chloroform. This is not possible for vitreous cryosections. In the technique described in this chapter, the modicum of tension imposed on a ribbon of cryosections by the micromanipulator induces some stretching of the sections. Images from sections prepared in this manner are improved, but not free of compression artifact (Ladinsky et al., 2006).

Crevassing is caused when sections are cut and bend to conform to the angle of the knife as they travel down the blade. Crevasses, not unlike their glacial namesake, appear as a series of fractures on the upper surface of the section, perpendicular to the cutting direction, that penetrate approximately 20% of the section's volume (Al-Amoudi et al., 2005). Some have observed that crevassing can be eliminated if sections are made especially thin (Zhang et al., 2004). An alternative interpretation is that the depth of crevasses is a constant percentage of the section volume, meaning that it is quite obvious in thicker sections but less so in thinner ones. Theoretically, decreasing the included angle of the cryodiamond knife, thereby lessening the angle by which the section must bend as it comes off the blade, should reduce crevassing. Thus, a 25° cryodiamond knife should produce sections with less crevassing than a 45° knife. Extrapolating on this idea, knives made from single carbon nanotubes, which have zero included angle, have been considered as a way for producing artifact-free cryosections (Singh et al., 2009). Although carbon "nanoknives" have been fabricated, they have not yet reached a design level that is practical for use in cryo-ultramicrotomy.

19. Conclusion

Cryo-electron microscopy of vitreously frozen biological material is now a powerful method that is yielding new insights into cellular structure that could not be attained by any other means. It is limited, however, by several factors, chief among which is sample thickness. Modern cryo-ultramicrotomes and cryodiamond knives, in conjunction with methods for using these tools, are making cryosectioning of frozen-hydrated samples a routine technique in several cryo-EM laboratories. As the technology continues to improve, the method will become popular among many more structural cell biologists. This particular method, which uses a micromanipulator to refine the hand of the skilled cryomicrotomist, is but one of several effective techniques that are making vitreous cryosectioning an important addition to the electron microscopist's toolbox.

ACKNOWLEDGMENTS

The author wishes to thank the following people for advice and support: J. Richard McIntosh (University of Colorado, Boulder), Kent McDonald (University of California, Berkeley), Helmut Gnaegi (Diatome, Ltd.), Jacques Dubochet (University of Lausanne, Switzerland), Jason Pierson (Netherlands Cancer Institute, Amsterdam), Lu Gan, Alasdair McDowall, Grant Jensen and Pamela Björkman (California Institute of Technology).

REFERENCES

Adrian, M., Dubochet, J., Lepault, J., and McDowall, A. W. (1984). Cryo-electron microscopy of viruses. *Nature* **308**, 32–36.

Al-Amoudi, A. J., Dubochet, H., Luthi, Gnaegi W., and Studer, D. (2003). An oscillating cryo-knife reduces cutting-induced deformation of vitreous ultrathin sections. *J. Microsc.* **212**, 26–33.

Al-Amoudi, A., Norlen, L. P. O., and Dubochet, J. (2004a). Cryo-electron microscopy of vitreous sections of native biological cells and tissues. *J. Struct. Biol.* **148**, 131–135.

Al-Amoudi, A., Chang, J.-J., Leforestier, A., McDowall, A., Salamin, L. M., Norlen, L. P. O., Richter, K., Blanc, N. S., Studer, D., and Dubochet, J. (2004b). Cryo-electron microscopy of vitreous sections. *EMBO J.* **23**, 3583–3588.

Al-Amoudi, A., Studer, D., and Dubochet, J. (2005). Cutting artifacts and cutting process in vitreous sections for cryo-electron microscopy. *J. Struct. Biol.* **150**, 109–121.

Bouchet-Marquis, C., Dubochet, J., and Fakan, S. (2005). Cryoelectron microscopy of vitrified sections: A new challenge for the analysis of functional nuclear architecture. *Histochem. Cell Biol.* **125**, 43–51.

Dubochet, J. (1995). High-pressure freezing for cryoelectron microscopy. *Trends Cell Biol.* **5**, 366–368.

Dubochet, J., Zuber, B., Eltsov, M., Bouchet-Marquis, C., Al-Amoudi, A., and Livolant, F. (2007). How to "read" a vitreous section. *Methods Cell Biol.* **79**, 385–406.

Echlin, P., Skaer, H. Le.B., Gardiner, B. O. C., Franks, F., and Asquith, M. H. (1977). Polymeric cryoprotectants in the preservation of biological ultrastructure. II. Physiological effects. *J. Microsc.* **110**, 239–255.

Franks, F., Asquith, M. H., Hammond, C. C., Skaer, H. Le.B., and Echlin, P. (1977). Polymeric cryoprotectants in the preservation of biological ultrastructure. I. Low temperature states of aqueous solutions of hydrophilic polymers. *J. Microsc.* **110**, 223–238.

Hsieh, C.-E., Leith, A., Mannella, C. A., Frank, J., and Marko, M. (2005). Towards high-resolution three- dimensional imaging of native mammalian tissue: Electron tomography of frozen-hydrated rat liver sections. *J. Struct. Biol.* **153**, 1–13.

Ladinsky, M. S., Pierson, J. M., and McIntosh, J. R. (2006). Vitreous cryo-sectioning of cells facilitated by a micromanipulator. *J. Microsc.* **224**, 129–134.

Matias, V. R., Al-Amoudi, A., Dubochet, J., and Beveridge, T. J. (2003). Cryo-transmission electron microscopy of frozen-hydrated sections of *Escherichia coli* and *Pseudomonas aeruginosa*. *J. Bacteriol.* **185**, 6112–6118.

Matzelle, T. R., Gnaegi, H., Ricker, A., and Reichelt, R. (2003). Characterization of the cutting edge of glass and diamond knives for ultramicrotomy by scanning force microscopy using cantilevers with a defined tip geometry, Pt.II. *J. Microsc.* **209**, 113–117.

McDonald, K. E., and Auer, M. (2006). High-pressure freezing, cellular tomography, and structural cell biology. *Biotechniques* **41**, 137.

McDonald, K. L., Morphew, M., Verkade, P., and Muller-Reichert, T. (2007). Recent advances in high-pressure freezing: Equipment- and specimen-loading methods. *Methods Mol. Biol.* **369**, 143–173.

McDowall, A. W., Chang, J.-J., Freeman, R., Lepault, J., Walter, C. A., and Dubochet, J. (1983). Electron microscopy of frozen hydrated sections of vitreous ice and vitrified biological samples. *J. Microsc.* **131**, 1–9.

McEwen, B. F., Marko, M., Hsieh, C.-E., and Mannella, C. (2002). Use of frozen-hydrated axonemes to assess imaging parameters and resolution limits in cryoelectron tomography. *J. Struct. Biol.* **138**, 47–57.

Michel, M., Gnaegi, H., and Muller, M. (1992). Diamonds are a cryosectioner's best friend. *J. Microsc.* **166**, 43–56.

Norlen, L. (2008). Exploring skin structure using cryo-electron microscopy and tomography. *Eur. J. Dermatol.* **18**, 279–284.

Pierson, J., Fernandez, J. J., Bos, E., Amini, S., Gnaegi, H., Vos, M., Bel, B., Adolfsen, F., Carrascosa, J. L., and Peters, P. J. (2010). Improving the technique of vitreous cryo-sectioning for cryo-electron tomography: Electrostatic charging for section attachment and implementation of an anti-contamination glove box. *J. Struct. Biol.* **169**, 219–225.

Richter, K. (1994). Cutting artefacts on ultrathin cryosections of biological bulk specimens. *Micron* **25**, 297–308.

Richter, K., Gnaegi, H., and Dubochet, J. (1991). A model for cryosectioning based on the morphology of vitrified ultrathin sections. *J. Microsc.* **163**, 19–28.

Singh, G., Rice, P., Mahajan, R. L., and McIntosh, J. R. (2009). Fabrication and characterization of a carbon nanotube-based nanoknife. *Nanotechnology* **20**, 1–6.

Skae, H. Le.B., Franks, F., Asquith, M. H., and Echlin, P. (1977). Polymeric cryoprotectants in the preservation of biological ultrastructure. III. Morphological aspects. *J. Microsc.* **110**, 257–270.

Tokuyasu, K. T. (1971). A technique for ultramicrotomy of cell suspensions and tissues. *J. Cell Biol.* **57**, 551–565.

Tokuyasu, K. T. (1986). Application of cryoultramicrotomy to immunocytochemistry. *J. Microsc.* **143**, 139–149.

Zhang, P., Bos, E., Heymann, J., Gnaegi, H., Kessel, M., Peters, P. J., and Subramanium, S. (2004). Direct visualization of receptor arrays in frozen-hydrated sections and plunge-frozen specimens of *E. coli* engineer to overproduce the chemotaxis receptor Tsr. *J. Microsc.* **216**, 76–83.

CHAPTER NINE

SITE-SPECIFIC BIOMOLECULE LABELING WITH GOLD CLUSTERS

Christopher J. Ackerson,* Richard D. Powell,[†] *and* James F. Hainfeld[†]

Contents

Abstract

Site-specific labeling of biomolecules *in vitro* with gold clusters can enhance the information content of electron cryomicroscopy experiments. This chapter provides a practical overview of well-established techniques for forming bio-molecule/gold cluster conjugates. Three bioconjugation chemistries are covered: Linker-mediated bioconjugation, direct gold–biomolecule bonding, and

* Colorado State University, Department of Chemistry, Fort Collins, Colorado, USA
[†] Nanoprobes, Incorporated, Yaphank, New York, USA

Methods in Enzymology, Volume 481   © 2010 Elsevier Inc.
ISSN 0076-6879, DOI: 10.1016/S0076-6879(10)81009-2   All rights reserved.

coordination-mediated bonding of nickel(II) nitrilotriacetic acid (NTA)-derivatized gold clusters to polyhistidine (His)-tagged proteins.

1. INTRODUCTION

Electron cryomicroscopy comprises a set of techniques brimming with promises and possibilities as varied as atomic resolution protein structure on noncrystalline biomolecules (Cong *et al.*, 2010) and visual proteomics (Nickell *et al.*, 2006). The suite of electron cryomicroscopy techniques fails to achieve its theoretical limits in large part because of various contrast problems: poor signal-to-noise ratio (SNR) constitutes a substantial practical limitation for all applications of electron cryomicroscopy to "unstained" biological materials.

Site-specific labeling with gold cluster tags has been used to skirt the SNR limitations inherent to electron cryomicroscopy techniques. Gold clusters have amplitude contrast in addition to phase contrast in the electron microscope, and so are easily distinguished in images and in image processing from biological molecules which have only phase contrast. While in principle, clusters or nanoparticles comprised of any heavy metal could be used as an electron contrast marker, gold is adopted most widely because it is comparatively inert, insensitive to molecular oxygen, forms nanoparticles comparatively easily because of the low surface energy of Au (Alvarez *et al.*, 1997), and because the chemistries for surface functionalization of gold nanoparticles are comparatively well understood and controlled in aqueous solution.

This chapter focuses on methods for site-specific gold cluster labeling of biomolecules. Here, we define "biomolecules" as including proteins, nucleic acids, lipids, and biologically active small molecules. Gold clusters are here defined as a cluster of gold atoms containing multiple gold–gold bonds which are rendered kinetically metastable by a protecting or stabilizing ligand layer. They are differentiated from colloidal gold in that they have molecule-like properties and often well-defined molecular formulae, and sometimes, crystal structure. This is in contrast to colloidal gold where "monodisperse" is usually defined as ± 10% or lower coefficient of variation in core size, and the gold cores are generally comprised of multiply twinned crystallites with many different crystal structures present in the same preparation. Figure 9.1 shows a selection of well-established gold clusters, with structures shown if presently determined. A representation of 5 nm colloidal gold is included for size comparison.

Just as there are many different modalities of cryomicroscopy, there are multiple approaches to making gold cluster/biomolecule conjugates for

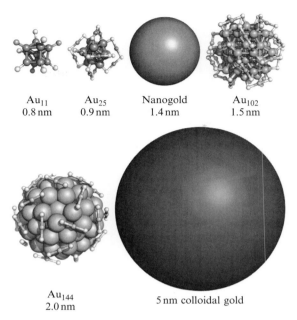

Figure 9.1 Representative inorganic gold clusters. The organic components of the ligand layer of each cluster are not shown. Orange, yellow, purple, and green represent gold, sulfur, phosphorous, and chlorine, respectively. The images of Au_{11}, Au_{25}, and Au_{102} are created from X-ray crystallographic coordinates. The image of Au_{144} is from a density functional theory model (Lopez-Acevedo *et al.*, 2009). A 5-nm diameter colloidal gold particle is shown for comparison. (See Color Insert.)

electron cryomicroscopy (Table 9.1; Fig. 9.2). Generally, the methods can be categorized according to the nature of the interface between biomolecule and gold cluster. There are three commonly used nanoparticle/biomolecule interfacial chemistries: Electrostatic, linker-mediated bonding, and direct bonding (Fig. 9.2).

This chapter focuses on the linker-mediated and direct bioconjugation strategies that make stable covalent bonds between well-defined gold cluster compounds and biomolecules. This introduction provides an overview of the bioconjugate chemistries shown in Fig. 9.2. These various clusters can be used to form gold cluster-cysteine, -amine, or Ni(II)-His_6 cross-links to biomolecules (Fig. 9.2 and Table 9.1).

Linker-mediated bioconjugate chemistries, as depicted in the upper right and lower right panels of Fig. 9.2, have been used for decades to reliably make gold cluster bioconjugates that may be molecularly defined. Given the widespread and easy use of these clusters, their use will be discussed in detail. The best known examples of gold nanoparticles used

Table 9.1 Properties of selected gold cluster labels

Gold cluster	Bioconjugate reactive group(s) of cluster	Labeled biomolecule moiety	Nominal diameter
Undecagold	Monoamino-, maleimido, *Sulfo*-NHS-,	$-SH$, $-NH_2$	0.8 nm
Nanogold	Amino-, maleimido, *Sulfo*-NHS,	$-SH$, $-NH_2$	1.4 nm
Ni-NTA–Nanogold	Ni(II) nitrilotriacetic acid (NTA)	Polyhistidine, particularly His_6	1.8, 5 nm
MPCs: Au_{25}, Au_{38}, Au_{68}, Au_{102}, Au_{144}	Au	$-SH$	0.9, 1.1, 1.3, 1.5, 2.0 nm

for linker-mediated biological labeling are undecagold (Bartlett *et al.*, 1978; Cariati and Naldini, 1971; Safer *et al.*, 1982) and the commercial product Nanogold (Hainfeld and Furuya, 1992).

Nanogold and undecagold are available in nominally monofunctional forms (Nanoprobes, Inc., Yaphank, NY). The monomaleimido and mono-*Sulfo*-NHS forms of Nanogold and undecagold have been used widely for site-specific labeling of cysteine and primary amine-containing residues respectively within proteins. This bioconjugate chemistry has been used to label antibodies and their fragments (Hainfeld and Furuya, 1992; Hainfeld and Powell, 1997; Ribrioux *et al.*, 1996), other proteins (Rappas *et al.*, 2005; Yonekura *et al.*, 2006), peptides (Segond von Banchet *et al.*, 1999), oligonucleotides (Alivisatos *et al.*, 1996; Hamad-Schifferli *et al.*, 2002; Park *et al.*, 2005; Xiao *et al.*, 2002), and lipids that assemble into liposomes (Hainfeld *et al.*, 1999a). The biological (Hainfeld and Powell, 1997, 2000) and nanobiotechnological (Hainfeld *et al.*, 2004) aspects of covalent gold labeling with Nanogold and undecagold have been reviewed previously.

Metal cluster labels can also be targeted to genetically encoded tags. A 1.8-nm Nanogold cluster label, stabilized and solubilized using a different ligand shell, has been functionalized with a derivative of nitrilotriacetic acid (NTA). Upon incubation with nickel(II) salts, the Nanogold cluster forms a nickel(II) complex with two vacant adjacent binding sites, which binds strongly to histidine residues (Hainfeld *et al.*, 1999b). Consequently, this structure binds very strongly to polyhistidine (His) tags such as hexahistidine (His_6), which are commonly used as affinity purification tags. A 5-nm analog has recently been introduced (Dubendorff *et al.*, 2010; Reddy *et al.*, 2005).

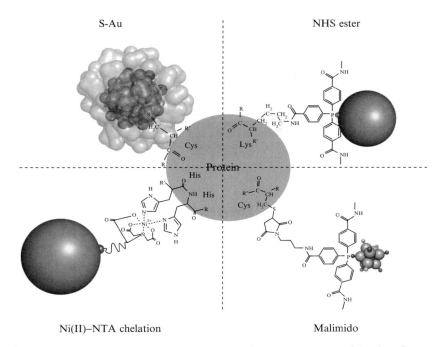

Figure 9.2 Schematic of the bioconjugation chemistries presented in this chapter, which are summarized in Table 9.1. In the upper left corner is shown a "direct" biomolecule/gold cluster conjugate where a thiol moiety such as a cysteine residue bonds directly to the gold cluster core of $Au_{102}(p$-mercaptobenzoic acid$)_{44}$ through a sulfur–gold bond. The crystallographically determined ligand layer of the cluster is rendered as a partially transparent surface in that image, illustrating the steric constraint of this linkage. Shown in upper and lower right panels are the bioconjugates formed by NHS and malimido derivatives of tris (aryl) ligands which protect clusters such as Nanogold and undecagold. Shown at lower left is an Ni^{2+}-mediated interaction between a polyhistidine tag and a nitroloacetic acid (NTA)-derivatized gold cluster. Note that all stable water-soluble gold clusters and colloids have a *full* organic ligand shell like that shown for Au_{102} in this diagram, but these ligand layers are sometimes not as well characterized as the cluster core.

It is also possible to form bonds between cysteine residues within proteins and gold clusters, on the basis of gold–thiol bonding. This strategy eliminates the organic linker between the protein and gold core, which is advantageous for highest resolution localization of cluster labels on proteins. This labeling strategy also takes advantage of a set of molecularly defined, and in some cases, structurally defined (Heaven *et al.*, 2008; Jadzinsky *et al.*, 2007) gold cluster compounds whose gold atom nuclearities include Au_{25}, Au_{38}, Au_{68}, Au_{102}, and Au_{144}. This strategy has been used to form gold cluster conjugates with proteins (Ackerson *et al.*, 2006, 2010; Aubin-Tam and Hamad-Schifferli, 2005; Aubin-Tam *et al.*, 2009), peptides (Krpetic *et al.*, 2009; Levy *et al.*, 2004), oligonucleotides (Ackerson *et al.*, 2005), and bioactive small molecules (Bowman *et al.*, 2008). This labeling strategy is more complicated than the linker-mediated strategies, because the site of label and chemical composition of the surface layer of the gold cluster influences both the yield of the labeling reaction and the biological function of the cluster/protein conjugate (Aubin-Tam *et al.*, 2009). Fortunately, guidelines for choosing sites for gold cluster labeling exist to guide the design of these experiments.

The bioconjugation strategies presented in this chapter are chosen because they are robust and have been repeated in many laboratories. Other *in vitro* labeling means have been used to make nanoparticle/protein bioconjugates. Tetrairidium clusters have been used to derivatize virus capsids for high-resolution EM and single particle analysis (Cheng *et al.*, 1999). Although it is too small to be directly visualized in a standard TEM, and hence is suited mainly to single particle analysis, this cluster provides very high resolution and minimal structural perturbation. Larger platinum clusters 0.8–2.5 nm in diameter, stabilized by 1,10-phenanthroline derivatives, have been used as covalent Fab′ labels (Powell *et al.*, 1999). However, covalent gold labels are preferred because they afford better synthetic control of particle size distribution, solubility, and stability, and more specific biomolecule conjugation.

Colloidal gold nanoparticles, defined here as particulate gold with diameters between 5 and 250 nm, have found widespread use in biological electron microscopy. Most commonly, they are used as fiducial markers in electron tomography and in the gold colloid/antibody conjugates widely used in the immunohistochemistry (Handley, 1989a,b). In contrast to the discrete molecular particle sizes that have recently emerged for gold nanoparticles approximately 3 nm in diameter or smaller, these colloidal particles can be prepared with average core diameters that comprise a more or less continuous distribution. From a physicochemical standpoint, when compared to the molecular or molecule-like gold clusters that are the focus of this chapter, these particles are generally larger, comprised of multiply twinned crystallites, and defined as monodisperse when the coefficient of variation of particle diameter within a preparation is 10% of the average diameter or less. From a bioconjugate chemical standpoint, gold colloids generally bind to biomolecules through electrostatic, rather than covalent, interactions.

A consequence of the electrostatic nature of the colloidal gold–protein bond is that site-specific biomolecule labeling with gold colloids is not routine. For instance, colloidal gold antibody conjugates are usually defined in terms of "average number of antibodies per colloid" for large colloids or "average number of colloids per antibody" for small colloids, as opposed to a specific labeled residue within a protein as is the case for the conjugates that are the focus of this chapter. Other consequences of the electrostatic interaction typical in gold colloid/biomolecule conjugates include dissociation of the colloid which can result in competition between labeled and unlabeled probe giving lower labeling (Kramarcy and Sealock, 1990), and the residual affinity of colloidal gold for proteins which sometimes results in nonspecific binding (Behnke *et al.*, 1986). To avoid some of these problems, additional macromolecules are sometimes used for stabilization, such as carbowax (polyethylene glycol) or bovine serum albumin. These, however, can add to probe size and hinder and target access. Thus, while colloidal gold conjugates have greatly enhanced the microscopy of fixed and plastic-embedded biological samples, their utility in the higher resolution experiments typical in electron cryomicroscopy is more limited.

There are also emerging strategies, and substantial excitement about these strategies, for producing electron contrast in a clonable manner, with the goal of producing a "Green Fluorescent Protein" for electron microscopy. Many efforts have focused on adapting metallothionein, a member of the metal binding class of phytochelatin proteins (Diestra *et al.*, 2009; Mercogliano and DeRosier, 2007). Metallothioneins function within cells as metal scavengers and are highly expressed when cultured cells are exposed to toxic metal ions. Mercogliano and DeRosier showed that recombinant concatenated metallothioneins can sequester a sufficient number of gold atoms to observe metal–nanoparticle-like densities in *in vitro* electron cryomicroscopy experiments. Subsequent *in vivo* experiments by other research groups have demonstrated some promise for such strategies.

Proteins with unusual electron density have also been used to localize areas of proteins (Alcid and Jurica, 2008). These strategies also show promise, but are not yet widely adopted.

2. GENERAL CONSIDERATIONS

Of the site-specific gold cluster bioconjugation strategies discussed here, the bioconjugate chemistry chosen will depend on the goal of the cryomicroscopy experiment. For intermediate resolution goals, the linker-mediated bioconjugate strategies incorporated in Nanogold and undecagold (Bartlett *et al.*, 1978; Cariati and Naldini, 1971; Safer *et al.*, 1982) deserve the highest consideration because of their commercial availability and ease of use. For the highest resolution goals, direct bonding strategies should be considered, because gold clusters of varying sizes with the exactly known molecular formulae of Au_{25}, Au_{38}, Au_{68}, Au_{102}, and Au_{144}, and in some cases, exactly known structure, can be affixed directly and rigidly to a specific site within the target biomolecule.

This chapter contains discussions and detailed protocols for labeling and purification of gold cluster bioconjugates with the direct, NHS–ester, malimido-, and Ni(II)-mediated bioconjugate chemistries shown in Fig. 9.2. While the reaction conditions for formation of the bioconjugates presented are specific for the particular bioconjugate chemistry, the assays of bioconjugate product, for instance, by gel electrophoresis, UV/VIS spectroscopy, or electron microscopy and purification conditions may be considered more general, applicable to any of the gold cluster bioconjugates presented in this chapter with minor adaptations.

With regard to purification in particular, colloidal gold conjugates are usually separated by "pelleting" in a centrifuge or ultracentrifuge, followed by resuspension. However, most of the gold cluster labels discussed in this chapter are not amenable to such a simple purification procedure. For instance, Nanogold is sufficiently small that it does not pellet at $100,000 \times g$. Although centrifugation is inefficient for purification of gold cluster bioconjugates, a number of general strategies have proved successful for purification of bioconjugates, and several specific separation strategies are discussed in the context of purification of specific bioconjugates. The purification strategies presented for particular bioconjugate strategies, however, may work well for other gold cluster bioconjugates. For instance, we include a detailed discussion on gel filtration chromatography strategies for Nanogold conjugate purification, including recommended reaction stoichiometries, to make the purification steps easier. These strategies are certainly also adaptable for purification of monolayer protected cluster (MPC) bioconjugates. Thus, when working through the purification of a novel gold cluster bioconjugate, a "mix and match" approach with the purification strategies presented in each subsection may be considered.

2.1. Stoichiometric labeling

In some applications, 100% labeling is desired, for example, to count the number of target sites on a protein or complex. Although this is possible, there are several factors that may interfere with the desired result: (a) it should be appreciated that these are chemical reactions that may not yield 100% efficiency; (b) reactions may generate some cross-reactivity with other chemical functional groups; (c) the gold particle may bind "nonspecifically" to the protein being studied; and (d) a particular gold cluster may have more than one reactive group. Thus, in any labeling experiment where subsequent electron cryomicroscopy might require stoichiometric labeling, appropriate controls must always be included. These include, for instance, competing or blocking agents, removal of the target site (e.g., cleavage of His tag), and use of nonfunctionalized gold particles or (for labeling at thiol sites obtained by disulfide reduction) nonreduced biomolecules.

3. MONOLAYER PROTECTED CLUSTER LABELING OF BIOMOLECULES

Beginning in the 1990s was a great increase in research in "nanotechnology," resulting in a great number of discoveries relating to nanoparticles in general and gold nanoparticles in particular. Just as an example, there were, according to Web of Science, 1060, 1299, and 1390 original research papers on the topic of "gold nanoparticle" in the years 2007, 2008, and 2009. Some of this research has involved developing, characterizing, and applying novel gold cluster/biomolecule conjugates that may be of direct interest for application in electron cryomicroscopy.

Among the research that contributes to improvements in gold nanoparticle/protein conjugates includes the work of several groups investigating the interactions of nanoparticle preparations with biological molecules (Bowman *et al.*, 2008; Tkachenko *et al.*, 2003; Verma *et al.*, 2004). Some research groups have put forth strategies for monofunctionalization of gold nanoparticles (Huo and Worden, 2007; Liu *et al.*, 2006; Worden *et al.*, 2004a,b). Another group has investigated the direct coupling of His_6 tags to gold nanoparticle surfaces, presenting spectroscopic evidence for His–gold bonds and also slow denaturation of the protein conjugated to the gold surface (Kogot *et al.*, 2008, 2009). Multidentate strategies for bioconjugation have also been put forth (Krpetic *et al.*, 2009). Of greatest interest to electron cryomicroscopy is the work of the Kornberg and Hamad-Schifferli groups, where gold–protein bonds through cysteine thiolate functionality have been formed. These results will form the basis for the methods that follow.

The gold nanoparticles that are amenable to the protein labeling strategies that are described herein come from a series of now well-defined thiolate MPCs that have been investigated for the past ~ 15 years (Sardar *et al.*, 2009). MPCs are composed of a compact crystalline gold core in which every gold surface atom is bonded to an organothiolate ligand (Fig. 9.3) and may generally be formulated as $Au_n(SR)_m$. These compounds are now understood to occur in a "magic number series" resulting from the complete filling of superatomic electron orbitals (Walter *et al.*, 2008). This means that MPCs of the generic formulae $Au_{25}(SR)_{18}$, $Au_{38}(SR)_{22}$, $Au_{68}(SR)_{34}$, $Au_{102}(SR)_{44}$, and $Au_{144}(SR)_{60}$ are understood to have special stability, analogous to the stability of a noble gas atom.

MPCs may be functionalized in well-known ligand exchange reactions (Scheme 9.1; Hostetler *et al.*, 1999; Song and Murray, 2002; Templeton *et al.*, 1998, 2000). The kinetics, rate, and extent of a ligand exchange reaction depend upon the nature of the incoming and outgoing ligands, as well as the charge on the cluster core. The incoming ligand can be any molecule presenting a thiol, including modified nucleic acids, small

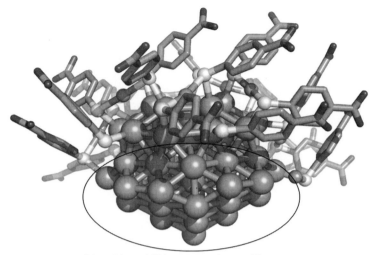

Ligand layer hidden to reveal crystalline core

Figure 9.3 An image generated from Au_{102}(p-mercaptobenzoic acid)44 ligand protected cluster X-ray crystal coordinates. Some of the ligand layer is hidden to reveal the
crystalline core which is fcc (bulk) gold. Orange is gold, yellow is sulfur, gray is carbon,
and red is oxygen. (See Color Insert.)

$$Au_mSR_n + pR'SH \rightleftharpoons Au_mSR_{(n-p)}(SR')_p + pSR$$

Scheme 9.1

molecule drugs, and cysteine-containing proteins and peptides, allowing the
formation of biomolecule/MPC conjugates.

This protein–gold bonding can lead to a remarkably rigid interface between
the protein and the metal cluster. For instance, Au_{144}(p-mercaptobenzoic
acid)$_{60}$ clusters can be conjugated to a protein such that the majority of Au_{144}
clusters have a positional displacement (mobility) of less than 3 Å relative to the
center of mass of the protein (Sexton and Ackerson, 2010). This means that
the *in vitro* conjugated cluster/protein conjugate is defined both with respect
to metal cluster formula and position of the metal relative to the protein with a
"precision" that approaches that of naturally occurring metalloproteins.

The primary caution when making directly bonded gold cluster/protein
conjugates is that forming such conjugates can perturb the structure and
function of the labeled protein (Aubin-Tam and Hamad-Schifferli, 2005,
2008; Aubin-Tam *et al.*, 2009; Kogot *et al.*, 2009). The nature of this
perturbation depends upon the ligand layer used to protect the clusters

and the location of the cysteine within the protein that is labeled. Because diminishment or destruction of protein structure will complicate or make impossible any electron cryomicroscopy analysis, the methods section contains a detailed discussion of design of the conjugates. A secondary consideration is labeling yield, which also depends upon the ligand layer of the cluster and the site of label. Generally, when labeling yield is good, protein structure is retained, but this is not always true.

3.1. Design considerations

The careful design of an MPC/protein conjugation experiment can help insure a resulting MPC/protein conjugate that behaves in the expected manner. There are three essential parameters to consider when designing the bioconjugation process. These are the position of the labeled cysteine residue within the larger protein, the choice of size of the gold cluster core, and the choice of ligand protecting the gold cluster core. Because of the anticipated audience of this chapter, we will limit the discussion to choosing the position of the labeled cysteine residue, as this single parameter has the greatest effect on the structure and function of the resulting gold cluster bioconjugate.

When the cysteine position is chosen well, the $Au_{144}(p\text{-mercaptobenzoic acid})_{60}$ does not diminish protein function and will label in high yield. The synthesis of this cluster is also reasonably straightforward in a laboratory equipped for standard molecular biology procedures. Other MPCs (Table 9.1) may also be adapted in the design and bioconjugation process, depending upon the desired result. These compounds are generally harder to synthesize as discrete products in good yield, and while some have been well investigated for protein labeling, they are less well established.

The "direct" labeling of a protein with a gold cluster as shown in Fig. 9.2, upper right panel, takes advantage of the spontaneous bonding that can occur by a ligand exchange reaction (Scheme 9.1) between a gold cluster and the thiol functional group of cysteine. Many, if not most, cysteine residues within proteins, however, are not amenable to direct labeling with MPCs. In our experience, only solvent exposed surface cysteine residues can be labeled, and in some cases, for instance, if the cysteine residue sits in a concave region of protein surface, not even solvent exposed residues can be labeled (Ackerson et al., 2010). Cysteine residues that are disulfide bonded or buried within the interior of proteins cannot be labeled.

If the X-ray structure of the target protein is known, then it is possible to examine the structure for the presence of any unpaired cysteine residues. If they are present, it is quite easy to test empirically if the target protein is label-able and whether the resulting structure is perturbed, which can be accomplished by taking a circular dichroism (CD) spectrum of the labeled protein. If the experimental goal can be accomplished with an adventitious naturally occurring cysteine residue, then more careful design is unnecessary.

If the structure of the target protein is unknown, but cysteine residues are present, in the primary sequence, then it is also recommended that a test should be done empirically (*vide infra*) for "label-ability" and maintenance of structure.

In most cases, adventitious cysteine residues will not exist. Standard molecular cloning protocols can be used to make serine for cysteine mutations for native cysteine residues that form undesired conjugates, and a single cysteine residue may be inserted into a position that allows the binding of a gold cluster to the region of the protein optimal for the goal of the broader experiment.

The goal of the broader experiment can only provide an approximate placement of the cysteine residue. Two other factors must be taken into account.

The first factor is that the cysteine residue that is labeled will become an integral component into the monolayer of the cluster which is labeled, and thus it must protrude from the protein a sufficient distance to penetrate the monolayer of the MPC. Solvent-accessible cysteine residues that are in recessions or cavities in a protein surface will usually not form conjugates. The second factor is that some cysteine residue positions, when labeled, will cause the protein to partially or completely denature, while other labeled cysteine positions leave protein structure intact. The studies of the Hamad–Schifferli group on labeling different cysteine positions of cytochrome *c* provide the greatest insight in choosing positions for labeled cysteine residues within proteins (Aubin–Tam *et al.*, 2009). Hamad–Schifferli concludes that to preserve protein structure, flexible and loosely folded motifs should be labeled. Labeling of nucleation centers of protein folding should be avoided. We have not done the same systematic study that Hamad–Schifferli has done, but we do note that some residue positions, when labeled, appear to diminish or abolish the affinity of a labeled antibody for its target.

In our own work, we have labeled cysteine residues that are both naturally occurring and engineered into proteins. We have the most experience with labeling cysteine residues appended to an engineered C-terminal extension of the NC10 single chain Fv antibody fragment (Ackerson *et al.*, 2010). In the constructs that we made, the labeled cysteine residue is separated from the normal C-terminus of the protein by genetically engineered linkers of 0–7 amino acid residues. These mutants were originally generated when the gold nanoparticles that we made did not label cysteine residues engineered elsewhere in the scFv and labeled cysteine residues on C-terminal linkers with never more than 50% conjugate yield (Ackerson *et al.*, 2006). We have overcome the yield and position limitations with subsequently described conjugation chemistries (Ackerson *et al.*, 2010).

With a sufficiently long linker, we observe no diminishment of practical protein function as measured by ELISA assay, and we also observe that with a three amino acid residue extension to the C-terminus of the protein, we observe a remarkably rigid interface between protein and cluster (Sexton and Ackerson, 2010). Thus, the diminishment of protein structure and presumably, function that sometimes occurs when labeling cysteine residues can certainly be avoided by labeling a C- or N-terminal extension, if the C- or N-terminal of the target protein is appropriately accessible. We have noticed that sometimes expression of proteins in recombinant over-expression systems (i.e., *Escherichia coli*) can be complicated by a C-terminal cysteine. Thus, we generally include a Glycine residue after the Cysteine residue so that the Cysteine is not the terminal residue. With a sufficiently long linker, the penetration of ligand layer problems is obviated as well. The primary drawback to C- or N-terminal labeling is that the label may not be positioned at a site of particular interest within the protein. For identifying subunits within a multicomponent protein complex, however, C- or N-terminal labeling may work very well and still produce a tight interface between gold cluster and protein.

In summary, designing a protein with a single label-able cysteine residue requires consideration of multiple factors. If it is compatible with the goal of the experiment, then the most facile position of a cysteine may be on a short N or C-terminal extension of the protein. To make a site-directed gold label within the "framework" of the protein, it is best to locate the labeled cysteine on a protruding and loosely folded motif.

3.2. Protocols

The protocol presented will describe how to site-specifically label a designed cysteine-containing protein with an Au_{144} cluster. Because synthesis of Au_{144} clusters is straightforward, the protocol will be divided into two sections, one describing particle synthesis, and a second describing particle bioconjugation.

3.3. Synthesis of $Au_{144}(pMBA)_{60}$

Generally, when gold clusters are synthesized, the result is a mixture containing a wide distribution of particle sizes, with particularly stable clusters (i.e., Magic Number clusters, *vide supra*) being more abundant, but generally within a large range of cluster core sizes. The $Au_{144}(p\text{-mercaptobenzoic acid})_{44}$ synthesis that we have discovered is rare in that it produces in a single synthetic step a uniform cluster. Other compounds in the "magic number series" such as Au_{102} or Au_{68} are synthetically more elusive and are thus less appropriate for synthetic discussion in this chapter.

Reagents

We have found that the vendor of 4–mercaptobenzoic acid (*p*MBA) matters in the synthesis, but have sourced the other components in the synthesis from many different vendors without detriment. Reagents required for this synthesis include 4–mercaptobenzoic acid (90%; *p*MBA), sourced from TCI America, hydrogen tetracholoroaurate trihydrate (HAuCl$_4$), sodium borohydride(NaBH$_4$), reagent grade methanol, nanopure or comparable water, and ammonium acetate.

1. Dissolve 0.002 mol of HAuCl$_4$ in 100 mL of methanol.
2. Dissolve 0.0068 mol of *p*MBA in 48 mL of water, to this add 3.2 mL of 10 *M* NaOH. Check that pH is above 13, and if so, add water to a final volume of 80 mL. If pH is not above 13, then add NaOH until pH is above 13, and then add water to a final volume of 80 mL.
3. Mix solutions made in steps 1 and 2, and stir at least overnight in a sealed vessel. This can be allowed to stir for many days if desired.

This makes a gold(I)-*p*-mercaptobenzoic acid compound, and the solution should appear clear at this point.

1. Dilute the gold(I)–*p*MBA made above to a final [Au] of ∼500 *μM* and a final water/methanol ratio of 27% water, 73% methanol. For instance, in a ∼1 L scale, take 50 mL of the gold/*p*MBA solution above, and add it to 260 mL of MeOH and 740 mL of water.
2. To this mixture add a ∼4.5–fold molar excess of NaBH$_4$:Au(I). For instance, in the 1.05-L scale referenced earlier, add 10 mL of 0.25 *M* NaBH$_4$ (dissolved freshly in water). This is the hardest step to reproduce, because it is difficult to know the potency of a particular stock of NaBH$_4$. This is because in normal storage, NaBH$_4$ will slowly absorb water from the atmosphere and convert to sodium borate, releasing H$_2$ gas. Thus, it is advised to initially attempt this reaction on a small scale (scale down to ∼1.0 mL) and run five reactions with different equivalents of NaBH$_4$, going from a 1.5× excess to a 10× excess. Assay these small-scale reactions and then use the optimized of NaBH$_4$ for scaling up.
3. Seal the reaction vessel and let it stir for 18 h at room temperature. The extended reduction time seems important for getting monodisperse product.
4. Once the reaction is complete, the product can be precipitated by addition of excess methanol. For instance, in the 1.05-L scale, addition of an additional 1.0-L of methanol should completely precipitate the product.

The precipitated product may be collected by centrifugation or filtration over a medium frit filter. If it is by centrifugation, limit the g-force that the particles are exposed to no more than $8000 \times g$. While $8000 \times g$ may be tolerated for hours, it is generally not necessary to centrifuge for more than 3 or 4 min, as the precipitated metal clusters will rapidly collect at the bottom of the centrifuge tube. Higher centrifugal forces can cause the particles to sinter.

After collection by filtration or centrifugation, the particles may be redissolved in 10 mM ammonium acetate, pH 8.0. For the 1.05-L volume referenced earlier, 50 mL is an appropriate volume of buffer to use. The product should then be subjected to two additional methanol precipitation and redissolve rounds to remove any unreacted starting materials. For the 50 mL redissolved volume, it is appropriate to add 50 mL methanol to provoke precipitation.

3.4. Assay of $Au_{144}(pMBA)_{60}$ synthesis

The most straightforward assay is to dry the product on a glow-discharged continuous carbon TEM grid and examine the product in the electron microscope. The apparent diameter of the product should be 2.0 nm, and it should appear very uniform, sometimes forming large ordered crystalline arrays. If the product appears polydisperse and larger than 2.0 nm, the culprit is probably too large of an excess of $NaBH_4$ in the synthetic step. If the product appears sufficient for the goals of the experiment at this stage, then it may not be necessary to do further analysis. If it is important to confirm molecular or near molecular dispersity of the product, then forming crystals and analyzing by mass spectroscopy can be convenient methods for this.

Product quality may be assessed by attempting to form crystals. Hexagonal plate crystals measuring between 100 and 300 μm in maximum dimension should form from a solution of the product placed in a 250-μM NaCl solution that is allowed to slowly evaporate. A microscope suitable for crystallography will be required to examine these crystals.

The product may also be amenable to assessment by MALDI-mass spectroscopy, although depending upon the instrument often data is not acquired. MALDI matrices of 2,5-dihydroxybenzoic acid and DCTB have been successful in our lab and others for generating good spectra (Chaki et al., 2008; Dass, 2009). The product may be quantitated using its weak absorption peak at 510 nm for which we have calculated a Beer's law extinction coefficient of $\varepsilon_{510} = 4.34 \times 10^5 \, M^{-1} cm^{-1}$.

3.5. Conjugation of $Au_{144}(pMBA)_{60}$ clusters

To conjugate the Au_{144} clusters described, it is first necessary to prepare the protein by reducing any oxidized or disulfide cysteine residues to thiol form. Thiol-based reductants, commonly used in protein handling, such as β-mercaptoethanol, mercaptoethylamine, and dithiothreitol (DTT), can participate in ligand exchange, altering or even destroying the cluster core in an "etching" process (Schaaff and Whetten, 1999) when they are used in amounts commonly used in protein chemistry (i.e., 1 mM). If these reductants are used, they must be completely removed, for instance, by gel filtration chromatography. We have found, however, that the phosphine-based cysteine reductant Tris-carboxyethyl phosphine (TCEP) can be used prior to the gold labeling reaction without the need for removal.

Reagents

- Protein to be labeled, with cysteine residue as described (*vide supra*), 1.0 mg or more in solution at a concentration of 1.0 mg/mL or greater.
- TCEP
- $pMBA$-protected Au_{144} clusters from previous protocol.

1. Add TCEP to the protein to a final concentration of 300 μM.
2. Add a fivefold molar excess of Au_{144} to the protein/TCEP mixture. Incubate at 37 °C for 1 h. Labeling may be complete in much shorter time periods (as short as 5 min) if the protein being used will not tolerate 37 °C for 1 h.
3. Prior to purifying the protein from excess Au_{144}, check the yield of the labeling reaction by running an SDS-PAGE gel. Note that reducing agents such as β-mercaptoethanol should not be included in these gels, as such reagents will disrupt the protein–gold bond. A coffee-colored band corresponding to the labeled protein should run gel-shifted about 8 kDa relative to the protein, as shown in Fig. 9.4. A successful labeling reaction will appear quantitative in this assay. The intensity of gold-labeled bands may be enhanced with an autometallographic reaction, as described in the Nanogold labeling section of this chapter.

3.6. Separation of $Au_{144}(pMBA)_{60}$ conjugates

Purification of the conjugate from free Au_{144} depends upon the nature of the labeled protein. Purification generally aims to accomplish removal of three undesired components of the labeling reaction: Excess Au_{144}, unlabeled protein, and Au_{144} conjugated to more than one target protein.

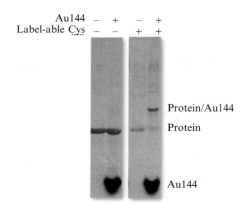

Figure 9.4 Coomassie-stained SDS–polyacrylamide gels showing the gel shift and color change in a protein band that results from gold cluster labeling, in the far right lane. The third lane from the right shows no gel-shifted band for a protein without label-able cysteine residues.

Purification can be achieved in a single step by native gel electrophoresis and glycerol gradient centrifugation for which standard protocols easily found elsewhere will work well. In each case, free Au_{144} is very well separated from Au_{144} conjugated to protein, and the protein that is conjugated to Au_{144} retains the coffee-like color of Au_{144}, making identifying the labeled protein in a gel or gradient possible by simple visual inspection. These methods provide single-step purification, but do not scale well, although this is often not problematic given the small material requirements of electron cryomicroscopy.

We have had success with other methods of purification. If the protein has an affinity tag, the protein can usually be purified away from excess gold by affinity chromatography. We have used FLAG and TAP tags successfully for this purpose. While this will remove free Au_{144}, there may be unlabeled protein retained and also Au_{144} with two or more proteins bound retained. We have found that ion exchange chromatography with a Q-column will generally produce excellent separation between unlabeled and labeled protein because of the high negative charge of the mercaptobenzoic acid ligand layer. We have used gel filtration chromatography to separate Au_{144} with two or more proteins from Au_{144} with only one protein.

In some instances, Au_{144} may become flocculent inside of chromatography columns, discoloring the column and possibly fouling it. This can be minimized by adding glycerol to the column running buffer to a concentration of 5%. If a column becomes discolored or fouled, the Au_{144} clusters may be effectively dissolved and removed with an injection of 200 mM glutathione, which will etch any flocculant gold nanoparticles on the column and solubilize them.

4. Nanogold Labeling of Proteins

The use of metal cluster compounds functionalized using synthetic organic ligands, which can be tailored to impart desired solubility properties or functionalized for selective cross-linking to specific functional groups on biomolecules, provides gold clusters with an extensive conjugation chemistry. The 0.8-nm undecagold cluster, containing 11 gold atoms, and the 1.4-nm Nanogold cluster label, are stabilized with organic ligands terminated with water-soluble functional groups. These compounds are typically prepared with a small number of amines, then chromatographed over an anionic resin to isolate the monoamino forms. These are then converted to maleimido-derivatives, which may be used to label thiol groups (cysteine residues) or *Sulfo*-NHS-derivatives, which may be used to label primary aliphatic amino groups (lysine residues or *N*-terminal amines). Both NHS- and malimido-bioconjugate chemistries are shown schematically in Fig. 9.2. In addition, the monoamino-functionalized clusters may be activated and cross-linked to a variety of biomolecules or other types of probes using homo- and heterobifunctional cross-linkers.

These reagents have been used to label a variety of proteins (Rappas *et al.*, 2005; Yonekura *et al.*, 2006) and peptides (Kelly and Taylor, 2005; Segond von Banchet *et al.*, 1999) for electron cryomicroscopy, enabling the preparation of probes that have been used for the molecular localization of protein subunits (Opalka *et al.*, 2003) and the reconstruction of a variety of protein complexes (Jeon and Shipley, 2000a,b; Volkmann *et al.*, 2001). Using maleimido-Nanogold, antibody IgG molecules and Fab' fragments may be labeled site-specifically at hinge thiol sites, generated by the selective reduction of IgG molecules or $F(ab')_2$ fragments with a mild reducing agent such as DTT or mercaptoethylamine hydrochloride (MEA); this positioning minimizes interference with target binding (Hainfeld and Furuya, 1992; Hainfeld and Powell, 1997). Even smaller Fv fragments may be labeled at naturally occurring lysine residues using *Sulfo*-NHS-Nanogold (Ribrioux *et al.*, 1996).

4.1. Conjugation of the Nanogold label

Nanogold and undecagold are supplied in with a choice of three different reactive functional groups, intended for different labeling applications (Fig. 9.2); each contains close to one reactive functional group. The starting material used to prepare all three is a "Monoamino" form of each cluster, which is separated from a statistical mixture of gold clusters containing different numbers of amines using ion exchange chromatography (Hainfeld, 1989). This is then converted to the maleimido-form using an excess of a bifunctional reagent that introduces a maleimide group. The activated undecagold or Nanogold reagents are separated from excess cross-linker by gel filtration over a desalting gel (GH25, Chisso Corporation). The reagents are eluted with the same buffer in which labeling is usually conducted: 0.02 M sodium phosphate, pH 6.5 with 150 mM sodium chloride, and 1 mM disodium dihydrogen EDTA to chelate transition metal ions, which might otherwise catalyze oxidation of the thiol-labeling sites. The highest selectivity of maleimides for thiols is between pH 6.0 and 7.0; at higher pH, competing hydrolysis and reactions with amines decrease specificity. Immediately after separation, the reagents are diluted to 30 nmol/mL (Nanogold) or 50 nmol/mL (undecagold) in the same buffer, then flash-frozen and lyophilized before being packaged for use. *Sulfo-N*-hydroxysuccinimido (NHS)-Nanogold is similarly activated and separated, but using a cross-linking reagent that introduces a *Sulfo*-NHS group; the activated Nanogold or undecagold is separated by gel filtration, but in this case is eluted with 0.02 M sodium phosphate at pH 7.5 with 150 mM sodium chloride. This pH is within the optimum range (7.5–8.2) for reaction of the *Sulfo*-NHS group with primary aliphatic amines.

This means that the only preparation required for use is reconstitution in degassed, deionized water to give a solution in a buffer that is optimized for labeling. When labeling thiol-containing compounds, the amenability of the thiol to maleimide labeling should be checked carefully before reaction. In many proteins, such as antibody IgG molecules, the target thiols are in the form of disulfides. If so, these should be carefully reduced before use. Thiol-based reducing agents such as DTT or MEA at concentrations between 10 and 50 mM are usually appropriate for selective reduction of antibody hinge thiols leaving intrachain thiols intact; generally, we have found that lower concentrations (20 mM) work well for goat antibodies, while mouse and, in particular, rabbit antibodies may require higher concentrations of reducing agent. Reduction is typically carried out for 1 h at room temperature in 0.1 M sodium phosphate buffer at pH 6.0. The reduced protein is then separated from excess reducing agent by gel filtration over a desalting gel (GH25, Chisso) eluted with the buffer in which the labeling reaction is to be conducted, usually 0.02 M sodium phosphate, pH 6.5, with 150 mM sodium chloride and 1 mM disodium dihydrogen EDTA. Because the reducing agent is used in relatively large excess, even residual traces can react with the maleimido-Nanogold, and it is therefore critical that it be completely separated from the reduced protein. Dialysis does not provide an acceptable level of purification: gel filtration is required to ensure complete removal.

A second consideration for labeling is accessibility of the target function-ality to the relatively large Nanogold cluster. Nanogold has a molecular weight (MW) close to 15,000, and its diameter, including its coordinated ligands, is 2.5–2.7 nm; this may hinder conjugation if the target group is located in a deep cleft. Therefore, it is usually worth checking the structure of the conjugate biomolecule before labeling. In addition, searching for parallel labeling reactions (such as fluorescent or enzymatic labeling) is highly recommended since the conditions used for successful labeling can provide insights into the conditions that may work well with Nanogold. Suitability for labeling may be assessed using ^{14}C labeling or colorimetrically (Grassetti and Murray, 1967). Buffers should be thoroughly degassed and plastic implements used to handle and transfer specimens, since transition metal ions, especially in the presence of atmospheric oxygen, can mediate the oxidation or reoxidation of thiols to disulfides.

Maleimido Nanogold and *Sulfo*-NHS-Nanogold, as sold, will react in a monovalent (1:1) manner with most biomolecules. While accurate analytical methods to "prove" monofunctionality are not readily available, monovalent reactivity has been demonstrated for a reasonably wide range of conjugates. To maximize the yield of 1:1 conjugate, and also to provide an easy separation, the relative sizes of the molecule to be labeled and the Nanogold particle should be compared before the reaction is carried out. In most labeling reactions, the products will be separated by gel filtration or size exclusion chromatography. Therefore, a reaction stoichiometry should be used that gives the greatest size difference between the conjugate and the reagent used in excess. For labeling proteins or other macromolecules that are larger than Nanogold, an excess of Nanogold reagent should be used; this can range from 2:1 up to 5:1, with the smaller ratios used where the reagents are similar in size. For small proteins or peptides that are similar in size to Nanogold or smaller, a stoichiometric excess of the protein or peptide should be used; a twofold excess is recommended for proteins close to Nanogold in mass, and a larger excess for those that are smaller, up to 10-fold for small peptides or small molecules. This strategy is to enable size separation, for example, by gel filtration column chromatography. If the protein is large, it can be easily separated from unreacted free Nanogold due to the size difference. For peptides smaller than Nanogold, the Nanogold can be separated from the peptide, but labeled Nanogold may not separate well from free Nanogold. By using an excess of the peptide, most of the Nanogold will be labeled, and it is then just necessary to separate excess peptide. A typical labeling procedure is as follows:

1. Dissolve protein in labeling buffer, at a concentration of 1.0 mg/mL or higher if possible.
2. Reconstitute the Nanogold reagent in deionized water. Brief vortexing may help dissolve any residual precipitate.
3. Combine the Nanogold reagent with the protein or biomolecule to be labeled. Gentle agitation for 40 min to 1 h may be used to ensure

reaction; otherwise, and afterwards, reaction mixtures should be incubated at 4 °C overnight.

4. Separate next day.

4.2. Separation of Nanogold conjugates

Gel filtration chromatography usually provides a successful method for purification of Nanogold bioconjugates. For conjugate purification, a gel should be used with an MW separation range that extends from below the smaller of the two reagents (Nanogold or conjugate biomolecule) to significantly above that of the conjugate, in order to allow the separation of multimers or small aggregates from monomeric 1:1 conjugates. For example, when labeling antibody Fab' fragments, Superose-12 or Superdex-75 (GE Healthcare), which have MW separation ranges of 10,000–300,000 and 500–70,000 respectively, are appropriate. An example of the separation that can be obtained is shown in Fig. 9.5: note that the conjugate product is separated from both larger and smaller impurities.

Larger proteins and protein complexes may be separated using gels with higher MW ranges such as Superose-6 (GE Healthcare) or A-5m (Bio-Rad), which have MW separation ranges of 5000–1,000,000 or 10,000–5,000,000, respectively. Smaller peptide conjugates may be separated using Superdex-75. For separating conjugates prepared with molecules that are significantly smaller than Nanogold, such as small peptides or substrate analogs, a gel such as Superdex Peptide/PG30, which has a separation range from 100 to 7000 with an exclusion limit of 10,000, often works well. It should be noted that because the majority of the mass of Nanogold is contributed by heavy gold atoms, its hydrodynamic radius is smaller than that of a protein of comparable MW: on many media, it will elute in a position similar to a protein of about 8000 MW. Before injection onto the column, the reaction mixture should be concentrated to a volume of 5% of the column or less using a membrane centrifuge filter with an MW cutoff below the MW of the conjugate: this will ensure maximum chromatographic resolution.

In some situations, gel filtration may not be the most appropriate method for separation. Alternative chromatographic methods may provide better separation. Hydrophobic interaction chromatography may be used to separate labeled from unlabeled proteins with similar MWs in situations where this is required (Hainfeld, 1989). Reverse-phase chromatography may be used to separate labeled oligonucleotides: a butyl column, eluted with a gradient of 0–70% acetonitrile in 0.1 M triethylammonium acetate buffer at pH 7, was found to separate labeled and unlabeled oligonucleotides from unconjugated Nanogold.

Caution should be exercised when using other separation methods, as the behavior of Nanogold or its conjugates is not always predictable. Gel electrophoresis may be used to separate Nanogold conjugates in a manner similar to

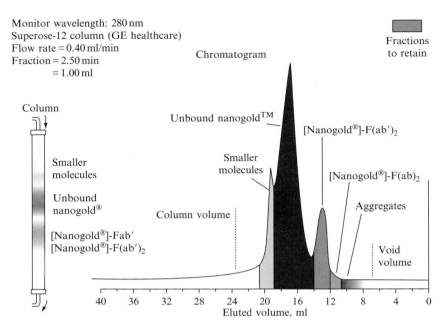

Figure 9.5 Chromatogram showing the separation of Nanogold-labeled Fab′ from larger species, excess Nanogold, and smaller species by gel filtration over Superose-12 gel filtration media. Column volume = 16 mL (50 cm length × 0.67 cm diameter). (See Color Insert.)

that shown in Fig. 9.5; however, as discussed for MPC conjugates, nonreducing gels should be used, and the shift arising from the Nanogold has been found to be significantly less than that predicted from its mass (Hainfeld and Furuya, 1995). When using gel electrophoresis, it is recommended that the product is run in duplicate gel lanes so that one may be developed using a protein-specific colorimetric stain such as Coomassie blue to visualize protein-containing bands, and the other with an autometallographic reagent such as LI Silver, which visualizes the bands containing Nanogold (Note that silver stain reagents for proteins, which use a different chemistry, are not appropriate for developing gels containing gold cluster bioconjugates.) A gold cluster/protein bioconjugate is indicated by a band that develops with both types of stains. Dialysis may also be used, but the membrane should be tested before use to ensure that it does not bind to Nanogold.

4.3. Assay of labeling and activity

The extent of labeling may be calculated from the UV/visible absorption spectrum of the conjugate. Gold labeling of most proteins is best calculated using the extinction coefficients at 280 nm, where proteins usually absorb strongly and their extinction coefficients are known, and at 420 nm, where Nanogold absorbs strongly but most proteins do not. If your protein does not absorb at 420 nm, use the following method to calculate labeling:

1. Since the absorption at 420 nm arises solely from the Nanogold, use the extinction coefficient of Nanogold at 420 nm (110,000) to calculate the concentration of Nanogold:

$$\text{Nanogold concentration} = A_{420\,\text{nm}}/110,000 \text{ mol/L}$$

2. Use the absorption at 420 nm and the extinction coefficients of Nanogold at 280 and 420 nm to calculate the absorption at 280 nm due to the Nanogold:

$$A_{280\,\text{nm}} \text{ of Nanogold} = A_{420\,\text{nm}} \times 300,000/110,000$$

3. Subtract this from the measured absorption at 280 nm. The difference is the absorption due to protein:

$$A(\text{protein}) = A(\text{measured}) - A(\text{Nanogold})$$

4. Use the absorption arising from protein to calculate the protein concentration:

$$\text{Protein concentration} = A(\text{protein})/E(\text{protein})$$

where E is the extinction coefficient of the protein at 280 nm. Convert optical densities to extinction coefficients using the MW as follows:

$$E = \text{OD}(1\%) \times \text{MW}(\text{protein})/10$$

5. The number of Nanogold labels per protein is simply the concentration of Nanogold divided by the concentration of protein:

$$\text{Labeling} = \text{Nanogold concentration}/\text{protein concentration}$$

The activity of the labeled conjugate may be checked by dot blots. We have developed an optimized detection procedure that maintains the very high sensitivity these conjugates provide, combined with a greatly reduced background and enhanced signal clarity. Best results are usually obtained using a procedure that incorporates (a) gold enhancement (Powell and Hainfeld, 2002) rather than silver enhancement to develop the signal after application of the Nanogold conjugate; in most cases, this will provide substantially lower background or nonspecific signal; (b) 0.1% Tween 20 (detergent) in the buffers used for blocking, antibody incubation, and washing; this will dramatically reduce background binding; and (c) inclusion of 1% nonfat dried milk (such as the material sold in supermarkets and food stores) as an additive in the incubation buffer in which the Nanogold conjugate is applied to the blot, and 5% nonfat dried milk in the blocking buffer used to block the membrane before application of probes. The suggested procedure is as follows.

Antigen application

1. Prepare antigen solutions with a series of dilutions (0.01, 0.001, 0.0005, 0.0001, 0.00005, 0.00001, and 0.000005 mg/mL) using PBS, pH 7.4.
2. Pipet 1 μL of the aforementioned solutions on to a dry nitrocellulose membrane (0.2 μm pore size); prepare two duplicates as a negative control: (a) Negative control 1: No antigen, No antibody; and (b) Negative control 2: No antigen with Nanogold conjugate incubation.
3. Air-dry for 30 min.

Blocking

1. Immerse membranes in 8 mL of TBS–Tween 20 (20 mM Tris pH 7.6, with 150 mM NaCl and 0.1% Tween 20) for 5 min.
2. Block membranes in 8 mL of TBS–Tween 20 containing 5% nonfat dried milk for 30 min at room temperature.

Binding of Nanogold conjugate

1. Dilute Nanogold conjugate in TBS–Tween 20 containing 1% nonfat dried milk to about 0.05 nmol/mL (about 4 μg/mL for antibodies; 1:20 dilution, 300 μL conjugate + 5.30 mL TBS-gelatin containing 1% nonfat dried milk).
2. Incubate the membranes in 8 mL of diluted conjugate solution for 30 min at room temperature.
3. Incubate the control membrane in 8 mL of TBS–Tween 20 containing 1% nonfat dried milk for 30 min at room temperature.

Autometallographic detection

1. Wash membranes three times for 3 min each in 8 mL of TBS–Tween 20. Wash membranes thoroughly in 8 mL of deionized water (4 × 3 min). Make sure strips are washed separately according to what they are incubated in (strips incubated in one lot of a conjugate are washed in a separate dish from strips that are incubated in TBS–Tween 20 with 1% nonfat dried milk without conjugate; strips incubated in different lots are washed separately).
2. Perform Gold Enhancement using GoldEnhance EM (Nanoprobes Product No. 2113) according to instructions (mix solutions A and B, wait 5 min, then add C and D).
3. Record the number of observed spots and time when the spots appear. Record the time at which background appears on the control.
4. After 15 min, remove the enhancement solution. Rinse membranes with water (3 × 3 min) and air-dry for storage.

4.4. Electron cryomicroscopy with incorporated Nanogold conjugates

Many electron cryomicroscopy experiments require the incorporation of a Nanogold-labeled component into the structure that is being observed, rather than labeling of the assembled structure, and this can require a modified approach to microscopy. Because of the variety of systems and probes that can be used for such experiments, a single best procedure is difficult to identify; however, that described by Kelly and Taylor (2005), who used Nanogold-labeled β1-integrin to identify the binding site on α-actinin, provides a useful starting point and demonstrates the resolution that is possible with this approach. The cytoplasmic domain of β1-integrin is a 41 amino acid peptide. In order to identify its binding site, the authors synthesized the cytoplasmic domain with a histidine tag at its N-terminus. The binding of this peptide to a lipid monolayer containing a chelated nickel (II) atom mimics the native environment at the cytoplasmic leaflet of the plasma membrane. Preliminary results using a binding assay indicated the presence of two binding sites on the β1-integrin: electron cryomicroscopy studies were then pursued in order to localize these. It was synthesized with two different cysteine modifications: in the first, a cysteine was inserted in the middle of the sequence just after T757 (C758), and in the second, a cysteine was added at the C-terminus (C779). Monomaleimido Nanogold was conjugated to the cysteines by mixing the peptides with the target molecule (without reducing agents) for approximately 16 h at room temperature. Conjugates were chromatographically purified by gel filtration using

Sephadex G-25 (Sigma), and labeling efficiency was determined from the UV/visible spectrum of the conjugate.

1. Nickel-modified lipid monolayers were set up in wells 5 mm in diameter and 1 mm deep milled into Teflon blocks. Dilaurylphosphatidylcholine (DLPC) was used as a filler lipid, with different percentages of dimyristoylphosphatidyl cholinesuberimidenitriloacetic acid nickel(II) salt (nickel-chelating lipid), and different amounts of DDMA to facilitate crystallization.

2. Monolayers were cast upon aqueous buffer containing His-tagged β1-integrin peptide.

3. α-Actinin was then injected into the aqueous phase. Arrays were obtained using 0.0114 nmol of α-actinin and 0.148 nmol of the β1-integrin peptide, using 20 mM Tris buffer (pH 7.5) with 50 mM NaCl and 1 mM MgCl$_2$. Crystals formed overnight at 4 °C.

4. Specimens were recovered from the monolayer using hydrophobic reticulated carbon films on 300 Mesh copper grids, and were either negatively stained with 2% aqueous uranyl acetate or plunge frozen without stain in liquid ethane for cryomicroscopy. Low-dose EM data were collected on a Philips CM300-FEG, a 300-kV electron microscope equipped with a field emission gun. Frozen hydrated specimen grids were transferred to a cryoholder and examined at a temperature of -180 °C. Images were recorded at magnifications of 3,040,000 under low-dose conditions.

5. The 2D arrays of the β1-integrin–α-actinin complex were examined with and without the gold label. Averaged projections were calculated for each specimen along with a difference map to determine the relative position of the gold-labeled β1-integrin peptide. The β1-integrin peptide with gold label at two sites gives difference peaks consistent with binding of the β1-integrin cytoplasmic domain to α-actinin between the first and second of the four 3-helix motifs in the central rod domain (at the R1–R2 junction). The observed differences in binding for the two gold-labeled peptides were attributed to the asymmetrical arrangement of the R1–R4 domain with respect to the lipid monolayer, steric factors because of packing of adjacent molecules within the 2D array, and flexibility of the C-terminal region of the β1-integrin cytoplasmic domain.

A comparison of their results with the model of α-actinin (Liu *et al.*, 2004) is shown in Fig. 9.6. A principle advantage of electron cryomicroscopy is the lack of specimen processing, so the goal is to minimize the use of counterstains and processing steps. However, negative stains are helpful for defining the edges of protein particles, particularly where particle averaging is planned.

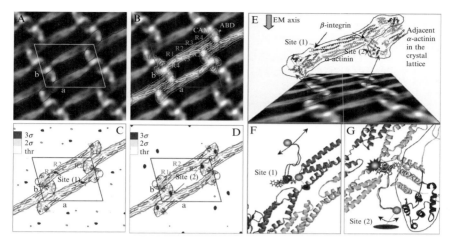

Figure 9.6 Use of Nanogold-labeled β1-integrin to identify the binding site on α-actinin (Kelly and Taylor, 2005). (Left) (A) 2D projection map of the β1-integrin: α-actinin arrays. α-Actinin dimers are indicated by the red outline; the unit cell is indicated in yellow. (B) Docking of the α-actinin atomic model (Liu *et al.*, 2004) into the EM projection map. P2-symmetry-related dimers were also generated. (C) 2D difference map between unlabeled β1-integrin:α-actinin arrays, and arrays labeled with the Nanogold attached to C758, that is, in the middle of the integrin peptide sequence. The binding in this case occurs at site (1) on α-actinin. The most significant differences are at the level of 3σ (blue). (D) 2D difference map between unlabeled β1-integrin:α-actinin arrays and arrays with Nanogold labeling at C779. Labeling in this case occurs at site (2) on α-actinin. The density levels in both difference maps were truncated at a threshold corresponding to the values of 2σ (two times the standard deviation) for each map, respectively. 2σ peaks are colored sky-blue in (C) and (D). (Right) (E) Model of the unlabeled β1-integrin bound to both potential sites on α-actinin between the first and second spectrin repeats. The view direction chosen to show the atomic model is not perpendicular to the arrays, but rather at an angle that is approximately parallel with the local twofold axis of the R1–R4 domain. The effect is accentuated by the foreshortened appearance of the projection image shown below the model. This direction is maintained in (F) and (G), but because the local twofold symmetry of R1–R4 is only approximate, the integrin peptides are not seen in identical orientations. Arrows in (F) and (G) indicate the general degree of disorder in the gold label when attached to C779. (F) Integrin binding at site (1), while (G) refers to the integrin binding site (2) related by the local twofold of the R1–R4 domain. (F) The Nanogold label at C758 in the integrin peptide (red) fits into the α-actinin crystal lattice in an ordered fashion to give rise to a significant signal seen in the first difference map. The Nanogold label at C779 in the peptide (magenta) is too disordered to detect at a significant level in the second difference map. W755 is shown in stick rendering here and in (G). (G) Integrin cytoplasmic domain binding at site (2). When the Nanogold label is bound to C758 in the peptide (red), the fit is very tight because the inserted amino acid shifts the position of W755 toward the neighboring α-actinin molecule. No difference density is detected for this peptide when bound to site (2) suggesting that the peptide cannot bind under these conditions within the 2D lattice. However, the Nanogold probe positioned at C779 (green) of the integrin peptide fits with less restriction into the α-actinin lattice. The red line in (G) delineates an adjacent α-actinin molecule in the 2D lattice. Blue disks in (F) and (G) indicate the approximate location of the difference peak if the model were projected in a direction perpendicular to the plane of the 2D arrays (gold sphere = Nanogold). (See Color Insert.)

5. Ni(II)-NTA–Nanogold Labeling of His-Tagged Proteins

Studies on single His–tag–NTA-Ni(II) interactions have determined values for the dissociation constant between 7×10^{-8} and 7×10^{-7} M (Kröger *et al.*, 1999; Nieba *et al.*, 1997). Recombinant proteins are often expressed with His tags so that they can be purified over chromatographic resins derivatized with NTA-Ni(II) groups (Hochuli *et al.*, 1988). In the context of chromatographic resins in which multiple NTA-Ni(II) groups can chelate to a single polyhistidine tag, the dissociation constant has been estimated to be 10^{-13}, significantly tighter binding than antibody–antigen (Schmitt *et al.*, 1993). Ni(II)-NTA–Nanogold contains multiple NTA-Ni (II) chelates, and therefore a similar value is reasonable, making Ni(II)-NTA–Nanogold a sensitive and specific probe for His-tagged proteins.

This probe has an important advantage over antibody and protein probes: the targeting entity is much smaller even than the Fab′ antibody fragment. Therefore, the distance from the center of the Nanogold particle to the His-tag site is much shorter, estimated to be about 2.6 nm (Fig. 9.7). This provides significantly higher resolution, which is sufficient to obtain specific structural information on the position and orientation of protein subunits within a protein complex. Its initial use to localize the PsbH subunit in the photosynthetic reaction center Photosystem II (PSII; Buchel *et al.*, 2001; Bumba *et al.*, 2005) provides an excellent illustration of the power of this approach to locating specific proteins within multisubunit protein complexes. To identify the location of the His-tagged PsbH subunit within PSII, Bumba *et al.* (2005) used dimeric PSII complexes from cyanobacterium *Synechocystis* sp. PCC 6803 (obtained by solubilization of thylakoid membranes with 1% dodecyl-maltoside and affinity chromatography over an Ni(II) chelate gel). These were applied to glow-discharged grids, excess buffer was removed using filter paper, and the grid was then incubated upside-down on a droplet of NTA-Ni(II)–Nanogold solution (170 nM in 50 mM MES, pH 6.5) for 10 min at 4 °C. Labeling was terminated by removing the grid from the droplet, rinsing thoroughly with water, and staining with 1% uranyl acetate. A typical EM image showed dispersed particles with uniform size and shape, almost free of contaminants: these are dimeric PSII particles, mostly in top-view projections. All the projections had the same handedness and no mirror images were detected, indicating preferred orientation of the PSII dimers with their stromal side to the carbon support film. Not all the complexes were labeled, and this was attributed to the location of the PsbH His-tag on the stromal side of the complex, which renders the labeling site somewhat inaccessible. The short *N*-terminus of the *Synechocystis* PsbH protein enabled precise

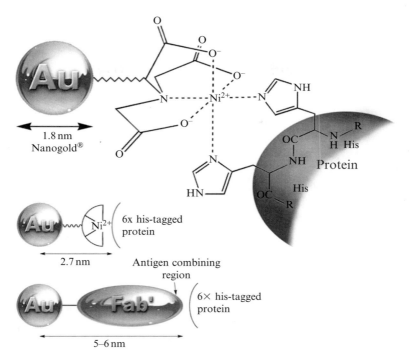

Figure 9.7 Ni(II)-NTA–Nanogold, showing binding mechanism (top) and size comparison with Nanogold-labeled Fab′.

identification of the location of the His site within the PSII complex. A careful comparison of the location of the gold clusters in *Synechocystis* PSII with those in *Chlamydomonas reinhardtii* revealed that gold label in *Synechocystis* preparation is slightly shifted with respect to the longer edge of the complex. The location of the PsbH subunit is in good agreement with the assignment of the PsbH subunit in the model of Ferreira *et al.* (2004), and suggests that a single transmembrane helix close to the CP47 subunit corresponds to the PsbH protein.

This reagent has since become quite widely used for the structural characterization of protein complexes (Adami *et al.*, 2007; Balasingham *et al.*, 2007) and a viral capsid (Chatterji *et al.*, 2005) by high-resolution electron cryomicroscopy. Ni(II)-NTA–Nanogold has been used in the structural characterization of protein complexes by single particle analysis (Collins *et al.*, 2006; Young *et al.*, 2008), with electron tomography and single particle analysis to determine the structure of the *Saccharomyces cerevisiae* gamma-tubulin small complex (gamma-TuSC) at 25-Å resolution (Kollman *et al.*, 2008), and even as a heavy atom derivative in the assignment of the 3D structure of $Ca_v3.1$ (Walsh *et al.*, 2009).

Larger Ni-NTA-derivatized gold nanoparticles have been prepared and used for derivatization of His-tagged proteins, and afford similar resolution combined with higher EM visibility. A 3.9-nm analog was used to assemble *Mycobacterium tuberculosis* 20S proteasomes tagged with 6×-histidine into 2D arrays, thus improving electron cryomicroscopic resolution by 6–8 Å compared to analysis of single particles (Hu *et al.*, 2008), and a 4.4-nm Ni-NTA–gold label was used to label adenovirus serotype 12 (Ad12) knob protein for STEM observation (Hu *et al.*, 2007; Briñas *et al.*, 2008). A 5-nm NTA-Ni(II)–gold probe is now available commercially (Dubendorff *et al.*, 2010; Reddy *et al.*, 2005).

Ni(II)-NTA–Nanogold is usually used to tag subunits or components of a specimen, which are then assembled or modified to give the functional unit under investigation; this is then examined by electron cryomicroscopy. An important consideration in the preparation of Ni(II)-NTA–Nanogold-labeled conjugates is that each particle contains multiple NTA-Ni(II) groups, and therefore care should be taken to avoid cross-linking and formation of multimers. This can usually be achieved by control of binding stoichiometry: the use of a moderate excess of Ni(II)-NTA–Nanogold (5–10-fold) to label a protein with a single His tag will usually ensure single labeling with minimal cross-linking. For proteins with multiple His tags, larger excesses may be appropriate. A protocol for this procedure usually includes the following elements:

1. Mix the His-tagged protein or peptide with Ni(II)-NTA–Nanogold in a binding buffer of 20 mM Tris, 0.15 M NaCl, pH 7.6 with 0.1% (w/v) Tween 20 and 1% (w/v) nonfat dried milk. Including Tween 20 and nonfat dried milk helps reduce the nonspecific binding of Ni-NTA–Nanogold to the proteins. The binding buffer must be free from thiols such as β-mercaptoethanol or DTT, or chelators such as EDTA or citrate which can remove Ni(II) ions and produce low binding, However, 5–20 mM imidazole may be added to reduce nonspecific binding.

2. Incubate for 5–30 min at room temperature. For most applications, 5–10 min incubation is sufficient to obtain positive staining.

3. The Nanogold-labeled His-tagged protein can be separated chromatographically in the same manner as proteins labeled with maleimido-Nanogold or *Sulfo*-NHS-Nanogold. However, because the mode of binding is different, different buffers are appropriate for elution and reaction. 50 mM Tris, 300 mM NaCl, 10–200 mM imidazole, and 0.1% Tween 20 adjusted to pH 7.6 are recommended for elution, reaction, and washing.

4. Mix the Nanogold-labeled His-tagged protein or peptide with the other components of the system under study, and allow to assemble or assume natural functionality.

5. Process for electron cryomicroscopy according to the usual procedure used for unlabeled specimens.

The wash buffer may be optimized for specific applications by adjusting the imidazole and sodium chloride concentrations. Because Ni(II)-NTA–Nanogold is charged, high salt concentrations (up to 1 M) may be helpful in preventing nonspecific interactions. Increasing the imidazole concentration may also help reduce interactions of the negatively charged NTA-Ni (II) entity with protein components. However, it can also decrease the binding to the target proteins: the optimum concentration may be found by trial and error.

The extent of labeling may be calculated in the same manner as for conjugates labeled with maleimido-Nanogold or *Sulfo*-NHS-Nanogold, using the extinction coefficients at 280 and 420 nm. However, the values for the extinction coefficients are different to those for maleimido-Nanogold or *Sulfo*-NHS-Nanogold, and can vary from lot to lot. The values that should be used for calculation with each lot are given in a specification sheet that is supplied with the product.

6. Conclusions

Site-specific labeling of biomolecules with gold clusters offers a flexible, adaptable approach to enhancing information content in electron cryomicroscopy of biological samples. The development of site-specific labeling methods have the potential to bring to electron microscopy many of the advantages that the new generations of fluorophores developed in the past three decades have brought to light and fluorescence microscopy. These advantages are particularly relevant to cryo-EM where specimens are observed at high resolution with significantly less processing and perturbation than with conventional EM methods, and a wealth of molecular information is available to be harvested.

REFERENCES

Ackerson, C. J., *et al.* (2005). Defined DNA/nanoparticle conjugates. *Proc. Natl. Acad. Sci. USA* **102,** 13383–13385.

Ackerson, C. J., *et al.* (2006). Rigid, specific, and discrete gold nanoparticle/antibody conjugates. *J. Am. Chem. Soc.* **128,** 2635–2640.

Ackerson, C. J., *et al.* (2010). Synthesis and bioconjugation of 2 and 3 nm-diameter gold nanoparticles. *Bioconjug. Chem.* **21,** 214–218.

Adami, A., Garcia-Alvarez, B., Arias-Palomo, E., Barford, D., and Llorca, O. (2007). Structure of TOR and its complex with KOG1. *Mol. Cell* **27,** 509–516.

Alcid, E. A., and Jurica, M. S. (2008). A protein-based EM label for RNA identifies the location of exons in spliceosomes. *Nat. Struct. Mol. Biol.* **15,** 213–215.

Alivisatos, A. P., Johnsson, K. P., Peng, X., Wilson, T. E., Loweth, C. J., Bruchez, M. P., Jr., and Schultz, P. G. (1996). Organization of "Nanocrystal Molecules" using DNA. *Nature* **382,** 609–611.

Alvarez, M. M., Khoury, J. T., *et al.* (1997). Critical sizes in the growth of Au clusters. *Chem. Phys. Lett.* **266,** 91–98.

Aubin-Tam, M. E., and Hamad-Schifferli, K. (2005). Gold nanoparticle cytochrome c complexes: The effect of nanoparticle ligand charge on protein structure. *Langmuir* **21,** 12080–12084.

Aubin-Tam, M. E., and Hamad-Schifferli, K. (2008). Structure and function of nanoparticle-protein conjugates. *Biomed. Mater.* **3,** 034001.

Aubin-Tam, M. E., *et al.* (2009). Site-directed nanoparticle labeling of cytochrome c. *Proc. Natl. Acad. Sci. USA* **106,** 4095–4100.

Balasingham, S. V., Collins, R. F., Assalkhou, R., Homberset, H., Frye, S. A., Derrick, J. P., and Tonjum, T. (2007). Interactions between the Lipoprotein PilP and the Secretin PilQ in *Neisseria meningitidis. J. Bacteriol.* **189,** 5716–5727.

Bartlett, P., *et al.* (1978). Synthesis of water-soluble undecagold cluster compounds of potential importance in electron microscopic and other studies of biological systems. *J. Am. Chem. Soc.* **100,** 5085–5089.

Behnke, O., Ammitzbøll, T., Jessen, H., Klokker, M., Nilausen, K., Tranum-Jensen, J., and Olsson, L. (1986). Non-specific binding of protein-stabilized gold sols as a source of error in immunocytochemistry. *Eur. J. Cell Biol.* **41,** 326–338.

Bowman, M. C., *et al.* (2008). Inhibition of HIV fusion with multivalent gold nanoparticles. *J. Am. Chem. Soc.* **130,** 6896–6897.

Briñas, R. P., Hu, M., Qian, L., Lymar, E. S., and Hainfeld, J. F. (2008). Gold nanoparticle size controlled by polymeric Au(I) thiolate precursor size. *J. Am. Chem. Soc.* **130,** 975–982.

Buchel, C., Morris, E., Orlova, E., and Barber, J. (2001). Localisation of the PsbH subunit in photosystem II: A new approach using labelling of His-tags with a Ni(2+)-NTA gold cluster and single particle analysis. *J. Mol. Biol.* **312,** 371–379.

Bumba, L., Tichy, M., Dobakova, M., Komenda, J., and Vacha, F. (2005). Localization of the PsbH subunit in photosystem II from the *Synechocystis* 6803 using the His-tagged Ni–NTA Nanogold labeling. *J. Struct. Biol.* **152,** 28–35.

Cariati, F., and Naldini, L. (1971). Trianionoheptakis(triarylphosphine)undecagold cluster compounds. *Inorg. Chim. Acta* **5,** 172–174.

Chaki, N. K., *et al.* (2008). Ubiquitous 8 and 29 kDa gold:alkanethiolate cluster compounds: Mass-spectrometric determination of molecular formulas and structural implications. *J. Am. Chem. Soc.* **130,** 8608–8610.

Chatterji, A., Ochoa, W. F., Ueno, T., Lin, T., and Johnson, J. E. (2005). A virus-based nanoblock with tunable electrostatic properties. *Nano Lett.* **5,** 597–602.

Cheng, N., Conway, J. F., Watts, N. R., Hainfeld, J. F., Joshi, V., Powell, R. D., Stahl, S. J., Wingfield, P. E., and Steven, A. C. (1999). Tetrairidium, a 4-atom cluster, is readily visible as a density label in 3D cryo-EM maps of proteins at 10–25 Å resolution. *J. Struct. Biol.* **127,** 169–176.

Collins, R. F., Beis, K., Clarke, B. R., Ford, R. C., Hulley, M., Naismith, J. H., and Whitfield, C. (2006). Periplasmic protein–protein contacts in the inner membrane protein Wzc form a tetrameric complex required for the assembly of *Escherichia coli* group 1 capsules. *J. Biol. Chem.* **281,** 2144–2150.

Cong, Y., Baker, M. L., *et al.* (2010). 4.0-A resolution cryo-EM structure of the mammalian chaperonin TRiC/CCT reveals its unique subunit arrangement. *Proc. Natl. Acad. Sci. USA* **107,** 4967–4972.

Dass, A. (2009). Mass spectrometric identification of Au(68)(SR)(34) molecular gold nanoclusters with 34-electron shell closing. *J. Am. Chem. Soc.* **131,** 11666–11667.

Diestra, E., *et al.* (2009). Visualization of proteins in intact cells with a clonable tag for electron microscopy. *J. Struct. Biol.* **165,** 157–168.

Dubendorff, J. W., Lymar, E., Furuya, F. R., and Hainfeld, J. F. (2010). Gold labeling of protein fusion tags for EM. *Microsc. Microanal.* **16** (Suppl. 2): Proceedings CD866.

Ferreira, K. N., Iverson, T. M., Maghlaoui, K., Barber, J., and Iwata, S. (2004). Architecture of the photosynthetic oxygen-evolving center. *Science* **303,** 1831–1838.

Grassetti, D. R., and Murray, J. F., Jr. (1967). Determination of sulfhydryl groups with 2, 2'- or 4, 4'-dithiodipyridine. *Arch. Biochem. Biophys.* **119,** 41–49.

Hainfeld, J. F. (1989). Undecagold-antibody method. *In* "Colloidal Gold: Principles, Methods, and Applications, Vol. 2," (M. A. Hayat, ed.), pp. 413–429. Academic Press, San Diego.

Hainfeld, J. F., and Furuya, F. R. (1992). A 1.4-nm gold cluster covalently attached to antibodies improves immunolabeling. *J. Histochem. Cytochem.* **40,** 177–184.

Hainfeld, J. F., and Furuya, F. R. (1995). Silver-enhancement of Nanogold and undecagold. *In* "Immunogold-Silver Staining: Principles, Methods and Applications," (M. A. Hayat, ed.), pp. 71–96. CRC Press, Boca Raton, FL.

Hainfeld, J. F., and Powell, R. D. (1997). Nanogold technology: New frontiers in gold labeling. *Cell Vis.* **4,** 408–432.

Hainfeld, J. F., and Powell, R. D. (2000). New frontiers in gold labeling. *J .Histochem. Cytochem.* **48,** 471–480.

Hainfeld, J. F., Furuya, F. R., and Powell, R. D. (1999a). Metallosomes. *J. Struct. Biol.* **127,** 152–160.

Hainfeld, J. F., Liu, W., Halsey, C. M. R., Freimuth, P., and Powell, R. D. (1999b). Ni-NTA–gold clusters target his-tagged proteins. *J. Struct. Biol.* **127,** 185–198.

Hainfeld, J. F., Powell, R. D., and Hacker, G. W. (2004). Nanoparticle molecular labels. *In* "Nanobiotechnology," (C. A. Mirkin and C. M. Niemeyer, eds.), pp. 353–386. Wiley-VCH, Weinheim, Germany (Chapter 23).

Hamad-Schifferli, K., Schwartz, J. J., Santos, A. T., Zhang, S., and Jacobson, J. M. (2002). Remote electronic control of DNA hybridization through inductive coupling to an attached metal nanocrystal antenna. *Nature* **415,** 152–155.

Handley, D. A. (1989a). The development and application of colloidal gold as a microscopic probe. *In* "Colloidal Gold: Principles, Methods and Applications," (M. A. Hayat, ed.), pp. 1–11. Academic Press, San Diego, CA (Chapter 1).

Handley, D. A. (1989b). Methods for synthesis of colloidal gold. *In* "Colloidal Gold: Principles, Methods and Applications," (M. A. Hayat, ed.), pp. 12–22. Academic Press, San Diego, CA (Chapter 2).

Heaven, M., *et al.* (2008). Crystal structure of the gold nanoparticle [N(C(8)H(17))(4)][Au (25)(SCH(2)CH(2)Ph)(18)]. *J. Am. Chem. Soc.* **130,** 3754–3755.

Hochuli, E., Bannwarth, W., Döbeli, H., Gentz, R., and Stber, D. (1988). Genetic approach to facilitate purification of recombinant proteins with a novel metal chelate adsorbent. *Bio/Technology* **6,** 1321–1325.

Hostetler, M., *et al.* (1999). Dynamics of place-exchange reactions on monolayer-protected gold cluster molecules. *Langmuir* **15,** 3782–3789.

Hu, M., Qian, L., Brinas, R. P., Lymar, E. S., and Hainfeld, J. F. (2007). Assembly of nanoparticle-protein binding complexes: From monomers to ordered arrays. *Angew. Chem. Int. Ed. Engl.* **46,** 5111–5114.

Hu, M., Qian, L., Briñas, R. P., Lymar, E. S., Kuznetsova, L., and Hainfeld, J. F. (2008). Gold nanoparticle-protein arrays improve resolution for cryo-electron microscopy. *J. Struct. Biol.* **161,** 83–91.

Huo, Q., and Worden, J. G. (2007). Monofunctional gold nanoparticles: Synthesis and applications. *J. Nanopart. Res.* **9**, 1013–1025.

Jadzinsky, P. D., *et al.* (2007). Structure of a thiol monolayer-protected gold nanoparticle at 1.1 angstrom resolution. *Science* **318**, 430–433.

Jeon, H., and Shipley, G. G. (2000a). Vesicle-reconstituted low density lipoprotein receptor: Visualization by cryoelectron microscopy. *J. Biol. Chem.* **275**, 30458–30464.

Jeon, H., and Shipley, G. G. (2000b). Localization of the N-terminal domain of the low density lipoprotein receptor. *J. Biol. Chem.* **275**, 30465–30470.

Kelly, D. F., and Taylor, K. A. (2005). Identification of the beta1-integrin binding site on alpha-actinin by cryoelectron microscopy. *J. Struct. Biol.* **149**, 290–302.

Kogot, J. M., *et al.* (2008). Single peptide assembly onto a 1.5 nm Au surface via a histidine tag. *J. Am. Chem. Soc.* **130**, 16156–16157.

Kogot, J. M., *et al.* (2009). Analysis of the dynamics of assembly and structural impact for a histidine tagged FGF1-1.5 nm Au nanoparticle bioconjugate. *Bioconjug. Chem.* **20**, 2106–2113.

Kollman, J. M., Zelter, A., Muller, E. G., Fox, B., Rice, L. M., Davis, T. N., and Agard, D. A. (2008). The structure of the gamma-tubulin small complex: Implications of its architecture and flexibility for microtubule nucleation. *Mol. Biol. Cell* **19**, 207–215.

Kramarcy, N. R., and Sealock, R. (1990). Commercial preparations of colloidal gold-antibody complexes frequently contain free active antibody. *J. Histochem. Cytochem.* **39**, 37–39.

Kröger, D., Liley, M., Schiwek, W., Skerra, A., and Vogel, H. (1999). Immobilization of histidine-tagged proteins on gold surfaces using chelator thioalkanes. *Biosens. Bioelectron.* **14**, 155–161.

Krpetic, Z., *et al.* (2009). A multidentate peptide for stabilization and facile bioconjugation of gold nanoparticles. *Bioconjug. Chem.* **20**, 619–624.

Levy, R., *et al.* (2004). Rational and combinatorial design of peptide capping ligands for gold nanoparticles. *J. Am. Chem. Soc.* **126**, 10076–10084.

Liu, J., Taylor, D. W., and Taylor, K. A. (2004). A 3-D reconstruction of smooth muscle - actinin by cryoEM reveals two diVerent conformations at the actin binding region. *J. Mol. Biol.* **338**, 115–125.

Liu, X., *et al.* (2006). Monofunctionat gold nanopartictes prepared via a noncovalent-interaction-based solid-phase modification approach. *Small* **2**, 1126–1129.

Lopez-Acevedo, O., Akola, J., *et al.* (2009). Structure and bonding in the ubiquitous icosahedral metallic gold cluster Au-144(SR)(60). *J. Phys. Chem. C* **113**, 5035–5038.

Mercogliano, C. P., and DeRosier, D. J. (2007). Concatenated metallothionein as a clonable gold label for electron microscopy. *J. Struct. Biol.* **160**, 70–82.

Nickell, S., *et al.* (2006). A visual approach to proteomics. *Nat. Rev. Mol. Cell Biol.* **7**, 225–230.

Nieba, L., Nieba-Axmann, S. E., Persson, A., Hämäläinen, M., Edebratt, F., Hansson, A., Lidholm, J., Magnusson, K., Karlsson, Å.F., and Plückthun, A. (1997). BIACORE analysis of histidine-tagged proteins using a chelating NTA sensor chip. *Anal. Biochem.* **252**, 217–228.

Opalka, N., Beckmann, R., Boisset, N., Simon, M. N., Russel, M., and Darst, S. A. (2003). Structure of the filamentous phage pIV multimer by cryo-electron microscopy. *J. Mol. Biol.* **325**, 461–470.

Park, S. H., Yin, P., Liu, Y., Reif, J. H., LaBean, T. H., and Yan, H. (2005). Programmable DNA self-assemblies for nanoscale organization of ligands and proteins. *Nano Lett.* **5**, 729–733.

Powell, R. D., and Hainfeld, J. F. (2002). Silver- and gold-based autometallography of Nanogold. *In* "Gold and Silver Staining: Techniques in Molecular Morphology," (G. W. Hacker and J. Gu, eds.), pp. 29–46. CRC Press, Boca Raton, FL (Chapter 3).

Powell, R. D., Halsey, C. M. R., Liu, W., Joshi, V. N., and Hainfeld, J. F. (1999). Giant platinum clusters: 2 nm covalent metal cluster labels. *J. Struct. Biol.* **127,** 177–184.

Rappas, M., Schumacher, J., Beuron, F., Niwa, H., Bordes, P., Wigneshweraraj, S., Keetch, C. A., Robinson, C. V., Buck, M., and Zhang, X. (2005). Structural insights into the activity of enhancer-binding proteins. *Science* **307,** 1972–1975.

Reddy, V., Lymar, E., Hu, M., and Hainfeld, J. F. (2005). 5 nm Gold–Ni-NTA binds His Tags. Microsc. Microanal., 11 (Suppl. 2). (R. Price, P. Kotula, M. Marko, J. H. Scott, G. F. Vander Voort, E. Nanilova, M. Mah Lee Ng, K. Smith, P. Griffin, P. Smith, and S. McKernan, eds.), Cambridge University Press, 1118CD.

Ribrioux, S., Kleymann, G., Haase, W., Heitmann, K., Ostermeier, C., and Michel, H. (1996). Use of Nanogold- and fluorescent-labeled antibody Fv fragments in immunocytochemistry. *J. Histochem. Cytochem.* **44,** 207–213.

Safer, D., et al. (1982). Biospecific labeling with undecagold: Visualization of the biotin-binding site on avidin. *Science* **218,** 290–291.

Sardar, R., Funston, A., et al. (2009). Gold nanoparticles: Past, present, and future. *Langmuir* **25,** 13840–13851.

Schaaff, T., and Whetten, R. (1999). Controlled etching of Au:SR cluster compounds. *J. Phys. Chem. B* **103,** 9394–9396.

Schmitt, J., Hess, H., and Stunnenberg, H. G. (1993). Affinity purification of histidine-tagged proteins. *Mol. Biol. Rep.* **18,** 223–230.

Segond von Banchet, G., Schindler, M., Hervieu, G. J., Beckmann, B., Emson, P. C., and Heppelmann, B. (1999). Distribution of somatostain receptor subtypes in rat lumbar spinal cord examined with gold-labelled somatostatin and anti-receptor antibodies. *Brain Res.* **816,** 254–257.

Sexton, J. Z., and Ackerson, C. J. (2010). Determination of rigidity of protein bound Au_{144} clusters by electron cryomicroscopy. *J. Phys. Chem. C.* Published on Web July 14. DOI: 10.1021/jp101970x.

Song, Y., and Murray, R. (2002). Dynamics and extent of ligand exchange depend on electronic charge of metal nanoparticles. *J. Am. Chem. Soc.* **124,** 7096–7102.

Templeton, A., et al. (1998). Gateway reactions to diverse. Polyfunctional monolayer-protected gold clusters. *J. Am. Chem. Soc.* **120,** 4845–4849.

Templeton, A., et al. (2000). Monolayer-protected cluster molecules. *Acc. Chem. Res.* **33,** 27–36.

Tkachenko, A. G., et al. (2003). Multifunctional gold nanoparticle-peptide complexes for nuclear targeting. *J. Am. Chem. Soc.* **125,** 4700–4701.

Verma, A., et al. (2004). Recognition and stabilization of peptide alpha-helices using templatable nanoparticle receptors. *J. Am. Chem. Soc.* **126,** 10806–10807.

Volkmann, N., Amann, K. J., Stoilova-McPhie, S., Egile, C., Winter, D. C., Hazelwood, L., Heuser, J. E., Li, R., Pollard, T. D., and Hanein, D. (2001). Structure of Arp2/3 complex in its activated state and in actin filament branch junctions. *Science* **293,** 2456–2459.

Walsh, C. P., Davies, A., Butcher, A. J., Dolphin, A. C., and Kitmitto, A. (2009). Three-dimensional structure of CaV3.1: comparison with the cardiac L-type voltage-gated calcium channel monomer architecture. *J. Biol. Chem.* **284,** 22310–22321.

Walter, M., et al. (2008). A unified view of ligand-protected gold clusters as superatom complexes. *Proc. Natl. Acad. Sci. USA* **105,** 9157–9162.

Worden, J. G., et al. (2004a). Monofunctional group–modified gold nanoparticles from solid phase synthesis approach: Solid support and experimental condition effect. *Chem. Mater.* **16,** 3746–3755.

Worden, J. G., et al. (2004b). Controlled functionalization of gold nanoparticles through a solid phase synthesis approach. *Chem. Commun.* 518–519.

Xiao, S., Liu, F., Rosen, A. E., Hainfeld, J. F., Seeman, N. C., Musier-Forsyth, K., and Kiehl, R. A. (2002). Self assembly of metallic nanoparticle arrays by DNA scaffolding. *J. Nanopart. Res.* **4,** 313–317.

Yonekura, K., Yakushi, T., Atsumi, T., Maki-Yonekura, S., Homma, M., and Namba, K. (2006). Electron cryomicroscopic visualization of PomA/B stator units of the sodium-driven flagellar motor in liposomes. *J. Mol. Biol.* **357,** 73–81.

Young, M. T., Fisher, J. A., Fountain, S. J., Ford, R. C., North, R. A., and Khakh, B. S. (2008). Molecular shape, architecture, and size of P2×4 receptors determined using fluorescence resonance energy transfer and electron microscopy. *J. Biol. Chem.* **283,** 26241–26251.

CHAPTER TEN

How to Operate a Cryo-Electron Microscope

Jingchuan Sun* *and* Huilin Li*,†

Contents

Abstract

We describe the basic principles for imaging frozen-hydrated specimens in a transmission electron microscope and provide a step-by-step guide to a new user, from starting up and aligning the microscope, to loading a cryo-grid into

* Biology Department, Brookhaven National Laboratory, Upton, New York, USA
† Department of Biochemistry & Cell Biology, Stony Brook University, Stony Brook, New York, USA

Methods in Enzymology, Volume 481
ISSN 0076-6879, DOI: 10.1016/S0076-6879(10)81010-9

the specimen holder and inserting the holder into the microscope, to setting up the low dose mode for imaging, and eventually to shutting down the microscope. The procedure is based on a JEOL TEM; however, it is applicable to microscopes from other manufacturers with small modifications. We also give several tips on how to minimize specimen exposure before taking pictures; how to minimize the specimen drift and charging during exposure; how to estimate the thickness of the vitreous ice, and how to quickly estimate the electron exposure dose. Despite recent advances in instrumentation, the microscopist's patience and attention to detail may still be the key to acquiring a high quality cryo-electron micrograph.

1. INTRODUCTION

Electrons interact with matter strongly. This is reflected by their several thousand times higher cross section than that of X-rays (Henderson, 1995). The strong interaction is the primary reason why EM can reveal structural information of individual biological macromolecules. However, the biological specimens are susceptible to electron beam–induced damage due to their nonconductive nature and the weak chemical bonds that hold atoms together in biological structures. Cryo-electron microscope (EM) was invented primarily to avoid negative staining, enabling direct visualization of biological structures in the high vacuum of the microscope (Dubochet et al., 1988). However, the low temperature also significantly reduces specimen damage (see the Rubinstein chapter): at room temperature, the dose tolerance of the protein sample is about 2 e/$\mathring{A}^2$; at liquid nitrogen (LN_2) temperature, beam tolerance increases by almost an order of magnitude to 10–20 e/$\mathring{A}^2$ (Glaeser and Taylor, 1978); and at liquid helium temperature, there is another gain of 2–4 times, resulting in a beam tolerance of 40 e/$\mathring{A}^2$ (Fujiyoshi et al., 1991; Knapek et al., 1982). However, at the extreme low temperature of liquid helium, the density of vitreous ice increases, resulting in reduced specimen contrast, thus potentially canceling out the gain in beam tolerance.

In a nutshell, a cryo-EM is a regular TEM equipped with a cryo-specimen holder. A twin-blade anticontaminator, inserted either from the side of the microscope column, or as a built-in device within the objective lens pole pieces, is necessary for cryo-work. The twin-blades sandwich the specimen and are maintained at a temperature about 10 °C lower than the specimen, thus acting as a cold trap to reduce contamination to the specimen. Higher end and dedicated cryo-EMs may feature more expensive hardware and software, such as liquid helium cooling capability, in-column or postcolumn energy filters, large format (4 K × 4 K or 8 K × 8 K) ultrasensitive CCD cameras, and fully automated single particle or tomographic tilt series collection (see chapter by Carragher and Potter).

The use of column anticontaminator and the twin-blades anticontaminator alleviates but does not entirely prevent specimen from being contaminated: at extremely low temperature (-170 °C), the specimen itself is also a trap to the ever-present residual gases in the microscope column. Thus, the cryo-microscopist should be aware that once a specimen is inserted into the microscope, it is being gradually contaminated, even before the electron beam is turned on.

In the first half of the chapter, we give several protocols on how to operate a cryo-EM. At Brookhaven National Laboratory, we use a JEOL JEM 2010F TEM (200 kV) equipped with a Gatan 626 cryo-specimen holder and a Gatan 4 K CCD camera. However, the instructions are general enough that they should be useful to all cryo-EM users, with some adaptation to the specifics of the particular instrument being used. In the second half of the chapter, we outline a few practical precautions to problems encountered on a daily basis of a typical cryo-EM session.

2. BASIC OPERATIONS OF CRYO-EM

2.1. Microscope startup and cryo-EM grid insertion procedure

- Read the logbook and check if the previous session was normal. Check column vacuum ($<3 \times 10^{-5}$ Pa) and gun vacuum (0.1×10^{-6} Pa). Make sure high tension (HT) is READY.
- Power-on the CCD camera controller. It takes about 1 h to cool down and to stabilize the camera temperature. Lower temperature reduces the number of thermal electrons in CCD chip, thus minimizes the noise level in the final images.
- Fill up the anticontamination device (ACD) of the microscope with LN_2. Refill after boiling, and refill every 4 h afterward.
- Check that HT is at 180 kV. Set the target HT to 200 kV in the HT ramp-up program, set step 0.1 kV, time 30 min, and then start the program. The HT ramp-up program automatically raises the voltage to the target value, and relieves the user of the tedious task.
- Remove the specimen holder from microscope. Make sure the holder is centered (specimen position X, Y, Z, and Tilt-X and Tilt-Y are zeroed). Unplug cable from holder, close the cryo-shutter blade to prevent the EM grid from dropping into the column during holder removal. Flip the goniometer vacuum pump switch to AIR position. First pull the holder straight out until stopped, then turn ~60° counterclockwise until stopped, pull outward again until stopped, turn ~30° counterclockwise until stopped, stop and wait until the vacuum release sound is stopped and the green light goes off, then pull out the holder. This is a relatively simple procedure, yet most new users make mistake here that can cause

the column vacuum to crash. The trick is to do it one step at a time, and try not to rush and combine steps. A common mistake is to pull out the holder at the end of the last 30° turn without waiting for the vacuum in the prepump chamber to be fully released.

- Evacuate the sample holder Dewar on the holder pumping station. Make sure the vacuum reaches $\sim 10^{-5}$ Torr. This takes 30–60 min. An insufficiently evacuated Dewar will 'sweat' when LN_2 is poured into due to moisture condensation on the cold exterior surface, and will not be able to reach the desired temperature range (-170 to -180 °C).
- Examine the sample holder, particularly the O-rings, and remove any hairs or dust particulates. If necessary, regrease O-rings using a small amount of high vacuum grease, about half the size of a rice grain.
- Examine the cryo-workstation to make sure it is dry and there are no used EM grids in the chamber. This includes removing the aluminum workstation from the work chamber and inspecting for used grids at the bottom of the chamber. Insert specimen holder into cryo-workstation. Open the shutter, press the clip-ring tool vertically onto the grid-retaining clip ring, rotate clockwise about half a turn, and lift the clip-ring tool vertically to remove clip ring from the holder. The clip-ring should now be attached to the tip of the clip-ring tool. Make sure to locate and remove the used EM grid; otherwise it may stick to the holder, make its way into the microscope, and eventually drop into the objective lens.
- Cool down the cryo-workstation and the specimen holder. Only clean LN_2 devoid of ice contamination should be used to avoid contaminating the EM grid or causing excessive LN_2 bubbling in the specimen holder Dewar. Put on the workstation plexiglass cover and the specimen holder Dewar cap only after LN_2 has boiled. Let the workstation and sample holder cool down for ~ 10 min, or measure the temperature with controller to make sure that the specimen holder tip has reached -190 °C.
- Transfer a cryo-grid box from the large LN_2-cooled storage Dewar (34 L capacity) to a wide-mouth cylindrical LN_2 Dewar (2 L capacity). Use a large and long pair of tweezers to transfer a grid box into cryo-workstation.
- Make sure the LN_2 level in the cryo-workstation covers the grid box. Add more LN_2 if needed. Loosen the box retaining screw with a screwdriver. Insert a pair of sharp-tipped tweezers through one hole, and insert the clip ring tool with a mounted clip through the other hole in the plexiglass cover, and let both cool down for 2 min.
- Make sure the LN_2 level is slightly above the grid seat at the tip of the cryo-holder. Wait if the level is too high or add a bit more LN_2 if the level is too low. Use the precooled tweezers to transfer one EM grid from the box to the grid seat in the holder tip, and then press the precooled clip-ring tool vertically onto the grid, and turn counterclockwise half a turn,

then pull out vertically. Check with the tweezers that the clip ring is properly seated and the EM grid is not distorted, then close the shutter and replenish LN_2. Tighten the grid box cap screw and transfer it back into the storage Dewar. Loading cryo-grid is an important step and the user should make sure: (1) The tweezers, the clip-ring tool, and the clip ring have been cooled before touching the cryo-grid; (2) the cryo-grid has been properly fixed by the clip ring and the grid is not distorted during the process. The grid should be discarded if it is distorted at this step.

- Take the cryo-workstation with the holder to the microscope. Pump out most of LN_2 from the specimen holder Dewar with a LN_2 bubbler, a rubber plug with an inserted plastic tube. This is to prevent LN_2 from pouring onto the operator when specimen holder is rotated while being inserted into the column.
- Flip the goniometer airlock switch from 'AIR' to 'PUMP.' Take out the specimen holder from the cryo-workstation, align the holder airlock pin with the goniometer slot, and push straight in until stopped. Hold the holder rod initially to prevent it from rotating under its own weight. After the airlock opens and the green light is on, allow for additional 5–10 min pumping time to reduce the amount of ice being brought into the column. Turn the holder slowly clockwise $\sim 30°$ until stopped, make sure to hold it firmly so that it is sucked in slowly by vacuum until stopped. Continue to turn $\sim 60°$ clockwise until stopped, and then let holder in slowly and completely. Add LN_2 to the specimen holder Dewar. The LN_2 level in the Dewar should be maintained below and not touching the copper bar; otherwise disturbance in LN_2 will be transmitted via the copper bar to the specimen. Check LN_2 level in the cryo-holder Dewar every 30 min and add more when the level is too low.
- Wait for the microscope column vacuum to recover ($\sim 1–3 \times 10^{-5}$ Pa). Open the specimen holder shutter. Wait 3 min before opening the gun valve (V1) to obtain electron beam.

2.2. Basic microscope alignment

- Turn on electron beam. For EM with a field emission gun (FEG), open the column isolation valve (V1); for EM with a LaB_6 gun, slowly ramp up the filament heating current to saturation in about 3 min. Find beam by varying objective magnification (MAG) and/or the second condenser lens focus (C2).
- Select the desired SPOT SIZE (3) and beam convergence (α-SELECTOR, 3).
- *Gun alignment*. At MAG ~ 50 k×, focus beam (C2), activate Anode Wobbler, press GUN Deflection button, adjust GUN DEF X, Y to

minimize swipe of the illumination disk, and GUN SHIFT to center the disk. Turn off Anode Wobbler and Gun Deflection.

- *Condenser alignment.* At MAG ∼50 k×, center beam with GUN SHIFT X, Y at SPOT SIZE 1, and center beam with BEAM SHIFT X, Y at SPOT SIZE 5, and repeat a few times until beam position is stable when SPOT SIZE is varied.
- Insert and select a proper condenser aperture size (e.g., the third one, 40 μm). At MAG ∼50 k×, center the aperture by spreading beam with C2 to half the screen size, center the illumination disk by adjusting the mechanical X, Y shift of the C2 aperture; focus the beam (C2), and center the beam with BEAM SHIFT X, Y. Repeat several times.
- If the illumination disk is elliptic, press COND STIG button, adjust DEF X, Y so that the beam disk is round as you go over and under the probe focus. (DEF X and Y are multiple functional knobs; their function depends on which functional button is selected: it can be OBJ STIG, CON STIG, BRIGHT TILT, or others.)
- *Voltage centering.* This step is to minimize image movement when image focus is varied. At MAG ∼200 k×, either find, center, and focus a recognizable sample feature, or simply use the granular phase contrast of the carbon support film. Activate HT WOBBLER and BRIGHT TILT, adjust DEF X, Y so that the feature at the center of the small fluorescent screen remains stationary when viewed through the binoculars.
- Adjust beam tilt purity (also called the pivot point alignment. This is to ensure that there is no beam tilt when beam is shifted, and no beam shift when beam is tilted): At MAG 200 k×, focus and center beam, activate the CON DEF ADJ TILT button, flip the white toggle switch TILT to X position. This wobbles the beam by modulating the deflection lens current. Use SHIFT X and DEF X knobs to form a single spot. Repeat for TILT Y position. Because the condenser deflection shift is usually stable and does not require frequent adjustment, the following procedure is optional: spread beam, switch to DIFFRACTION mode, set CAMER LENGTH to 50 cm, focus diffraction spot, and adjust the intermediate lens stigmator to make the spot round, activate the COND DEF ADJ SHIFT button, set the toggle switch SHIFT to X position, use SHIFT X and DEF X knobs to form a single spot. Repeat for SHIFT Y.
- The optimal position of a modern computerized five-axis side entry goniometer stage should be calibrated by a service engineer. The user only needs to adjust specimen height to the ideal focal plane of the objective lens. Set objective lens current to the standard 5.05 A (this value is dependant on the specific type of objective lens installed). Adjust Z-HEIGHT to focus the sample. If specimen is not too far from the ideal position, focus electron beam on support carbon film, adjust Z-HEIGHT such that the diffuse diffraction rings converge into a single spot. When the specimen has been brought to the ideal focal plane of the objective

lens, set DV to 0. DV stands for deviation from ideal focal plane. During normal operation, use OBJ FOCUS knob to focus specimen, and DV shows how far the lens current is away from its optimal value (5.05 A). If the specimen is not very flat, the specimen height may need to be readjusted for different regions.

- Insert and center the first (largest) objective aperture. In diffraction mode, center the aperture by manually adjusting the aperture shift X and Y until the diffraction spot is at the center of the aperture image.

- Correct the objective lens astigmatism. This needs to be carried out every time when you plan to record an image. Set MAG to 250–400 k×. View specimen feature, for example, the amorphous carbon film granularity, through the binoculars. Focus objective lens to obtain minimum contrast image (near focus). Carefully discern the parallel striation of the carbon granularity: the striations turn 90° as the focus point is crossed. Press OBJ STIG button, adjust DEF X, Y to minimize the striation effect. When the objective lens astigmatism has been corrected, there should be virtually no feature at focus (minimal contrast), and sharp and granular feature with no preferred directionality on either side of focus. Correction of astigmatism can be confirmed by examining the power spectrum of a CCD image: in the absence of objective astigmatism, the concentric contrast transfer rings should be perfectly round.

2.3. Low dose mode setup

To minimize the specimen exposure, and to facilitate recording images of radiation sensitive materials, a modern electron microscope provides functionality to predefine and to recall several sets of electron-optical parameters that are intended for different tasks during a typical microscopy session, such as inspecting or searching the specimen area with a minimum electron dose (SEARCH mode), focusing the objective lens and correcting the objective astigmatism away from but near the area to be imaged (FOCUS mode), and recording diffraction or high-resolution images at the selected area, at a predefined MAG with the desired exposure time and electron beam intensity (PHOTO mode).

- To set up the SEARCH Mode, first choose the smallest possible condenser aperture size while maintaining enough beam intensity in the PHOTO mode, so that the aperture size does not have to be increased in the PHOTO mode. We use the third one (40 μm) out of the five available sizes in JEOL 2010F (150, 70, 40, 30, 10 μm). Select the smallest spot size (5), completely spread the beam (C2), switch the microscope to Diffraction mode, set Camera Length to 100 cm, overfocus the diffraction spot by turning the DIFF FOCUS knob clockwise from the focus point to form a disk about half the size of the small fluorescent screen.

Make sure that the disk, when focused, converges concentrically to the center of the small screen. This ensures that the specimen feature found at the center of the disk is centered in the PHOTO mode. Use PRJ SHIFT to center the disk, and BEAM SHIFT to reduce the nonconcentric disk swipe when the disk is over- or underfocused. This defocused diffraction mode is equivalent to a highly defocused MAG mode, and provides high contrast. In this mode, the level of DIFF defocus controls the MAG. Use of overfocusing is to maintain the sense of orientation; in an underfocused diffraction disk, the image moves in the opposite direction of specimen translation.

- To set up the FOCUS Mode, first define the beam/image shift distance at MAG $\sim$50 k$\times$ by moving the focused beam (BEAM SHIFT) from the screen center to a screen edge (e.g., the front edge), and then move the beam back to the screen center with IMAGE SHIFT. Once defined, do not change IMAGE SHIFT, and use only beam shift to move or to recenter electron beam in subsequent operations. Increase MAG to 200–400 k$\times$, form an illumination disk with desired size (4 cm) and brightness with C2 and spot size.

- To set up the PHOTO Mode, at the desired MAG (for example, 60 k$\times$), form (by overfocusing C2, that is, by turning the BRIGHTNESS knob clockwise after first reaching beam focus) and center (BEAM SHIFT) an illumination disk that is slightly larger than the recording media (negative films or CCD camera), adjust the beam brightness with C2 and SPOT SIZE (2 or 3), measure the electron current density with the small fluorescent screen inserted (10 pA/cm^2), and set the exposure time (1 s) for the targeted electron dose (20 e/Å^2). At a given aperture size and spot size, an overfocused beam provides better quality illumination (coherence) than an underfocused beam, because the virtual source size is smaller in the overfocused beam.

- To complete the setup process, go through the three modes sequentially several times, and readjust in each mode as described above until the beam is stable when mode is switched. It is important to change the mode sequentially and complete the cycle! For example, if the microscope is in FOCUS mode, and you wish to go to SEARCH mode, do not go back to SEARCH mode. Instead, go forward to PHOTO mode, then to the next SEARCH mode. Operating in sequence minimizes the magnetic hysteresis of the electromagnetic lens system, which in turn makes the beam position more stable.

- Check to ensure that the object in the center of the SEARCH disk is exactly centered in PHOTO mode. This is usually true if (1) the microscope is properly aligned, (2) there is no image shift in SEARCH or PHOTO mode, and (3) the SEARCH disk concentrically shrinks to or expands from the center of the screen when changing diffraction focus. However, for whatever reason, if a mismatch is found, identify and center a feature in PHOTO mode by moving the specimen drive, and center the same feature in SEARCH disk by adjusting the PRJ SHIFT X and Y. This will cause the SEARCH disk to be slightly off center on the fluorescent screen.

2.4. Target search and picture recording

- Depending on the specific application, the method for searching the target can be different. For the general single particle application, the specimen concentration and whether the particles are properly embedded in vitreous ice should be assessed by recording several CCD images. For 2D protein crystals embedded in a thin layer of tannic acid, trehalose, or glucose, the size of the crystals, the smoothness of their appearance, and the thickness of the embedding medium should be assessed, and the quality of the crystals can be evaluated by recording a shortly exposed electron diffraction pattern. For tomography, one or two initial exposures can be made to evaluate if the target is desirable, with good contrast, and its view is unblocked at high tilt angles. More details can be found in other chapters of this volume dealing with the specific approaches.
- When recording electron diffraction pattern of 2D protein crystal, a CCD camera is far more advantageous than the negative films because of the digital camera's liner response to electron dose and the large dynamic range (14 bits). In this case, the PHOTO mode should be set to diffraction. Before taking a diffraction pattern, it is crucial to ensure that the central diffraction spot is sharply focused, the area of exposure corresponds exactly to the center of the disk in SEARCH mode, and the beam size in PHOTO mode matches or is slightly larger than the particular crystal size. Find and center a distinguishing feature, such as a piece of dirt in SEARCH mode, and examine the feature in the PHOTO mode by defocusing the central diffraction spot. If the feature is not centered, move specimen drive to center it, and switch the scope to SEARCH mode, and adjust PROJ SHIFT X and Y to center the feature.
- Since electron diffraction pattern is not affected by slight specimen drift, one can take advantage of this by recording the pattern shortly after a crystal is identified, without waiting for the specimen drift to slow down, and by recording long exposure diffraction pattern with a weak beam intensity (20–40 s). Use of a weak beam and a long exposure time can minimize the blooming effect of the CCD camera, so that the beam stop does not have to be used. Blooming, or overflowing of electrons into the adjacent pixels, occurs when the number of electrons in a CCD pixel has exceeded its charge capacity.
- Contrary to electron diffraction, high-resolution imaging is affected by specimen movement. Before recording an image, the objective astigmatism

should be corrected, and the specimen drift should have slowed down to an acceptable level, say 1 nm/min. These are done in the FOCUS mode. The drift rate can be estimated either by eye through binoculars at high MAG (>400 k×), or assessed qualitatively by inspecting the power spectrum of a CCD image. The direction and position of the broken Thon rings are good indicators for the drift direction and rate. One can assess the drift rate quantitatively by recording two consecutive images in the same area and find the drift by computationally searching their cross-correlation peak, divided by the time interval between the two images.

• Set the underfocus value in the FOCUS mode. This value will be applied to the PHOTO Mode by the low dose program. Check to make sure if a compensatory value between FOCUS and PHOTO has been entered in the low dose program, and take that into consideration when setting the defocus. High-resolution images should always be recorded with an underfocused objective lens, that is, turning the FOCUS knob counterclockwise from the focus. This is because the objective lens usually has a positive spherical aberration value (C_s, 2 mm) and underfocusing the lens can partially counter the effect of this defect resulting in better images. Also note that the focus step size displayed on microscope monitor can be inaccurate. Perform a calibration by determining the defocus values of several CCD images of a carbon film, and compare that with the displayed values to obtain a correction factor.

• Depending on the application, the desirable underfocus value can be quite different. For tomography, this value is high, and can vary from 3 to 15 μm; for single particle work, underfocus ranges from 0.5 to 8 μm; and for 2D crystal, from 0.1 to 1 μm. When operating the microscope at higher tension (300–400 kV), a larger underfocus value is needed to produce comparable image contrast, because the electron at higher voltage has a shorter wavelength, and thus induces a smaller phase shift with the same underfocus value. For 2D crystallography and single particle work, one should also keep in mind that images should be recorded at different underfocus values, in order to 'fill up' the information gap caused by the microscope contrast transfer function.

• The choice of MAG in PHOTO mode should be carefully considered. This depends on the specimen, specific EM method being used, the picture-recording medium, and the targeted resolution. In general, electron tomography is used for studying larger objects at lower resolution, so MAG is set to a value from 20 k× to 50 k×. For single particle or 2D crystal work, MAG is recommended at 60 k× when recording on film, but higher on CCD camera (80–200 k×). Use of higher MAG for CCD cameras is due to their larger pixel size ($\sim$7–15 μm as compared to the $\sim$0.2 μm silver halide grain size of negative film), thus their worse modulation transfer function near the sampling (Nyquist) frequency.

- When ready (objective astigmatism corrected, underfocus value set, drift rate acceptable), raise the fluorescent screen, keep your arms and body away from the microscope, hold your breath, and click the EXPOSURE button in the low dose program. The program will automatically switch the microscope to PHOTO mode, and unblank electron beam for a preset time (1 s) to record a picture.

2.5. Cryo-grid replacement during an EM session

Some microscopes are conveniently designed so that several cryo-grids can be loaded at a time. But most are not. In any case, one frequently needs to replace the cryo-grid in the middle of an EM session. Taking out the specimen holder is essentially a reversed procedure of inserting the holder into the microscope.

- Before pulling out the specimen holder, first make sure that the cryo-workstation is cooled, and the interior of the chamber is free of contaminating ice. After the first use, one should remember to keep the cryo-workstation cooled by plugging the specimen holder side port, maintaining LN_2 inside the work chamber, keeping on the plexiglass cover and the two plugs, and covering the holes in the two plugs with, for example, a small Kimwipe box, to prevent ice contamination inside the chamber. Keep a second cryo-workstation in the EM room can be handy, because even with the best effort, the station will be coated with ice after several uses, making it difficult to load more grids. If the work chamber has ice contamination, the cryo-workstation should be warmed up to remove ice and moisture. Do not heat up the station with a hair dryer. The intense heat can damage the plastic casing of the chamber. A negative film dryer cabinet supplies warm airflow and is ideal for warming up the cryo-workstation.
- Remove the specimen holder from the column (see Section 2.1) and quickly insert it into the cooled cryo-workstation. First close the shutter to prevent the grid from dropping inside column. This also protects the cold grid seating area from ice contamination during the brief exposure of the specimen holder to the room air during transfer from the microscope column to the cryo-workstation. Use the nitrogen bubbler to pump out most of LN_2 from the holder Dewar; this is to avoid LN_2 pouring over the operator when the holder is rotated while being pulled out.
- Replenish LN_2 to the work chamber and the holder Dewar once the specimen holder is inserted in the cryo-workstation. Remove the existing grid from the holder seat and take it out from the chamber, and reload a new grid, and finally reinsert the holder into the column (see Section 2.1 for detail).

2.6. Microscope shut down procedure

- Center the specimen stage (translation X, Y, Z, and Tilt X and Tilt Y all zeroed).
- Quit the low dose program. Set scope to a low MAG (5 k×) to reduce the electric currents in the lens system when microscope is idle. Remove all apertures from the beam path to facilitate vacuum recovery: it is quite inefficient for the gas molecules to pass through the tiny holes in the aperture when inserted in the column.
- Shut off electron beam by closing the V1 value (FEG) or turn off the filament heating current (LaB_6). Cover the viewing window.
- Leave the cryo–holder shutter open. While making sure not to put much pressure on the Dewar, use the LN_2 bubbler to pump out LN_2 from the holder. Connect the heater control cable to the specimen holder, and turn on heater (30 °C). This heats up and dries the EM grid, and the grid is left in the column until next microscope session.
- Wear gloves for this operation to prevent introducing grease into column: if images were recorded on films, take out the film-receiving and the film-filled magazines from the desiccator, turn on the nitrogen gas cylinder, vent the camera chamber, replace the film magazines, and close the camera chamber door. Load into the microscope only films that have been desiccated. If it takes an unusually long time (>1 h) for the camera chamber vacuum to recover, reopen the camera door, run a finger slowly across the square-shaped O-ring at the back of the door to make sure it is properly seated. Inspect the O-ring for any dirt or lint and remove them if found. Regrease the O-ring if necessary.
- When the camera chamber vacuum has recovered (V2 open), insert electric heater into ACD, connect power supply cable, and turn on the ACD heater. This boils off the LN_2 in the device, heats up the cold copper braid inside the column, thus releasing the captured water vapor and other contaminating gases into the column that can be subsequently pumped out by diffusion pump during overnight vacuum recovery.
- Set the target HT to 180 kV, and start the HT program to ramp down the HT.
- Turn off CCD camera controller.
- Empty LN_2 in the small Dewar onto the cement outside the building, but not onto the floor in the hallway, to avoid freeze damage to the floor. Leave the Dewar uncapped, upside down, half-tilted and lean against a solid support, such as inside a box. This is to ensure that any residual LN_2 and water condensation drain properly, so the Dewar warms up quickly, and will be completely dry and ready for the next session.
- Clean up all work areas.
- Sign off logbook. Describe any problems encountered during the session.

3. SEVERAL PRACTICALS ON IMAGING FROZEN-HYDRATED BIOLOGICAL SPECIMENS

3.1. How to quickly find out if a newly inserted EM grid is heavily contaminated with crystalline ice

To quickly examine if a grid is suitable, first look at the grid at very low MAG (100×), to see if a significant portion of the grid is covered by ice of suitable thickness. To check if the grid is heavily contaminated by crystalline ice, switch the scope to DIFFRACTION mode, focus the central diffraction spot, and examine the diffraction rings from ice: vitreous ice produces fuzzy rings. If the rings are sharp or if strong diffraction spots are observed as the grid is moved about, the grid is heavily contaminated by polycrystalline or large pieces of crystalline ice, the grid should be replaced.

Do not waste valuable time on a contaminated EM grid. If it is deemed contaminated, the grid should be replaced immediately, until a good one is identified with minimum contamination and with a suitable ice thickness. The number of grids that have been inspected does not matter: one often hears people say that this is the fifth or eighth grid of the day, and if not taking pictures with that grid, the whole day will be wasted. But the truth is if one takes picture with the contaminated grid, one would be wasting many more days down the road. One should be uncompromising—keep replacing grids until a good one is found. If all the grids in stock have run out, one should go back to specimen preparation room and freeze a few new boxes of EM grids.

3.2. How to quickly estimate the electron dose at the exposure time used and at a particular MAG

To determine the electron dose, one needs to know exposure time (t), scope MAG (M), and electron current density at the specimen. While the exposure time and MAG displayed on the microscope are generally accurate enough, the current density needs to be determined with a Faraday cage. However, we find the current density displayed on the EM internal computer monitor, as measured via the small fluorescent screen, is a good approximation. The current density should be measured in an open area of the grid devoid of carbon or ice, at the electron-optical setting for image recording: at the MAG used for exposure, with the electron beam centered, spread to the intended size for recording the image (slightly larger than the negative film, that is, with the beam diameter slightly larger than the four marked corners on the large fluorescent screen). Since electric current (ampere) is defined as 1 coulomb electric charge (6.242×10^{18} electrons) transferred in 1 s, the exposure dose can be derived by multiplying the current density read off the small focusing screen with the exposure time:

$$
\begin{aligned}
\text{Dose } (e/\mathring{A}^2) &= \text{electron current density at specimen} \times \text{exposure time} \\
&= (\text{electron current density at screen} \times M^2) \\
&\quad \times \text{exposure time} \\
&\approx (1/16) \times (M')^2 \times C \times t
\end{aligned}
$$

where M' is MAG in the exposure mode in the unit of 10,000×, t the exposure time in second, and C the on-screen current density in pA/cm^2. For example, at MAG (M) of 60 k× (i.e., $M' = 6$) with the on-screen beam intensity of 10 pA/cm^2 and with 1 s exposure, the electron dose at the specimen is approximately $(1/16) \times 6^2 \times 10 \times 1$ e/$\mathring{A}^2 = 22.5$ e/$\mathring{A}^2$.

3.3. How to quickly estimate the thickness of vitreous ice

The desirable thickness of the vitreous ice depends on the specimen. Ideally, it is the thinnest ice that can totally embed the target molecules. In practice, the ice should be slightly thicker than the largest dimension of the mole-cules. A simple way to estimate the thickness of the ice is to tilt the specimen by 45°, drill a hole in the vitreous ice with the sharply focused electron beam, and tilt specimen back and measure the length of white trace of the hole, which is equal to the ice thickness. For example, if a white line is 1 cm long on the fluorescent screen at the MAG of 100 k×, the ice thickness in this region is 1 cm/100,000 = 100 nm thick.

If the microscope is equipped with an electron energy spectrometer, the ice thickness (T) can be calculated from the ratio of total integrated intensity (I_{tot}) and the zero-loss peak (I_0) of the electron energy loss spectrum, based on the assumption that inelastic scattering increases exponentially with specimen thickness (Egerton and Leapman, 1995):

$$
T = \Lambda \times \ln(I_{tot}/I_0)
$$

where Λ is electron inelastic mean free path in ice. Λ is approximately 200 nm at 120 kV (Grimm et al., 1996). Since Λ is a square root function of acceleration voltage ($U^{1/2}$), Λ is 260 and 320 nm at 200 and 300 kV, respectively. This is done by first recording an unfiltered CCD image of the ice region of interest, inserting the energy filter and choosing the 10 eV slit to record the filtered CCD image at the same region, and calculating the integrated intensities of these two images.

Zhang et al. (2003) suggested a simpler empirical method that does not require the electron energy spectrometer:

$$T = k \times \log(I_{ref}/I_{ice})$$

where I is the average value of a CCD image recorded at the empty hole (I_{ref}) or over ice (I_{ice}). k is a scaling factor, and at 120 kV and with a 40 μm objective aperture, is $\sim$750 nm. Such measurement is reported to be accurate within 15% (Zhang et al., 2003). The k value at higher acceleration voltage was not given. The CCD response is proportional to the exposed electron dose. Therefore, if a CCD camera is not available, one could in principle replace I_{ref}/I_{ice} with the beam attenuation ratio C_{hole}/C_{ice}, where C_{hole} is the current density of the beam measured in an open area, and C_{ice} is the same illumination at the ice region, as measured with the small fluorescent screen of the microscope.

3.4. How to minimize exposure to the specimen before taking a picture

The electron microscopist should be acutely aware that every electron spent on searching specimen area is a significant bit of signal taken away from the final image. There appears occasionally to be a misperception that the search mode is 'safe,' and one could take time to look for a good area. This is simply not true. The contribution of the electrons among the total allowed dose, say 20 e/Å^2, is not equal: the first electron is the best, and the 20th electron is the least useful. This is so because the sample is being continuously degraded: the first electron sees the best sample and carries the most information, and by the time the last electron arrives, the sample is already significantly damaged by the preceding electrons and this electron carries the least amount of information.

If one uses an illumination disk in the search mode that covers 10 holes of a quantifoil grid in diameter, assuming a weak beam intensity of 2.5 pA/cm^2 (on screen) at a low MAG of 5 k$\times$, and one spends 5 s to locate and center each suitable hole for imaging, the overhead cost, based on a single shot, is $1/16 \times 0.5^2 \times 5 \times 2.5 = 0.2$ e/Å^2. This is about 1% of the total useful dose (20 e/Å^2). However, if one takes picture one hole after another without skipping any holes, when a particular hole is reached, that hole would have already been exposed many times in the search mode with an accumulative dose of $\sim(\pi/4) \times 10^2/2 \times 0.2$ e/$\text{Å}^2 \sim 8$ e/Å^2, even before taking the picture of that ice hole. Taking picture in this manner, the overhead is not 1% but 40%. And remember this is the best 40% of the total exposure!

Search the EM grid systematically, that is, from one side to the other, to avoid going over the same area multiple times. The area of illumination should be minimized in search model. The electron beam intensity should also be minimized. This can be achieved by selecting the smallest spot size, and by using the smallest condenser aperture size that still provides strong enough beam intensity in PHOTO mode for 1 s exposure. One should spend as little time as possible during search, and blank the beam immediately after a suitable ice hole has been located and centered (GUN BLANK), or quickly switch to the FOCUS mode. One should also avoid taking pictures in every hole, even if they look perfect; skip every other one or two holes will mitigate the pre-exposure accumulative dose.

3.5. How to minimize the specimen drift and vibration during exposure

For cryo-EM, specimen drift is a major source of image quality degradation. When a new grid is first inserted into the microscope, it takes over 30 min for the temperature to stabilize. During the normal operation, if one notices a sudden increase in specimen drift rate, this is usually an indication that LN_2 in the holder Dewar is drying up or has dried up. If this ever happens, immediately check the specimen temperature by connecting the cable from the temperature controller to the holder. If the temperature is still below $-145\ °C$, refill the holder Dewar; if the temperature is unfortunately above $-145\ °C$, the grid needs to be replaced, because the vitreous ice may have transformed into crystalline ice at the raised temperature (the phase transition point at atmospheric pressure is $-137\ °C$). Therefore, it is very important to maintain the proper LN_2 level in the holder Dewar, and check the level regularly, say, every half an hour. A timer can be used if one tends to forget this.

During the session, drift is mainly caused by the frequent need to move specimen across a large distance after each exposure to locate next suitable area. The residual mechanical strain in the specimen holder drive can only be relieved over a certain period. The specially designed piezo-controlled specimen stage can counter a persistent drift, thus alleviates but does not eliminate the problem. An image drift compensator, an electronic device that introduces a continuous image shift in the direction opposite to the specimen drift, sometimes works effectively. However, the image drift compensator is not commercially available, so it has to be made in-house (Cattermole and Henderson, 1991).

Mechanical vibration transmitted by the side entry cryo-specimen holder is also a large source of image degradation. Specimen holder can transmit room noise and air draft into specimen vibration. The door of the EM room should be closed to minimize temperature gradient and noise from outside.

Liquid nitrogen bubbling inside the Dewar of specimen holder is another major source of image degradation. Ice contamination in liquid nitrogen is the main cause for bubbling. Thus, use of clean liquid nitrogen devoid of ice is very important. If the room is maintained quiet, the bubbling noise from the Dewar can actually be heard by a careful operator sitting in front of the scope. However, one should not always rely on his/her sense of hearing, and instead, it is more reliable to stand on a step stool, position a ear near but not touching the opening of the Dewar to listen to the inside of Dewar. A pen flashlight can help one to see the small bubbles. Nitrogen bubbling should be monitored frequently, with an interval of at least every half an hour. Liquid nitrogen bubbler can be used to pump out the ice particles sunk at the bottom of Dewar. Replenish the Dewar with clean liquid nitrogen afterward.

In addition to the patience to wait for the specimen to stop moving completely, and the diligence to eliminate nitrogen bubbling, the expose time is another important factor: at the same specimen drift rate, a shorter exposure time will incur less drift in the image. Balancing the requirement of beam intensity and the convenience of operation, 1 s exposure seems to be a good compromise. A shorter exposure would require stronger beam intensity, which in turn requires the use of a large condenser aperture. And using large condenser aperture will incur unnecessary beam damage to specimen during SEARCH, and switching the aperture size between PHOTO and SERACH mode is not feasible because of the aperture alignment issue.

3.6. How to minimize the beam-induced specimen charging and movement during exposure

Positive charging is caused by electrons in specimen being knocked out by the primary illumination electrons, and the lost electrons are not compensated due to the insulative nature of ice and biological specimen. Charging degrades image in two ways: it may act like an ill-defined mini-lens, causing uneven electron phase shift thus blurring the image; it also subjects specimen to mechanical stress and causes physical warp and distortion, consequently causing specimen movement.

Spot-scanning TEM where a focused beam is scanned across the specimen, was reported to reduce beam-induced specimen movement (Bullough and Henderson, 1987; Downing and Glaeser, 1986). Such method is frequently used in high-resolution imaging of two-dimensional protein crystals, but is rarely used in single particle application, primarily because the uneven intensity of the beam spots can overwhelm the specimen contrast.

A clean objective aperture made of gold was reported to reduce specimen charging (Miyazawa et al., 1999). The back scattered and secondary electrons from the clean aperture likely neutralize the positive charges on specimen. The so-called Unwin illumination, where the electron beam is slightly larger than the ice hole, and reaches the surrounding carbon film, was also reported to reduce charging and beam-induced specimen movement (Miyazawa et al., 1999).

The conductivity of the substrate appears to have a positive effect on specimen charging. Use of high purity carbon source (99.9999%), and evaporating the carbon film under high vacuum (better than 10^{-5} mbar) and avoiding carbon rod sparking, can produce carbon film with improved conductivity and smoothness. Preirradiation of the holey carbon grid by 100 kV electron beam in the projection chamber of a microscope with a total dose of 100 e/Å^2 was reported to reduce the charging and markedly improve image quality (Miyazawa et al., 1999). The use of a highly conductive support film, such as the amorphous TiSi film, instead of the carbon film, was proposed to be beneficial for cryo-EM (Rader and Lamvik, 1992). A metallic coating, either with gold or TiSi, on the carbon film was reported to significantly improve the high-resolution contrast of the organic monolayer paraffin crystals ($C_{44}H_{90}$) (Typke et al., 2004).

3.7. Safe handling of liquid nitrogen

LN_2 can cause cold burns. If proper care is exercised, it is relatively safe to work with LN_2. Wear safety glasses and thermal gloves when withdrawing LN_2 from the large storage tank. Wear long pants that cover the shoes, so LN_2 spills and splashes will not be trapped inside the shoes. Monitor the LN_2 level as well as the internal pressure every time you withdraw LN_2 from the tank, and ensure the pressure relief safety valve works properly. Do not overtighten the valve after dispensing of LN_2. This is a common mistake of novice user. If the valve is closed too tightly when cold, it will be very difficult to open it when it warms up and expands.

LN_2 boils immediately on contact with a warm object. So keep some distance when filling LN_2 into ACD, cryo-workstation, and specimen holder Dewar, until it calms down. Do not tiptoe and raise the small LN_2 Dewar above your head to refill ACD or the specimen holder Dewar; instead, one should stand on a step stool.

As LN_2 evaporates it reduces the oxygen level in the room and might act as an asphyxiant. So working in a confined room without airflow can be lethal. This is usually not a problem because EM room has a constant airflow that takes away the heat released from the microscope. However, if the air conditioning system is not reliable, an oxygen sensor should be installed.

4. CONCLUDING REMARKS

The procedures described in the chapter are based on a JEOL TEM, so there will be some modifications for FEI or other types of instruments. Users of those microscopes should refer to their operation manuals.

As cryo-EM includes electron crystallography, single particle, and electron tomography, the data collection strategies and routines might vary depending on the application. Furthermore, some microscopes are equipped with automation software. Users should refer to other chapters in this volume for details.

REFERENCES

Bullough, P., and Henderson, R. (1987). Use of spot-scan procedure for recording low-dose micrographs of beam-sensitive specimens. *Ultramicroscopy* **21,** 223–230.

Cattermole, D., and Henderson, R. (1991). An electronic image drift compensator for electron microscopy. *Ultramicroscopy* **35,** 55–57.

Downing, K. H., and Glaeser, R. M. (1986). Improvement in high resolution image quality of radiation-sensitive specimens achieved with reduced spot size of the electron beam. *Ultramicroscopy* **20,** 269–278.

Dubochet, J., Adrian, M., Chang, J. J., Homo, J. C., Lepault, J., McDowall, A. W., and Schultz, P. (1988). Cryo-electron microscopy of vitrified specimens. *Q. Rev. Biophys.* **21,** 129–228.

Egerton, R. F., and Leapman, R. D. (1995). Quantitative electron energy-loss spectroscopy. *In* "Energy-Filtering Transmission Electron Microscopy," (L. Reimer, ed.), p. 269. Springer Verlag, Heidelberg.

Fujiyoshi, Y., Mizusaki, T., Morikawa, K., Yamagishi, H., Aoki, Y., Kihara, H., and Harada, Y. (1991). Development of a superfluid helium stage for high-resolution electron microscopy. *Ultramicroscopy* **38,** 241–251.

Glaeser, R. M., and Taylor, K. A. (1978). Radiation damage relative to transmission electron microscopy of biological specimens at low temperature: A review. *J. Microsc.* **112,** 127–138.

Grimm, R., Typke, D., Bärmann, M., and Baumeister, W. (1996). Determination of the inelastic mean free path in ice by examination of tilted vesicles and automated most probable loss imaging. *Ultramicroscopy* **63**(3–4), 169–179.

Henderson, R. (1995). The potential and limitations of neutrons, electrons and X-rays for atomic resolution microscopy of unstained biological molecules. *Q. Rev. Biophys.* **28,** 171–193.

Knapek, E., Lefranc, G., Heide, H. G., and Dietrich, I. (1982). Electron microscopical results on cryoprotection of organic materials obtained with cold stages. *Ultramicroscopy* **10,** 105–110.

Miyazawa, A., Fujiyoshi, Y., Stowell, M., and Unwin, N. (1999). Nicotinic acetylcholine receptor at 4.6 A resolution: Transverse tunnels in the channel wall. *J. Mol. Biol.* **288,** 765–786.

Rader, S. S., and Lamvik, M. K. (1992). High-conductivity amorphous TiSi substrates for low-temperature electron microscopy. *J. Microsc.* **168,** 71–77.

Typke, D., Downing, K. H., and Glaeser, R. M. (2004). Electron microscopy of biological macromolecules: Bridging the gap between what physics allows and what we currently can get. *Microsc. Microanal.* **10,** 21–27.

Zhang, P., Borgnia, M. J., Mooney, P., Shi, D., Pan, M., O'Herron, P., Mao, A., Brogan, D., Milne, J. L., and Subramaniam, S. (2003). Automated image acquisition and processing using a new generation of 4 K × 4 K CCD cameras for cryo electron microscopic studies of macromolecular assemblies. *J. Struct. Biol.* **143,** 135–144.

CHAPTER ELEVEN

Collecting Electron Crystallographic Data of Two-Dimensional Protein Crystals

Richard K. Hite,* Andreas D. Schenk,* Zongli Li,*,† Yifan Cheng,‡ *and* Thomas Walz*,†

Contents

Abstract

Similar to X-ray crystallography, which requires three-dimensional (3D) crystals, electron crystallography is used to obtain structural information for proteins that form two-dimensional (2D) crystals. However, unlike data collection in X-ray crystallography, which is typically fast and straightforward, data collection in electron crystallography can take months to years and requires substantial expertise. In this contribution, we first discuss the proper preparation of 2D

* Department of Cell Biology, Harvard Medical School, Boston, Massachusetts, USA
† Howard Hughes Medical Institute, Harvard Medical School, Boston, Massachusetts, USA
‡ The W.M. Keck Advanced Microscopy Laboratory, Department of Biochemistry and Biophysics, University of California San Francisco, San Francisco, California, USA

Methods in Enzymology, Volume 481   © 2010 Elsevier Inc.
ISSN 0076-6879, DOI: 10.1016/S0076-6879(10)81011-0

crystals for electron microscopy, which, besides the quality of the 2D crystals, may be the most defining parameter for successful data collection. In the second part, we describe the procedures used to record high-resolution images and diffraction patterns.

ABBREVIATIONS

2D	two-dimensional
3D	three-dimensional
AQP	aquaporin
bR	bacteriorhodopsin
CCD	charge-coupled device
CTF	contrast transfer function
EM	electron microscopy
FEG	field emission gun
SA	selected area

1. INTRODUCTION

Electron crystallography was developed to determine the structure of bacteriorhodopsin (bR; Unwin and Henderson, 1975), a light-driven proton pump that forms crystalline arrays in the membrane of *Halobacterium salinarum*, known as purple membranes. By 1975, Henderson and Unwin had developed the methodology to a level that allowed them to calculate a density map of bR at 7 Å resolution, visualizing for the first time membrane-spanning α-helices and providing the first view of the organization of an integral membrane protein (Henderson and Unwin, 1975). After further developments in methods and instrumentation, electron crystallography finally produced a density map at 3.5 Å resolution and the first atomic model of bR (Henderson *et al.*, 1990). More atomic structures followed that were based on electron crystallographic density maps, including those of the plant light-harvesting complex 2 (Kühlbrandt *et al.*, 1994), the αβ tubulin dimer (Nogales *et al.*, 1998) and two members of the MAPEG (membrane associated proteins in eicosanoid and glutathione metabolism) family (Holm *et al.*, 2006; Jegerschöld *et al.*, 2008). Further developments in electron crystallography were, however, mostly driven by structural studies of members of the aquaporin (AQP) family (Schenk *et al.*, 2010), which resulted in atomic models for AQP1 (Murata *et al.*, 2000), AQP0

(Gonen *et al.*, 2004; Gonen *et al.*, 2005), and AQP4 (Hiroaki *et al.*, 2006; Tani *et al.*, 2009).

Electron crystallography is ideally suited for the structural analysis of membrane proteins (Hite *et al.*, 2007). It allows the structure of membrane proteins to be studied in a lipid bilayer, their native environment (reviewed in Fujiyoshi and Unwin, 2008; Raunser and Walz, 2009), rather than in detergent micelles, the most common membrane mimetic used in X-ray crystallography and nuclear magnetic resonance spectroscopy, which carry the risk of altering the protein structure. In two-dimensional (2D) crystals, membrane proteins remain fully functional, making it possible to use time-resolved techniques to capture short-lived intermediates for structural analysis (e.g., Berriman and Unwin, 1994; Subramaniam *et al.*, 1999). Another advantage of electron crystallography is that images provide phase information, so that the phases do not have to be determined by indirect methods as in X-ray crystallography. These and other advantages make electron crystallography an attractive approach for determining the structure of membrane proteins.

2. CHALLENGES

In theory, electron crystallography may be the ideal method to determine the structure of membrane proteins, but in practice it is far from easy to use this method to produce a density map at a resolution sufficient for building an atomic model. Like in X-ray crystallography, obtaining large and highly ordered crystals (three-dimensional (3D) crystals in X-ray crystallography and 2D crystals in electron crystallography) is the greatest challenge. However, while in X-ray crystallography, high-quality 3D crystals essentially guarantee that an atomic structure will be obtained in reasonably short order, in electron crystallography, collecting high-resolution data from 2D crystals remains very challenging and thus poses another major hurdle on the way to an atomic model of the crystallized protein.

Like all biological molecules, membrane proteins in 2D crystals are sensitive to radiation damage. Low-dose imaging techniques have been developed in electron microscopy (EM) to minimize the electron dose to which specimens are exposed. All instrumental alignments are routinely performed on a specimen area adjacent to the area of interest, and then the actual image or diffraction pattern is recorded with a very small electron dose (typically 10–20 electrons/$Å^2$) on the area of interest. Furthermore, the specimen is cooled to liquid nitrogen or even liquid helium temperature, which reduces the visible effects of radiation damage. However, even with the use of low-dose and cryo-techniques, each crystal can only be exposed to the electron beam once before radiation damage destroys most of the high-resolution structural information. Therefore, producing a complete

3D data set of Fourier terms extracted from images and/or diffraction patterns of untilted and tilted specimens means that data collection requires a large number of 2D crystals.

Obtaining the many images and/or diffraction patterns needed for structure determination is a time-consuming process because the yield of high-resolution images and diffraction patterns is very low. Recording images is particularly challenging, because, unlike diffraction, which is translation-invariant, the quality of images deteriorates with any movement of the specimen during image acquisition, which essentially blurs the information in the image. Hence, temperature fluctuations and acoustic and mechanical vibrations, all of which cause the specimen to move, have a deleterious effect on image quality. The overall environment in which the electron microscope is housed, and in particular the immediate environment of the specimen, that is, the specimen holder, thus has to be very stable to record high-quality images. Side-entry specimen cryo-holders contain a Dewar that holds the liquid nitrogen needed to cool the specimen during data collection (Hayward and Glaeser, 1980; Henderson *et al.*, 1991). The large, heavy Dewar is on the opposite side of the holder that carries the EM grid and is exposed to the environment. This design makes side-entry holders inherently somewhat unstable, both mechanically and thermally. Newer high-end electron microscopes are therefore now equipped with top-entry stages or stages that uncouple the specimen thermally and mechanically from the environment. It is always recommended to avoid moving or speaking during the actual recording of an image, because even such small disturbances can affect the image quality, especially when using side-entry holders. Such a requirement makes the image collection process even more tedious and physically challenging. In the newest models, the entire electron microscope is encased, uncoupling the entire instrument from the environment and greatly increasing the stability of the specimen. These models also allow the user to operate the instrument remotely, which decreases the likelihood of a user-induced disturbance and reduces the physical stress on the operator. Finally, electromagnetic stray fields that exceed the instrument's specifications can also not be tolerated, as they interfere with the electron optics of the microscope.

Even under ideal conditions, the signal of the high-resolution structure factors derived from an image very rarely reach one-tenth of what would be predicted for theoretically perfect imaging (Henderson, 1995). In most cases, the signal is even less than 5% of the theoretically possible value (Glaeser, 2008). As a result, the intensities of the highest resolution structure factors fall below the background noise and can thus not be measured. Early studies on images of 2D crystals showed that there were significant deviations in the crystal lattice in the images, which were explained by random, chaotic movements of the specimen during exposure to the electron beam

(Henderson and Glaeser, 1985). The exact cause of these beam-induced movements remains unclear. Until recently, the build-up of charge during illumination was thought to be responsible, but a newer theory states that the movements may also be due to structural changes resulting from beam damage to the proteins. The most effective way to minimize beam-induced movement, whatever the cause, is the use of spot scanning, a method in which the beam is focused to a small diameter and scanned over the specimen to record the image (Downing, 1991). Irradiating only small areas of the specimen at a time rather than the entire specimen area reduces the global beam-induced movements. Another method that has been shown to increase the quality of images is the "carbon sandwich" technique, in which 2D crystals are prepared between two layers of identical carbon film (Gyobu et al., 2004). Even though the exact reason why it improves image quality is under discussion, in some cases the carbon sandwich technique dramatically increased the yield of high-quality images recorded from tilted specimens.

Collecting images and electron diffraction patterns from highly tilted specimens is by far the most challenging task in electron crystallographic data collection. The reason is that the quality of data recorded from tilted specimens depends not only on all the factors that influence the quality of data recorded from untilted specimens, but it is also greatly affected by the flatness of the specimen. If the crystal is not perfectly flat, different regions of the crystal have different orientations relative to the electron beam, causing reflections in the direction perpendicular to the tilt axis to become increasingly blurred with increasing resolution until they completely disappear in the background. Studies on bR 2D crystals showed that a difference in tilt angle of 1° or more within the sampled region adversely affects the resolution perpendicular to the tilt axis (Baldwin and Henderson, 1984). While several parameters affect specimen flatness (Vonck, 2000), including the quality and material of the grids and the adherence of the crystals to the carbon support film, the flatness of the carbon film is by far the most important factor for producing flat specimens.

3. Images and Diffraction Patterns

Electron microscopes can be operated in two distinct modes to produce either images or diffraction patterns. To collect high-resolution image data, the 2D crystals do not have to be very large, areas of 500 nm × 500 nm can suffice to extract near-atomic resolution information, in particular because image processing makes it possible to computationally correct slight errors in crystal order (lattice unbending). By contrast, recording high-resolution diffraction patterns usually requires the crystalline areas to be at least 1 μm $\times$ 1 μm in size, and besides subtracting the background little can be done to computationally enhance the recorded information. The exact size of a crystalline area required to collect high-resolution data depends mostly on the number of unit cells that are exposed to the electron beam, which in turn depends on the unit cell dimensions and the number of layers making up the 2D crystal. The more unit cells contribute, the stronger the intensity of the reflections in diffraction patterns and Fourier transforms of images. For images, the size of the 2D crystal included in an image is usually limited by the size of the recording medium, which is typically EM film. In the case of electron diffraction, the size of the area of the 2D crystal that will contribute to the diffraction pattern can be selected by the choice of the selected area (SA) aperture. The aperture size that produces the highest quality diffraction patterns will depend on the nature of the 2D crystals. Even with very large 2D crystals, it can be beneficial to expose only smaller crystal areas, because long-range disorder becomes more significant with larger crystal areas and can deteriorate the quality of the recorded diffraction patterns.

Images and diffraction patterns provide different information. Fourier transforms of images can be used to measure both the phases and amplitudes of reflections, which are, however, affected by the contrast transfer function (CTF) of the electron microscope, which is a consequence of the electron optics of the objective lens (Fig. 11.1A). Correcting the phases of images from untilted specimens for the CTF can be accomplished by simple phase flipping and is rather straightforward. Correcting the phases of images from tilted specimens is, however, more involved as the defocus gradient across the image leads to splitting of the diffraction spots in the Fourier transform of the image (Philippsen *et al.*, 2007). Image amplitudes, on the other hand, are more difficult to correct for the CTF, and the amplitudes are also affected by the envelope function, which causes a gradual dampening of the amplitudes toward higher resolution (see Penczek, this volume). The fall-off of the amplitudes toward higher resolution can be crudely compensated for by applying a negative temperature factor to the amplitude data, which will result in a sharpening of the high-resolution features in the density map (Havelka *et al.*, 1995). Electron diffraction patterns record the reflection intensities, the squares of the amplitudes, much more accurately, because they are not affected by the CTF and envelope function (Fig. 11.1B). On the other hand, diffraction patterns contain only intensity but no phase information. The traditional way of producing a density map by electron crystallography thus consists of recording images to obtain phase information and diffraction patterns to measure the reflection intensities.

In certain situations, only images or diffraction patterns can or have to be collected. If 2D crystals cannot be grown to sufficiently large sizes to collect electron diffraction patterns, only images can be recorded. To account for the inaccuracies of the high-resolution amplitudes, more images are ideally

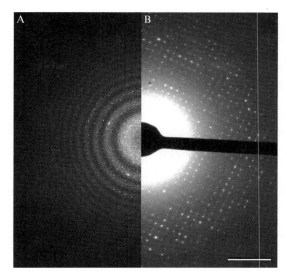

Figure 11.1 Electron crystallographic data of double-layered AQP0 2D crystals. (A) Fourier transform of an image of an untilted AQP0 2D crystal. (B) Diffraction pattern of an untilted AQP0 2D crystal. Scale bar is $(10 \text{ Å})^{-1}$.

included in the data set, further extending the duration of data collection. On the other hand, if 2D crystals allow collection of near-atomic resolution diffraction patterns and the atomic coordinates of a homologous protein are available, phases can be obtained computationally by molecular replacement. Since it is much easier and faster to collect high-resolution diffraction patterns than images, especially from highly tilted specimens, efforts are now underway in several groups to establish methods that relieve the dependency of electron crystallographic structure determination on image data and take more advantage of diffraction data.

4. Specimen Preparation

Possibly the most critical prerequisite for successful data collection is the preparation of the 2D crystals in a way that best preserves their crystalline order. Electron microscopes operate under high vacuum conditions, posing a problem for biological samples, which consist mostly of water. The prime objective of specimen preparation is thus to protect the samples from dehydration and the resulting structural collapse of the proteins. Two major methods have been developed to prepare 2D crystals for data collection, sugar embedding, and vitrification.

4.1. Sugar embedding

The first 2D crystals used for data collection were prepared by sugar embedding, a technique developed by Unwin and Henderson (1975) during their work on bR. Several sugars and sugar derivatives have been successfully used for the embedding of 2D crystals, including glucose (Henderson *et al.*, 1990), tannic acid (Kühlbrandt *et al.*, 1994), and trehalose (Kimura *et al.*, 1997).

The easiest sugar-embedding procedure is carried out at room temperature, typically using glucose. A protocol is provided in Section 4.1.1 and a diagram is shown in Fig. 11.2. The sugar-embedded specimen is transferred into the microscope and its quality can be assessed at room temperature. A good specimen has a large intact grid area (not too many grid squares with broken carbon film), a sufficient number of 2D crystals per grid square (10 or more), and an appropriate thickness of the sugar layer. Too thin of a sugar layer results in partial dehydration of the crystals, whereas too thick of

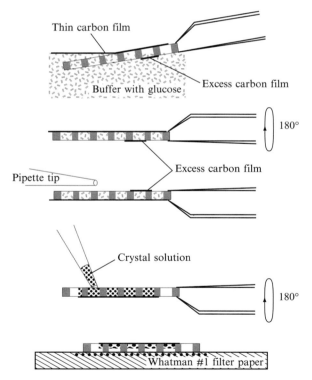

Figure 11.2 Room temperature glucose embedding of 2D crystals for electron crystallography. A detailed protocol is provided in Section 4.1.1.

a layer can interfere with adhesion to the carbon film causing problems with specimen flatness, both reducing the quality of the data that can be collected. The thickness of the sugar layer can be adjusted by changing the sugar concentration and/or the blotting time. These parameters may have to be adjusted regularly even for the same batch of 2D crystals, as the thickness of the sugar layer is affected by the relative humidity and temperature of the environment in which the specimen is prepared. Reproducibility can be improved by preparing sugar-embedded specimens in an environment with controlled humidity and temperature, such as a cold room. The quality of the sugar embedding can be tested by recording electron diffraction patterns, which should show sharp spots. Figure 11.3 shows representative electron diffraction patterns recorded of a fully embedded (Fig. 11.3A) and a partially embedded (Fig. 11.3B) double-layered AQP0 2D crystal. While both crystals had a similar appearance in the electron microscope when viewed in the overfocused diffraction search mode, the reflections in the pattern recorded of the only partially embedded crystal are smeared out and only extend to lower resolution. Because grids are initially not frozen in this preparation technique, it is easy to exchange specimens, which makes it possible to test many specimens until an ideal grid has been prepared. Once a good grid has been identified, it is cooled down in the microscope for actual data collection. In addition, glucose embedding usually produces very little ice contamination as the sample is only cooled down once it is in the microscope, where the vacuum protects the specimen from being contaminated when the temperature is being lowered.

A different sugar-embedding procedure involves plunge-freezing the specimen in liquid nitrogen or liquid ethane before it is transferred into the microscope. In this case, trehalose is typically used as the embedding sugar. A detailed protocol for this preparation method was published by

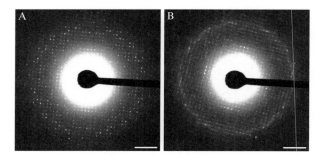

Figure 11.3   Electron diffraction patterns of untilted sugar-embedded, double-layered AQP0 2D crystals. (A) Well-embedded crystals produce sharp reflections in all directions. (B) Partially embedded crystals produce smeared reflections in all directions. Scale bars are $(10 \text{ Å})^{-1}$.

Fujiyoshi and coworkers (Hirai *et al.*, 1999). In contrast to the previously described method, which produces a specimen that is somewhat dried by the vacuum of the microscope, freezing the sample prior to its transfer into the microscope yields a "frozen-hydrated" sample. While milder to the crystals than the extensive drying used for glucose embedding, the thickness of the sugar layer has to be controlled very carefully by closely monitoring the sugar concentration and blotting time, as too thick of a sugar layer will create specimens that are not flat enough for data collection of highly tilted specimens. A second important parameter that has to be considered is the time between blotting and freezing; samples must be frozen as quickly as possible to prevent evaporation of the remaining buffer. Since the specimen is frozen prior to transfer into the microscope, the grids are more prone to contamination and specimen exchange is more difficult, allowing only a limited number of specimens to be tested in a day.

In a further development, Fujiyoshi and coworkers introduced the "carbon sandwich" technique, for which they published a detailed protocol (Gyobu *et al.*, 2004). A slightly modified version is included in Section 4.1.2 and diagramed in Fig. 11.4. In this method, the sugar-embedded crystals are sandwiched between two layers of carbon film, providing additional protection against dehydration. A major advantage of this preparation method is that it significantly reduces beam–induced specimen movement, which proved crucial in the structure determination of AQP4 (Hiroaki *et al.*, 2006)

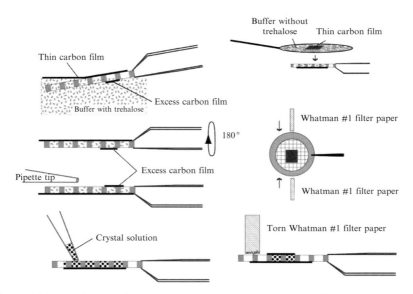

Figure 11.4 Carbon sandwich specimen preparation technique for electron crystallography. A detailed protocol is provided in Section 4.1.2.

and allowed structure determination of AQP0 to 1.9 Å (Gonen *et al.*, 2005). However, the carbon sandwich method is technically challenging and has not been successful for every specimen. It is particularly important that the sugar layer between the two carbon films is not too thick, as it can otherwise cause excessive electron scattering. Therefore, thin 300-mesh molybdenum grids are used to minimize the distance between the two carbon films. In addition, the samples are thoroughly blotted to remove excess sugar solution, but thorough blotting can cause breakage of the carbon film. While some breakage is unavoidable, selecting carbon films that are both strong and somewhat flexible can minimize this problem. For data collection, only specimen areas should be selected in which both carbon films are intact.

4.1.1. Protocol for room temperature glucose embedding

1. Pipette 2 ml of 1% glucose solution into a glass well (see Technical note 1).
2. Cut a small piece (~5 × 5 mm) from a sheet of mica coated with a thin carbon film and float the carbon film off the mica onto the glucose solution (see Technical note 2).
3. Pick up the carbon film with a glow-discharged (30 s at 45 mA) 300-mesh molybdenum grid and turn it over so that the carbon is below the grid.
4. Using a pipette tip turned sideways, gently move over the top surface of the grid that is not covered by the carbon film to remove any carbon debris.
5. Apply 1 μl of 2D crystal solution to the top of the grid that is not covered by the carbon film, agitate with a pipetter and incubate for 1 min (see Technical note 3).
6. Turn the grid over again and place it with the side not covered by the carbon film on a piece of filter paper (Whatman #1) for 10 s (see Technical note 4).
7. Pick the grid up from the filter paper and place it into a cryo-transfer holder at room temperature.
8. Insert the cryo-transfer holder into the microscope and record 5–10 diffraction patterns of the untilted specimen.
9. If the reflections visible in the diffraction patterns are sharp, cool the specimen and, once the temperature of the specimen has stabilized, begin data collection.

Technical notes:

1. The glucose solution should be prepared in the same buffer that was used to grow the crystals. The best glucose concentration has to be determined experimentally.

2. The age, thickness, and hydrophobicity of the carbon film can influence how well the crystals adhere, and several independently prepared pieces of carbon should be tested to identify the optimal carbon film.
3. The best volume of crystal solution to be used has to be determined experimentally. If the crystal density on the grid is too low, crystals can be allowed to sediment by gravity and the sediment can be used for grid preparation. If the crystal density on the grid is too high or the crystals stack, the solution can be diluted by adding the dialysis buffer that was used to grow the crystals.
4. The best blotting time has to be determined experimentally. The relative humidity of the environment in which the specimen is prepared can influence the thickness of the glucose layer. A controlled environment, such as a cold room or EM room, can improve the reproducibility of specimen preparation.

4.1.2. Protocol for trehalose carbon sandwich embedding

1. Pipette 2 ml of 4% trehalose solution into a glass well (see Technical note 1).
2. Pipette 2 ml of trehalose-free buffer (the same as the one in which the crystals are stored) into a second glass well.
3. Cut a small piece ($\sim 5 \times 5$ mm) from a sheet of mica coated with a thin carbon film and float the carbon film off the mica onto the trehalose solution.
4. Pick up the carbon film with a glow-discharged (30 s at 45 mA) 300-mesh molybdenum grid and turn it over so that the carbon is below the grid (see Technical note 2).
5. Using a pipette tip turned sideways, gently move over the top surface of the grid that is not covered by the carbon film to remove any carbon debris.
6. Apply 2 μl of 2D crystal solution to the top of the grid that is not covered by the carbon film, agitate with a pipetman, and incubate for 1 min (see Technical note 3).
7. Cut a second, smaller piece ($\sim 2 \times 2$ mm) from the same carbon-coated sheet of mica and float the carbon film off the mica onto the trehalose-free buffer.
8. Remove excess solution from the grid using a pipetman, but a small amount of liquid should remain on the grid (~ 0.5 μl).
9. Pick up the second piece of carbon film with a platinum loop (~ 8 mm diameter), remove excess buffer from the loop by gently blotting the side of the loop with a torn filter paper, and deposit the second piece of carbon film on the grid by touching the surface of the loop to the surface of the grid (see Technical note 4).

10. Blot off excess buffer by touching the side of the grid with two pieces of filter paper (not torn) from opposite directions (see Technical note 5).
11. Gently touch the top surface of the grid (side with the smaller piece of carbon film) at the edges with small pieces of torn filter paper to completely remove all excess buffer (see Technical note 6).
12. Plunge the grid into liquid nitrogen and place it into a pre-cooled cryo-transfer holder.

Technical notes:

1. The trehalose solution should be prepared in the same buffer that was used to grow the crystals. The best trehalose concentration has to be determined experimentally.
2. The molybdenum grid should be glow-discharged on both sides for at least 30 s at 45 mA.
3. The best volume of crystal solution to be used has to be determined experimentally. If the crystal density on the grid is too low, crystals can be allowed to sediment by gravity and the sediment can be used for grid preparation. If the crystal density on the grid is too high or the crystals stack, the solution can be diluted by adding the dialysis buffer that was used to grow the crystals.
4. Ensure that the orientation of the carbon film is not flipped during the loop transfer so that the surface of the carbon film that contacts the crystals is the same surface that contacted the mica. If the second piece of carbon film does not transfer from the loop to the grid, it can help to blot the loop while it remains in contact with the grid. If this is required, it is essential that motion be minimized as movement of the loop with respect to the grid may cause the carbon film to rupture.
5. The side blotting from opposite directions is performed to ensure that the small piece of carbon film remains in the center of the grid and should be done from both directions at the same time.
6. The blotting with torn pieces of filter paper is best done underneath a focused light source so that the amount of buffer remaining between the two layers of carbon film can be monitored visually. Blotting is complete when the grid squares appear to be gray when held to a light source. Once the specimen appears to be sufficiently dry, it should be plunged into liquid nitrogen immediately. Additional exposure to the environment can result in overdrying of the specimen and poorly embedded crystals.

4.2. Vitrification

In vitrification, the crystals are adsorbed to a carbon–coated EM grid, blotted, and quick-frozen in liquid ethane, which embeds the crystals in a layer of vitrified ice (Adrian *et al.*, 1984; Dubochet *et al.*, 1988).

The preparation of vitrified specimens used to be an art, but has been made much easier and more reproducible with the introduction of commercially available, semiautomated plunge freezing instruments such as FEI's Vitrobot and Gatan's Cryoplunge. Vitrification is a very mild specimen preparation method that generates little stress on the crystals. On the other hand, vitrified 2D crystals tend to be not very flat, causing problems for data collection from highly tilted specimens, which may be the reason why only a single atomic structure has been determined to date using vitrified 2D crystals (Ren *et al.*, 2001).

5. SPECIMEN FLATNESS

A second objective of sample preparation is to ensure that the crystals lie perfectly flat on the carbon support film so that high-quality images and diffraction patterns can be recorded from tilted specimens. The key to obtaining flat specimens is to use atomically flat carbon film, which is not trivial to prepare. To prepare atomically flat carbon films, Fujiyoshi and coworkers have developed a protocol, which is detailed in Section 5.1 (Fujiyoshi, 1998). Rods made of ultrapure carbon (impurity < 2 parts per million) are sharpened and installed in a carbon evaporator without mica. The evaporator is evacuated (vacuum has to be better than 1×10^{-4} Pa), and carbon is slowly evaporated until all sparking has subsided and a carbon cluster has formed around the tip of the carbon rod. Freshly cleaved mica is then placed in the evaporator, and the vacuum is restarted. Once the vacuum is stable, carbon is evaporated very slowly, taking utter care that no sparking occurs, which would otherwise cause carbon clusters to be deposited, ruining the flatness of the film. To this end, the current is increased very slowly while the vacuum is closely monitored. If the vacuum starts to fail, the filament is immediately turned off, and evaporation is only started again once the vacuum has fully recovered.

Cooling can also affect the flatness of the specimen. Carbon, used to prepare the specimen support film, and copper, the most common material used to make EM grids, have different linear thermal expansion coefficients. As a result, when the specimen is cooled down, the carbon film shrinks less than the EM grid, and the carbon film forms wrinkles, a phenomenon that has been dubbed "cryo-crinkling" (Booy and Pawley, 1993). To avoid this issue, EM grids are now used that are composed of molybdenum, a metal that has a similar linear thermal expansion coefficient as carbon. In addition to the material, the smoothness of the edges of the holes in the grid also affects the flatness of the carbon film (Fujiyoshi, 1998). The best grids for preparing 2D crystals are currently the molybdenum grids commercially available from JEOL.

One additional parameter that influences specimen flatness is the adherence of the crystals to the carbon support film. Partial adherence of the crystals to the carbon film, a problem more common to vitrified specimens than to sugar–embedded specimens, will result in bends or folds in the crystals, limiting the achievable resolution. The affinity of 2D crystals for the carbon support film, which must be determined empirically, can depend on the hydrophobicity of the carbon film. In some cases baking or aging the carbon support film, both of which increases its hydrophobicity, have aided in increasing the deposition of crystals to carbon films. For other crystals, addition of small amounts of detergents or other additives has been helpful in improving the adherence of crystals to carbon films. Folds and bends in crystals can also result from stacking of multiple crystals, so care must be taken that only crystals that are well separated from other crystals are used for data collection.

5.1. Protocol for spark-free preparation of thin carbon films

1. File one carbon rod to a pencil point at about a 45° angle and file the other carbon rod flat (see Technical note 1).
2. Mount the two carbon rods into the holders in a carbon evaporator; the rod with the pointed end should be mounted in the spring-loaded holder.
3. Secure the vacuum bell jar on the apparatus and start the vacuum pump (see Technical note 2).
4. When the vacuum has stabilized at a pressure of less than 1×10^{-4} Pa, turn on the filament (see Technical note 3).
5. Very slowly increase the current until the carbon begins to evaporate (see Technical note 4).
6. After the carbon has begun to evaporate, carefully watch for sparks (see Technical note 5).
7. Once the sparking has subsided, turn off the filament and vent the bell jar.
8. Place a piece of freshly cleaved mica inside the apparatus (see Technical note 6).
9. Replace the bell jar and restart the vacuum (see Technical note 7).
10. When the vacuum has stabilized at a pressure of less than 1×10^{-4} Pa, turn on the filament.
11. Very slowly raise the current until the carbon begins to evaporate.
12. Carefully watch for sparks during evaporation and for changes in pressure (see Technical note 8).
13. If the pressure rises, turn off the filament and wait for the vacuum to completely recover.
14. Once the carbon evaporation has been completed, store the carbon films in a desiccator until use.

Technical notes:

1. Pencil point-shaped carbon rods tend to work best as they break less during heating and provide better control over the speed of carbon evaporation.
2. Clean the inside of the bell jar regularly to reduce contamination.
3. If the pressure rises during pre-evaporation, reduce the current to zero, turn off the filament, and wait until the pressure stabilizes again at a vacuum of 1×10^{-4} Pa or better before continuing. It is essential that the carbon evaporates without an increase in vacuum pressure.
4. Carbon evaporation can be monitored by placing a frosted glass cover slip into the evaporator. One half of the frosted portion of the cover slip should be coated with a thin layer of vacuum grease to shield it from the evaporated carbon and allow comparison to the exposed half.
5. Sparks can be very small and difficult to see. The best way to see sparks is by looking down the axis of the carbon rods during evaporation while not looking directly at the very brightly lit carbon rods.
6. When mica is cleaved, it breaks along crystal planes and the freshly cleaved surfaces are free of contamination, creating atomically flat surfaces.
7. The initial carbon evaporation step is performed to remove any gas from the carbon rod and to deposit a carbon cluster at the interface between the two carbon rods. This process often produces sparks and thus poor quality carbon film, so it is done in the absence of mica. Once these tasks have been completed, it is possible to produce atomically flat carbon.
8. If sparks occur during evaporation, discard the mica and wait for the carbon rods to cool down prior to exchanging them as they become very hot during evaporation.

6. DATA COLLECTION

Collecting high-resolution data in electron crystallography, especially images, is very challenging and time-consuming, and it is thus important to use a high-performance electron microscope to make the best use of the time invested in data collection. The two most important parameters for high-resolution data collection are the coherence of the electron beam and the stability of the specimen stage. While electron microscopes equipped with LaB_6 or tungsten filaments have been used in the past, only instruments with a field emission gun (FEG), which generates electron beams with far superior intensity and coherence, should now be used for data collection. The acceleration voltage at which the microscope is operated is less essential for electron crystallography than for other applications, because 2D crystals tend to be quite thin. Nevertheless, acceleration voltages of 200 or 300 kV are typically used for electron crystallographic data collection. As discussed above, top-entry or uncoupled specimen stages are preferred over side-entry holders due to their higher stability. Additional components such as energy filters, phase plates, and multiple sample changers may be useful and beneficial in some cases, but are not critical for data collection.

To minimize the effects of beam damage, the specimen is cooled to cryogenic temperatures. Initially, specimens were cooled to liquid nitrogen temperature, but a careful analysis using purple membranes as the test specimen suggested that lowering the temperature further to liquid helium temperature provides additional cryo-protection (Fujiyoshi, 1998). A recent study performed on catalase 2D crystals, however, found no improvement at liquid helium temperature (Bammes et al., 2010). The benefits of cooling the specimen to liquid helium temperature thus remain controversial, but it is noteworthy that almost all atomic models of membrane proteins produced by electron crystallography to date are based on data collected at liquid helium temperature.

Photographic film is the traditional medium used to record images of 2D crystals. Technological advances have improved the quality of electronic detectors such as charge-coupled device (CCD) cameras to a point that they are now commonly used in the collection of images of single particle specimens. For 2D crystals, however, very large images at very fine sampling have to be recorded to allow extraction of near-atomic resolution information. Collecting images on film with subsequent digitization of the film thus remains the preferred way to record images of 2D crystals. By contrast, a recording device with a high dynamic range is essential to accurately measure intensities in electron diffraction patterns, because these vary over several orders of magnitude. As CCD cameras have a much larger dynamic range and better linearity, they are preferred over film to record diffraction patterns. Imaging plates, which also have a high dynamic range, are less effective for recording diffraction patterns, presumably because imaging plates cannot be normalized and thus have a highly variable background (Li et al., 2010). In addition to CCD cameras, other technologies are being developed to provide electronic detectors for EM (Faruqi and Henderson, 2007). The goal of these technologies is single electron detection, which would improve the quality of diffraction patterns recorded and may also make it possible to collect images of 2D crystals with such recording devices.

By removing most of the background created by inelastically scattered electrons, energy filters improve the signal-to-noise ratio of EM images, in particular the amplitude contrast (Yonekura et al., 2006). Energy filters are now often used for imaging in electron tomography of thick specimens and in single-particle EM of vitrified specimens. In the case of 2D crystals, the Fourier transforms, rather than the images, are analyzed. In Fourier transforms the information is focused into discrete spots, whereas the background noise is randomly distributed over the entire transform. Since the diffraction spots have a much higher intensity than the background, the reflections are easy to detect, and since only the values of the reflections are measured, the background created by inelastically scattered electrons is more or less eliminated as part of the image analysis. For electron diffraction, energy filters do greatly reduce the noise in the diffraction patterns caused by the inelastically scattered electrons. Studies on diffraction patterns of bR showed, however, that the reduction in background noise only improved the $R_{Friedel}$ values of the low resolution but not those of the high-resolution reflections (Yonekura *et al.*, 2002). Additionally, there are practical challenges to the use of energy filters for electron crystallography. The inclusion of a post-column energy filter prevents the use of film or large-area CCD cameras, greatly reducing the potential resolution of data that can be obtained and thus limiting their value in electron crystallographic data collection.

6.1. Microscope alignment

Each data collection session should begin by aligning the electron-optical system and correcting the astigmatism of the lenses as described in Sections 6.1.1 through 6.1.5. A properly aligned microscope will produce a more coherent beam, increasing the likelihood that the high-resolution structure factors can be accurately recovered (examples of diffraction patterns collected with an imperfectly aligned microscope are shown in Fig. 11.5). When collecting images, it is important that an objective aperture with a sufficiently large diameter is selected so that it will not limit the resolution of the recorded data, while also not being too large such that the image contrast is reduced.

The procedures described in Sections 6.1.1 through 6.1.5 are specific for an FEI Polara F30 transmission electron microscope and have to be adjusted for other instruments. These protocols assume that the general alignment of the electron microscope (gun shift and tilt, pivot points, rotation center), which does not necessarily have to be repeated prior to each data collection session, has already been performed.

6.1.1. Protocol for alignment of the condenser aperture

These steps are necessary for recording both images and diffraction patterns and should be performed prior to each data collection session.

1. In the low-dose exposure mode, remove the objective and SA apertures and chose the appropriate condenser lens aperture.
2. Set the spot size to the size that will be used for data collection.
3. Focus the beam to its minimum size by changing the beam intensity using the intensity knob.

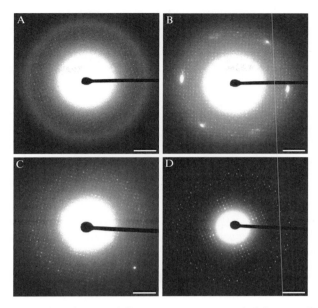

Figure 11.5 Imperfect electron diffraction patterns recorded from untilted, double-layered AQP0 2D crystals. (A) Diffraction pattern of a specimen that has been contaminated with a thin layer of ice. The ice layer produces a blurry ring at a resolution of ~3.5 Å. (B) Diffraction pattern of a specimen that contains crystallized glucose. Crystalline glucose produces a hexagonal diffraction pattern with strong reflections at a resolution of ~4 Å. (C) Diffraction pattern of a poorly focused specimen. Improper eucentric focusing can yield patterns, in which the Friedel mates differ significantly in intensity. (D) Diffraction pattern of a specimen recorded with a strongly astigmatic beam. Failure to correct diffraction astigmatism results in elongated, smeared reflections. Scale bars are $(10 \text{ Å})^{-1}$.

4. Move the beam to the center of the phosphor screen using alignment beam shift.
5. Slowly spread the beam until it fills the larger circle of the phosphor screen using the intensity knob.
6. If the beam is not centered on the circle, adjust the condenser aperture until it is centered.
7. Repeat steps 3–6 until the beam remains centered as the beam is spread.
8. Focus the beam to its minimum size by changing the beam intensity using the intensity knob.
9. Move the beam to the center of the phosphor screen using alignment beam shift.
10. Adjust the condenser astigmatism using the stigmator function until the beam remains spherical as the beam goes through the crossover point.

6.1.2. Protocol for centering the objective aperture

These steps are only necessary for recording images and should be performed prior to each data collection session.

1. Insert and align the appropriate condenser aperture (see Section 6.1.1).
2. Insert the specimen.
3. Insert the appropriate objective aperture (see Technical note 1).
4. Switch to diffraction mode and focus the beam to its minimum size using the focus knob.
5. Use the intensity knob to obtain a clear image of the objective aperture.
6. Move the objective aperture so it is centered around the beam.

   Technical notes:

1. The proper choice for the objective aperture should yield a sufficiently large diameter so that it will not limit the resolution of the recorded data, while also not being too large such that the image contrast is reduced.

6.1.3. Protocol for correction of diffraction astigmatism

These steps are only necessary for recording diffraction patterns.

1. Diffraction astigmatism must be corrected each time the z-height is adjusted by using the stigmator feature.
2. Move the beam out from behind the beam stop using diffraction shift.
3. Focus the beam to a sharp spot by adjusting the focus using the focus knob with a small focus step size.
4. Adjust the diffraction astigmatism by using the stigmator feature until the beam is spherical while the focus is slowly increased through the cross-over point.
5. Move the beam back behind the beam stop using diffraction shift.

6.1.4. Procedure for aligning the low-dose modes for imaging

1. Turn on low dose.
2. Align the electron–optical system as described in Sections 6.1.2 and 6.1.3.
3. Initially the microscope should be in search mode.
4. Switch to focus mode.
5. Switch to exposure mode.
6. Expose the specimen to the beam for ~ 30 s.
7. Switch to search mode and identify the exposed area, which should appear as a circular area that is lighter than the rest of the background.
8. Using the image shift feature, move the previously exposed area to the center of the TV-rate detector.
9. Move the stage to a new position and repeat steps 3–7 until no additional adjustments are necessary.

6.1.5. Procedure for aligning the low-dose modes for electron diffraction

1. Turn on low dose.
2. Align the electron–optical system as described in Sections 6.1.1 and 6.1.3.
3. Initially the microscope should be in search mode.
4. Switch to focus mode.
5. Switch to exposure mode.
6. Spread the beam using the focus knob and insert the SA aperture.
7. Adjust the aperture so that it is positioned in the center of the beam.
8. Focus the beam using the focus knob and place it behind the beam stop using the diffraction shift.
9. Switch to search mode.
10. Using the diffraction shift controls, move the image of the SA aperture to the center of the TV-rate detector.
11. Cycle through search mode to focus mode to exposure mode and back to search mode to establish a consistent hysteresis cycle and ensure that all modes are properly aligned.

6.2. Collecting images

Section 6.2.1 details the setup that is used in our laboratory to collect images with an FEI Polara F30 transmission electron microscope operated at 300 kV (the same protocol can also be used for data collection at other acceleration voltages and at liquid helium or liquid nitrogen temperatures). The parameters and the protocol are specific to the Polara instrument and must be adapted and optimized for different electron microscopes.

Images are recorded using low-dose procedures, which entails the use of three different instrument modes: search, focus, and exposure mode. In the search mode, the instrument is set to overfocused diffraction mode, using a camera length of ~ 3000 mm. This mode, which combines high contrast with low beam exposure of the specimen, is used to identify 2D crystals suitable for data collection, which is done with the help of a TV-rate camera. In the focus mode, the instrument is set to a high magnification, typically $\sim 200,000\times$, and the beam is positioned ~ 2 μm beside the specimen area to be imaged. The position used for focusing should be on the same height as the position used for imaging, and the offset of the focus position should therefore be set in the direction of the tilt axis. A second position for focusing can be set up on the opposite side of the specimen area, which can be used in case the first focusing position is blocked by an object or a grid bar. In the focus position, the astigmatism of the objective lens is corrected (if necessary) and the focus is set to a value between -400 and -1200 nm.

When imaging untilted crystals, the defocus should be varied from image to image within this range to change the positions of the zero transitions of the CTF, allowing the entirety of Fourier space to be sampled. Once the stage has settled and the specimen is no longer moving when viewed on the TV-rate camera, the plate camera is inserted and the microscope is switched to imaging mode, in which an image of the 2D crystal is recorded either by using an exposure with a flat beam for 1 s or by using spot scanning. The microscope is then returned to search mode to find the next 2D crystal to be imaged.

Once image collection is complete, the films are chemically developed, fixed and dried. The final step before data analysis is to examine the films with an optical diffractometer to identify and discard images that reveal specimen movement (loss of information in a certain direction in the optical Fourier transform). The optical Fourier transform also shows the diffraction spots, which reveal the quality of the imaged crystal and are used to identify the best-diffracting area of the crystal for digitization. The best images show isotropic reflections and have clear and symmetric Thon rings (see Fig. 11.1A).

6.2.1. Procedure for collecting images

The procedures are described for an FEI Polara F30 transmission electron microscope and have to be adjusted for other instruments.

Setup

- Apertures:
  - Condenser aperture: 100 μm (see Technical note 1)
  - Objective aperture: 70 μm (see Technical note 1)
  - SA aperture: removed
- Low dose
  - Search mode
    - Spot size 11 (smallest)
    - Diffraction mode
    - Strongly overfocused beam such that crystals can be identified on the TV-rate detector
    - Magnification of $\sim$4000$\times$
    - Camera length of $\sim$3000 mm
    - Beam intensity should be adjusted using the intensity knob such that crystals can be identified on the TV-rate detector while using the lowest dose possible
  - Focus mode
    - Spot size 2 (same as exposure mode)
    - Magnification of $\sim$200,000$\times$ (see Technical note 2)
    - Offset of 2 μm from the center of the beam in exposure mode in the direction parallel to the tilt axis

- Beam intensity should be adjusted such that the grains of the carbon film are clearly visible when viewed on the TV-rate detector near to focus
  ○ Exposure mode
    - Spot size 2 (see Technical note 1)
    - Magnification of ~50,000× (see Technical note 3)
    - Beam intensity is set by using the intensity knob to yield a dose of between 10 and 20 electrons/$\mathring{A}^2$ with a 1-s exposure over the entire film area (see Technical note 1)
- Recording device
  ○ Images are recorded on Kodak SO-163 film and developed for 12 min in full-strength Kodak D-19 developer at 20 °C
  ○ A TV-rate detector is used to identify crystals in search mode and to focus and correct the astigmatism in focus mode

Data collection

1. In search mode, locate an area of the grid with intact carbon and an appropriate density of crystals.
2. Use the wobbler tool to set the z-height of the stage.
3. Select the tilt angle for data collection (see Technical note 3).
4. Identify and center the beam on a crystal.
5. Switch to focus mode and set the defocus to a value between -400 and -1200 nm using the focus knob (see Technical note 5).
6. While in focus mode, correct any astigmatism of the objective lens (see Technical note 5).
7. Switch to exposure mode, insert the plate camera and wait 10 s (see Technical note 6).
8. Record the image using a 1-s flat beam exposure or with spot scanning (see Technical note 7).
9. Switch to search mode, identify another crystal for imaging and repeat.

Technical notes:

1. The condenser aperture and spot sizes for focus and exposure modes were chosen to yield an electron dose of between 10 and 20 electrons/$\mathring{A}^2$ in imaging mode. Other combinations can also be used to obtain the same electron dose. Smaller condenser aperture and/or spot size give a better spatial coherence of the illumination beam. The rule of thumb is to use the smallest possible condenser aperture and spot size combination that still yields a 10–20 electrons/$\mathring{A}^2$ dose in the imaging mode.
2. Once the tilt angle has been set, wait at least 15 min prior to beginning data collection and do not change the tilt angle until data collection is complete to prevent excessive drift of the stage. Images are generally recorded at tilt angles of 0°, 20°, 45°, 60°, and occasionally 70°.

3. Focusing should be conducted at the highest possible magnification of the SA mode to allow visualization of slight specimen drifts prior to image acquisition without changing the electron-optical system (which lenses are on or off) and potentially inducing hysteresis.

4. The magnification used for image acquisition in exposure mode is determined by the target resolution and finest possible step size of the scanner used for image digitization. Images should be recorded such that the pixel size at the specimen level is at least four times the target resolution and thus two times the Nyquist limit. This setup will ensure that the structure factors at the target resolution can be measured without ambiguity.

5. Vary the defocus from image to image over this range to sample the entire Fourier space. Images of highly tilted specimens taken without dynamic focusing should be recorded using a defocus of at least −1000 nm to ensure that the entire image area is underfocused.

6. Before collecting an image, wait in focus mode until the stage has stopped drifting. While this may take seconds to minutes depending on the stability of the stage, waiting is essential to record images with isotropic resolution. Steps 7 and 8 are generally performed using the exposure button, which switches the microscope to exposure mode, inserts the plate camera, waits 10 s and then records a 1-s exposure. If the exposure button is not used, the beam must be blanked until the image is recorded.

7. During image acquisition, the door should remain closed and no noise should be made.

8. If collecting tilted images, spot scan and dynamic focusing can be enabled to minimize beam-induced specimen movements and simplify subsequent image processing.

6.3. Recording electron diffraction patterns

As for imaging, the setup and protocol for collecting diffraction patterns detailed in Section 6.3.1 is again specific for the use of an FEI Polara F30 instrument and must be adapted and optimized for different electron microscopes.

Diffraction patterns are recorded using low-dose settings, but only the search and exposure modes are used during data collection. Focusing is done for each new grid square in exposure mode before starting with the actual data collection. First the z-height of the stage is set in search mode using the "wobbler" tool and then the image is focused in the exposure mode using the eucentric focus feature. In some microscopes, the z-height of the stage is fixed, preventing the use of eucentric focusing. In this case, focusing is also done in exposure mode at a position adjacent to the crystal using the method described above for recording images. Crystals are found in search mode, which is set up the same way as for collecting images. Once a crystal has been identified, an SA aperture is inserted. The size of the SA aperture is chosen so that only electrons that have passed through the crystal reach the detector. The use of an SA aperture is not absolutely necessary, but masking the beam that has not passed through the crystal increases the signal-to-noise ratio of the reflections and enables a more accurate measurement of the reflection intensities. In addition, the SA aperture can also reduce artifacts in the background intensity produced by the FEG. Diffraction patterns are recorded in exposure mode with the microscope operated in diffraction mode, using a low-intensity beam and long exposure times. Diffraction requires that the incident beam be absolutely parallel in order to produce focused diffraction spots. Furthermore, the beam should be sufficiently small to avoid damaging nearby crystals during recording. Such requirements can only be met by using a rather small spot size, resulting in a very weak intensity of the incident beam and the need for a prolonged exposure time. Because electron diffraction is invariant to specimen movement (as long as the crystal that is being exposed does not drift out of the SA aperture), long exposure times can be used for recording diffraction patterns without risking a reduction in quality. The exposure length is chosen to be as long as possible without saturating the CCD camera so that all reflections can be scaled together. Newer CCD cameras, which have a very high dynamic range, are treated differently as it would require an extremely long exposure to saturate them. For these cameras, a dosage of between 10 and 20 electrons/$\mathring{A}^2$ should be sufficient to measure every reflection. CCD cameras provide immediate feedback on the quality of the specimen. A poor specimen can thus be identified early in a session and exchanged for a better one, greatly speeding up the data collection as specimen preparations are not always consistent.

6.3.1. Procedure for recording diffraction patterns

The procedures are described for an FEI Polara F30 transmission electron microscope operated at 300 kV and have to be adjusted for other instruments.

Setup

- Apertures:
  - Condenser aperture: 100 μm (see Technical note 1)
  - Objective aperture is removed
  - SA aperture: 40 μm (2–5 μm at the specimen level)
- Low dose
  - Search mode
    - Spot size 11 (smallest)
    - Diffraction mode

- Strongly overfocused beam such that crystals can be identified on the TV-rate detector
- Magnification of ~4000×
- Camera length of ~3000 mm
- Beam intensity should be adjusted using the intensity knob such that crystals can be identified on the TV-rate detector while using the lowest dose possible
  - ○ Focus mode
    - Not used, but should be set up identically to the exposure mode
  - ○ Exposure mode
    - Spot size 11 (smallest) (see Technical note 1)
    - Diffraction mode
    - The beam should be focused to its minimum size in diffraction mode using the focus knob
    - Magnification of ~40,000×
    - Camera length of ~3000 mm (see Technical note 2)
    - The beam intensity should be spread using the intensity knob such that it illuminates an area with twice the diameter of the SA aperture in imaging mode
- Recording device
  - ○ Diffraction patterns are recorded on a CCD camera. Fresh dark and gain references should be prepared at the start of each data collection session.
  - ○ A TV-rate detector is used to identify crystals in search mode.
- Beam stop
  - ○ Insert the beam stop
  - ○ Take a series of images on the CCD camera to aid in placing the beam stop in the center of the camera

Data collection

1. In search mode, identify an area of the grid with intact carbon and a sufficient number of crystals.
2. Use the wobbler tool to set the z-height of the stage (see Technical note 3).
3. Set the tilt angle for data collection.
4. Switch to focus mode.
5. Switch to exposure mode.
6. While remaining in exposure mode, switch from diffraction to imaging
7. Use the eucentric focus feature to focus the image for the focus and exposure modes (see Technical note 2).
8. Center the beam on the beam stop by using alignment beam shift.
9. While remaining in exposure mode, switch back from imaging to diffraction.
10. Spread the beam using the focus knob, insert and center the SA aperture.

11. Remove the SA aperture and, if necessary, correct the diffraction astigmatism.
12. Focus the beam to its minimum size using the focus knob.
13. Center the beam behind the beam stop by using the diffraction shift.
14. Return to search mode.
15. Cycle through search mode to focus mode to exposure mode and back to search mode to ensure that all modes are properly aligned (see Technical note 4).
16. In search mode, identify a crystal and insert the SA aperture to mask out the background.
17. Blank the beam and switch to focus mode (see Technical note 5).
18. Switch to exposure mode, lower the screen, and ensure that the beam is behind the beam stop by quickly unblanking and reblanking the beam (see Technical note 6).
19. Raise the screen, remove the beam blank, and record a diffraction pattern using a 30-s exposure (see Technical note 1).
20. Return to search mode and identify another crystal repeat from step 16.
21. If moving to a new grid square, start from step 2.

Technical notes:

1. The condenser aperture size and the spot size for the focus and exposure modes were chosen to yield a parallel beam at a dosage that does not oversaturate the CCD camera and does not expose nearby crystals. For crystals that are larger or smaller, the condenser aperture and SA aperture sizes can be increased or decreased to ensure that the entire crystal is illuminated with a parallel beam while minimizing exposure of adjacent crystals. For CCD cameras with a dynamic range large enough that it is not saturated during prolonged exposures, a dosage of 30–40 electrons/$Å^2$ should be used.
2. The best camera length to be used depends on the quality of the crystals and can be determined empirically. Initially, a short camera length (~ 1500 mm) is used to record several diffraction patterns to determine the highest resolution at which reflections can be seen. Diffraction patterns are then recorded with increasingly longer camera lengths. Ideally, a camera length is chosen for data collection that produces diffraction patterns in which the highest resolution reflections are close to the edge of the recorded area. In this way, the spacing between the diffraction spots is maximized, simplifying subsequent processing of the diffraction patterns.
3. The z-height and eucentric focus should be set for each individual grid square.
4. Cycling through the three modes must be done each time the focus is changed to maintain a consistent hysteresis cycle and to ensure that they are aligned.

5. The focus mode is not used, but needs to be selected to preserve the hysteresis cycle.
6. Prolonged, direct exposure to the focused, intense beam can permanently damage the CCD array.

6.4. Collecting data of tilted specimens

To determine the 3D structure of a protein by electron crystallography, data have to be collected from tilted specimens. The procedure is the same as for collecting images and diffraction patterns of untilted specimens, except that the specimen is being tilted to a particular angle. The yield of good images and diffraction patterns usually drops with increasingly higher tilt angles, mostly because beam-induced specimen movement has a much more severe effect on images of tilted specimens and the increased influence of unevenness in the carbon support film.

When collecting diffraction patterns of tilted specimens, diffraction patterns should first be recorded of the untilted specimen to assess the quality of the preparation. Only if the diffraction patterns are of high quality, it is worth tilting the specimen and starting to collect data. A lack of specimen flatness results in diffraction patterns that show diffraction to high resolution in the direction parallel to the tilt axis and diffraction to lower resolution in the direction perpendicular to the tilt axis (Fig. 11.6). Larger crystals simplify the collection of diffraction patterns of highly tilted specimens, as the size of the crystals in projection is much smaller at high tilt angles. For the time-consuming task of collecting images of tilted specimens, the conditions should be as optimal as possible to ensure the highest chances of obtaining high-resolution images. Thus, even when collecting

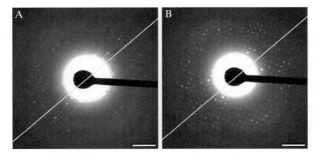

Figure 11.6 Electron diffraction patterns of highly tilted (70°), double-layered AQP0 2D crystals. (A) Diffraction pattern recorded of a crystal that is not flat. The reflections are blurred in the direction perpendicular to the tilt axis (white line) at low resolution and nonexistent at high resolution. (B) Diffraction pattern recorded of a flat crystal. The reflections are sharp in all directions. Scale bars are $(10 \text{ Å})^{-1}$.

images of tilted specimens, it is generally a good idea to first use electron diffraction to assess the quality of the preparation, since only well-prepared crystals have the potential to yield good images. For imaging tilted specimens, spot scanning not only increases the yield of good images but also offers the additional advantage of dynamic focusing. Since the beam is scanned in lines over the specimen area, the lines can be aligned with the tilt axis and the defocus value can be adjusted for each line of spots, thus compensating for the varying distance of the specimen from the electron source. As the defocus gradient within rows of spots is small enough to be neglected, images of tilted specimens recorded with spot scanning and dynamic focusing can be processed in the same way as images of untilted specimens, assuming a single defocus value for the entire image.

Each image or diffraction pattern of a tilted specimen represents a central section through Fourier space and cuts the reciprocal lattice lines, the Fourier transform of a 2D crystal, at specific positions. As data are collected from 2D crystals with different orientations relative to the tilt axis, a sufficiently high number of images and diffraction patterns collected at a single tilt angle would theoretically sample all positions on all lattice lines up to the chosen tilt angle. Due to practical limitations of the electron microscope and specimen holder designs, the highest tilt angle at which data can be collected with current instrumentation is 60–70°. The volume of Fourier space above this angle that is not sampled is known as the missing cone. The missing cone results in anisotropic resolution of the density map, with the resolution in the direction perpendicular to the specimen plane being lower than the in-plane resolution. Data collected at the highest tilt angle are the most valuable as they reduce the missing cone and contain the most information of the specimen in the z-direction. Ideally, all data would thus be collected of specimens tilted to 60° or even 70°. In practice, however, these data are the most difficult to collect, and data collected of specimens tilted to lower angles are necessary for accurate merging of the data set. Typically, data are collected at tilt angles of 0°, 20°, 45°, and 60°, and in rare cases also at 70°. As a rule of thumb, approximately half of the data should ideally be collected at the highest possible tilt angle. The rest of the data should be divided between 20° and 45°, with an emphasis on data collected at a 45° angle. Merging of diffraction patterns does not require as many diffraction patterns collected at low tilt angles, because diffraction is translation-invariant and no relative shift (phase origin) between the patterns has to be determined, as is the case for the merging of images. Ideally two-thirds to three-quarters of the data set should be collected at high tilt angles.

The number of images and diffraction patterns needed to fully sample Fourier space for a given target resolution can be roughly estimated using Eq. (11.1) (adapted from Radermacher, 1991):

$$N = 2\pi(D/dn)\sin\theta \qquad (11.1)$$

where N is the number of images, D is the unit cell size, d is the target resolution, n is the number of asymmetric units in the unit cell, and θ is the tilt angle.

The number resulting from this equation assumes that the images or diffraction patterns are evenly spaced in Fourier space, which is not the case for a real data set. Furthermore, images and diffraction patterns are often far from perfect and only contain partial information to a given resolution. In reality, the number of images and diffraction patterns to be collected at a certain tilt angle should thus be multiplied by a factor of 2 or even 3 to account for the deficiencies of the collected data.

ACKNOWLEDGMENTS

Electron crystallographic work in the Walz laboratory is supported by NIH grants P01 GM062580 (to S. C. Harrison) and R01 EY015107 and R01 GM082927 (to T. W.). Work in the Cheng laboratory is supported by NIH grants R01 GM082893 (to Y. C.) and P50 GM082250 (to A. Frankel). ADS is supported by a fellowship from the Swiss National Science Foundation (PA00P3_126253). T. W. is an investigator in the Howard Hughes Medical Institute.

REFERENCES

Adrian, M., Dubochet, J., Lepault, J., and McDowall, A. W. (1984). Cryo-electron microscopy of viruses. *Nature* **308,** 32–36.

Baldwin, J., and Henderson, R. (1984). Measurement and evaluation of electron diffraction patterns from two-dimensional crystals. *Ultramicroscopy* **14,** 319–335.

Bammes, B. E., Jakana, J., Schmid, M. F., and Chiu, W. (2010). Radiation damage effects at four specimen temperatures from 4 to 100 K. *J. Struct. Biol.* **169,** 331–341.

Berriman, J., and Unwin, N. (1994). Analysis of transient structures by cryo-microscopy combined with rapid mixing of spray droplets. *Ultramicroscopy* **56,** 241–252.

Booy, F. P., and Pawley, J. B. (1993). Cryo-crinkling: What happens to carbon films on copper grids at low temperature. *Ultramicroscopy* **48,** 273–280.

Downing, K. H. (1991). Spot-scan imaging in transmission electron microscopy. *Science* **251,** 53–59.

Dubochet, J., Adrian, M., Chang, J. J., Homo, J. C., Lepault, J., McDowall, A. W., and Schultz, P. (1988). Cryo-electron microscopy of vitrified specimens. *Q. Rev. Biophys.* **21,** 129–228.

Faruqi, A. R., and Henderson, R. (2007). Electronic detectors for electron microscopy. *Curr. Opin. Struct. Biol.* **17,** 549–555.

Fujiyoshi, Y. (1998). The structural study of membrane proteins by electron crystallography. *Adv. Biophys.* **35,** 25–80.

Fujiyoshi, Y., and Unwin, N. (2008). Electron crystallography of proteins in membranes. *Curr. Opin. Struct. Biol.* **18,** 587–592.

Glaeser, R. M. (2008). Retrospective: Radiation damage and its associated "information limitations" *J. Struct. Biol.* **163,** 271–276.

Gonen, T., Sliz, P., Kistler, J., Cheng, Y., and Walz, T. (2004). Aquaporin-0 membrane junctions reveal the structure of a closed water pore. *Nature* **429,** 193–197.

Gonen, T., Cheng, Y., Sliz, P., Hiroaki, Y., Fujiyoshi, Y., Harrison, S. C., and Walz, T. (2005). Lipid–protein interactions in double-layered two-dimensional AQP0 crystals. *Nature* **438**, 633–638.

Gyobu, N., Tani, K., Hiroaki, Y., Kamegawa, A., Mitsuoka, K., and Fujiyoshi, Y. (2004). Improved specimen preparation for cryo-electron microscopy using a symmetric carbon sandwich technique. *J. Struct. Biol.* **146**, 325–333.

Havelka, W. A., Henderson, R., and Oesterhelt, D. (1995). Three-dimensional structure of halorhodopsin at 7 Å resolution. *J. Mol. Biol.* **247**, 726–738.

Hayward, S. B., and Glaeser, R. M. (1980). High resolution cold stage for the JEOL 100B and 100C electron microscopes. *Ultramicroscopy* **5**, 3–8.

Henderson, R. (1995). The potential and limitations of neutrons, electrons and X-rays for atomic resolution microscopy of unstained biological molecules. *Q. Rev. Biophys.* **28**, 171–193.

Henderson, R., and Glaeser, R. M. (1985). Quantitative analysis of image contrast. *Ultramicroscopy* **16**, 139–150.

Henderson, R., and Unwin, P. N. (1975). Three-dimensional model of purple membrane obtained by electron microscopy. *Nature* **257**, 28–32.

Henderson, R., Baldwin, J. M., Ceska, T. A., Zemlin, F., Beckmann, E., and Downing, K. H. (1990). Model for the structure of bacteriorhodopsin based on high-resolution electron cryo-microscopy. *J. Mol. Biol.* **213**, 899–929.

Henderson, R., Raeburn, C., and Vigers, G. (1991). A side-entry cold holder for cryo-electron microscopy. *Ultramicroscopy* **35**, 45–53.

Hirai, T., Murata, K., Mitsuoka, K., Kimura, Y., and Fujiyoshi, Y. (1999). Trehalose embedding technique for high-resolution electron crystallography: Application to structural study on bacteriorhodopsin. *J. Electron Microsc.* **48**, 653–658.

Hiroaki, Y., Tani, K., Kamegawa, A., Gyobu, N., Nishikawa, K., Suzuki, H., Walz, T., Sasaki, S., Mitsuoka, K., Kimura, K., Mizoguchi, A., and Fujiyoshi, Y. (2006). Implications of the aquaporin-4 structure on array formation and cell adhesion. *J. Mol. Biol.* **355**, 628–639.

Hite, R. K., Raunser, S., and Walz, T. (2007). Revival of electron crystallography. *Curr. Opin. Struct. Biol.* **17**, 389–395.

Holm, P. J., Bhakat, P., Jegerschöld, C., Gyobu, N., Mitsuoka, K., Fujiyoshi, Y., Morgenstern, R., and Hebert, H. (2006). Structural basis for detoxification and oxidative stress protection in membranes. *J. Mol. Biol.* **360**, 934–945.

Jegerschöld, C., Pawelzik, S. C., Purhonen, P., Bhakat, P., Gheorghe, K. R., Gyobu, N., Mitsuoka, K., Morgenstern, R., Jakobsson, P. J., and Hebert, H. (2008). Structural basis for induced formation of the inflammatory mediator prostaglandin E2. *Proc. Natl. Acad. Sci. USA* **105**, 11110–11115.

Kimura, Y., Vassylyev, D. G., Miyazawa, A., Kidera, A., Matsushima, M., Mitsuoka, K., Murata, K., Hirai, T., and Fujiyoshi, Y. (1997). Surface of bacteriorhodopsin revealed by high-resolution electron crystallography. *Nature* **389**, 206–211.

Kühlbrandt, W., Wang, D. N., and Fujiyoshi, Y. (1994). Atomic model of plant light-harvesting complex by electron crystallography. *Nature* **367**, 614–621.

Li, Z., Hite, R. K., Cheng, Y., and Walz, T. (2010). Evaluation of imaging plates as recording medium for images of negatively stained single particles and electron diffraction patterns of two-dimensional crystals. *J. Electron Microsc.* **59**, 53–63.

Murata, K., Mitsuoka, K., Hirai, T., Walz, T., Agre, P., Heymann, J. B., Engel, A., and Fujiyoshi, Y. (2000). Structural determinants of water permeation through aquaporin-1. *Nature* **407**, 599–605.

Nogales, E., Wolf, S. G., and Downing, K. H. (1998). Structure of the $\alpha\beta$ tubulin dimer by electron crystallography. *Nature* **391**, 199–203.

Philippsen, A., Engel, H. A., and Engel, A. (2007). The contrast-imaging function for tilted specimens. *Ultramicroscopy* **107,** 202–212.

Radermacher, M. (1991). Three-dimensional reconstruction of single-particles in electron microscopy. *In* "Image Analysis in Biology," (D. P. Hader, ed.). CRC Press, Boca Raton.

Raunser, S., and Walz, T. (2009). Electron crystallography as a technique to study the structure of membrane proteins in a lipidic environment. *Annu. Rev. Biophys.* **38,** 89–105.

Ren, G., Reddy, V. S., Cheng, A., Melnyk, P., and Mitra, A. K. (2001). Visualization of a water-selective pore by electron crystallography in vitreous ice. *Proc. Natl. Acad. Sci. USA* **98,** 1398–1403.

Schenk, A. D., Hite, R. K., Engel, A., Fujiyoshi, Y., and Walz, T. (2010). Electron crystallography and aquaporins. *Methods Enzymol.* **483,** 91–119.

Subramaniam, S., Lindahl, M., Bullough, P., Faruqi, A. R., Tittor, J., Oesterhelt, D., Brown, L., Lanyi, J., and Henderson, R. (1999). Protein conformational changes in the bacteriorhodopsin photocycle. *J. Mol. Biol.* **287,** 145–161.

Tani, K., Mitsuma, T., Hiroaki, Y., Kamegawa, A., Nishikawa, K., Tanimura, Y., and Fujiyoshi, Y. (2009). Mechanism of aquaporin-4's fast and highly selective water conduction and proton exclusion. *J. Mol. Biol.* **389,** 694–706.

Unwin, P. N., and Henderson, R. (1975). Molecular structure determination by electron microscopy of unstained crystalline specimen. *J. Mol. Biol.* **94,** 425–440.

Vonck, J. (2000). Parameters affecting specimen flatness of two-dimensional crystals for electron crystallography. *Ultramicroscopy* **85,** 123–129.

Yonekura, K., Maki-Yonekura, S., and Namba, K. (2002). Quantitative comparison of zero-loss and conventional electron diffraction from two-dimensional and thin three-dimensional protein crystals. *Biophys. J.* **82,** 2784–2797.

Yonekura, K., Braunfeld, M. B., Maki-Yonekura, S., and Agard, D. A. (2006). Electron energy filtering significantly improves amplitude contrast of frozen-hydrated protein at 300 kV. *J. Struct. Biol.* **156,** 524–536.

CHAPTER TWELVE

Automated Data Collection for Electron Microscopic Tomography

Shawn Q. Zheng,* J. W. Sedat,† and D. A. Agard*

Contents

Abstract

A fundamental challenge in electron microscopic tomography (EMT) has been to develop automated data collection strategies that are both efficient and robust. UCSF Tomography was developed to provide an inclusive solution from target finding, sequential EMT data collection, to real-time reconstruction for both

* The Howard Hughes Medical Institute, University of California, San Francisco, California, USA
† The W. M. Keck Advanced Microscopy Laboratory, Department of Biophysics, University of California, San Francisco, California, USA

Methods in Enzymology, Volume 481
ISSN 0076-6879, DOI: 10.1016/S0076-6879(10)81012-2

single and dual axes. The predictive data collection method that is the cornerstone of UCSF Tomography assumes that the sample follows a simple geometric
rotation. As a result, the image movement in the *x*, *y*, and *z* directions due to
stage tilt can be dynamically predicted with the required accuracy (15 nm in *x–y*
position and 100 nm in focus) rather than being measured with additional
images. Lacking immediate feedback during cryo-EMT data collection can offset
the efficiency and robustness reaped from the predictive data collection and
this motivated the development of an integrated real-time reconstruction
scheme. Moderate resolution reconstructions were achieved by performing
weighted back-projection on a small cluster in parallel with the data collection.
To facilitate dual-axis EMT data collection, a hierarchical scheme for target
finding and relocation after specimen rotation was developed and integrated
with the predictive data collection and real-time reconstruction, allowing full
automation from target finding to data collection and to reconstruction of 3D
volumes with little user intervention. For nonprofit use the software can be
freely downloaded from http://www.msg.ucsf.edu/tomography.

1. INTRODUCTION

Electron microscopic tomography (EMT) is an important imaging
technology that fills the wide gap between X-ray crystallography
(X-RAY) and light microscopy (LM) in studying three-dimensional (3D)
biological structures. Not only does EMT expand our ability to view the
structures that fall outside the ranges of X-RAY and LM, but it also bridges
the corresponding gap in the chain of resolution and thus makes it possible
to perform hierarchical investigations of cell machinery at multiple scales.
EMT reconstructs the 3D structure from a series of two-dimensional (2D)
projections collected at various tilt angles by rotating the specimen around a
common tilt axis. These 2D projections are back-projected to form the 3D
structure. Unlike single particle microscopy that reconstructs 3D volumes
from projection images obtained from randomly oriented independent
particles on the premise that they are conformationally identical, EMT
acquires all the projections from the same object and is thus uniquely suited
to heterogeneous structures, such as many supramolecular assemblies, organelles, and cells. This versatility, however, was offset during the early stages
of EMT development by practical challenges that limited its extensive use.

The advent of computer-controlled electron transmission microscopes
and scientific grade slow scan CCD cameras infused great momentum into
the development of automated EMT. As a result, numerous implementations were developed that have dramatically eased the operation and
reduced the time for collecting EMT tilt series. EMT has now evolved
into a routine technology in structural analysis. Although the idea of EMT is
straightforward, practical difficulties arise from the mechanical imperfections

in the available goniometer tilt stages and the inability to precisely set the eucentric height of the specimen. The eucentric error causes the specimen to precess; significantly shifting in the x, y, and z directions during the course of data acquisition. When not too severe, the translational shift truncates the common area shared by each projection image and thus reduces the size of 3D volume that can be reconstructed. As the magnification increases, this adverse effect can be dramatically amplified to become disastrous, causing the object of interest to be shifted completely out of the field of view. In the meantime, the resulting z shift leads to severe focus changes, making it difficult to generate an accurate 3D reconstruction. Prior to automated EMT, manual operations were required to recenter the specimen and readjust the focus, making the data collection not only very inefficient and tedious but liable to high dose as well. Automated EMT endeavored to solve these problems by performing these otherwise manual operations by computer. The central idea is to accurately determine the x, y, and z shifts resulting from stage tilting and to then compensate for the x, y shifts using microscope beam and image shift coils and z shift (focus) by adjusting the objective lens current. Although the history of EMT can be traced back to 1968 (DeRosier and Klug, 1968; Hart, 1968), automated systems were only made available in the early 1990s (Dierksen et al., 1992; Downing, et al., 1992; Koster et al., 1992, 1993). These early systems achieved automation of stage tilting, translation, focusing, and image acquisition steps. During the data collection, auxiliary images were taken before and after each tilt and cross-correlated to measure the specimen translation in the x–y plane. Translation errors were then compensated by adjusting the beam and image shift coils. The z-shift, causing a change in focus, was also measured at each tilt angle by taking another set of additional images using the beam-tilt method (Koster et al., 1992). The objective lens current was then adjusted to compensate for the focus change induced by the z-shift. Through such an automation procedure, a dose saving of as much as 100-fold was achieved over manual operation. In addition, a significant amount of time was also saved (Braunfeld et al., 1994; Koster et al., 1992). However, the extra images taken to measure the specimen translation and focus change added to the time and about 11% to the total dose (Koster et al., 1997).

Cryo-EMT has two significant advantages over room temperature EMT. First of all, cryo-sample preparation preserves the genuine structures that may be otherwise modified by chemical fixation, dehydration, and freeze substitution and staining. Images acquired under cryo-conditions directly represent the interaction of electrons with the object and are thus free of the artifacts from sample preparation. These appealing advantages have stimulated continuous interests in cryo-EMT (Baumeister, 2002; Braunfeld et al., 1994; Koning and Koster, 2009; Leis et al., 2009; Rath et al., 1997; Steven and Aebi, 2003). Since structures are much more vulnerable to radiation damage under cryo-conditions, minimizing the extra dose involved in automated tracking and focusing procedures becomes critical, and several strategies have been proposed to achieve this goal. The earlier solution was to distribute the auxiliary exposures required for recentering and refocusing to areas outside the area containing the object to be reconstructed (Dierksen *et al.*, 1992, 1993; Grimm *et al.*, 1997; Rath *et al.*, 1997). While the primary area containing the object of interest receives no extra dose, users are required to define the supporting area typically along the tilt axis for auto-tracking and focusing. In addition, extra images are taken at each tilt step, reducing the data collection throughput. Ziese *et al.* (2002) suggested the possibility of precalibrating the image movement in the xy plane (image shifts) and the z direction (focus change) prior to data collection. Those calibration curves would then be applied during the data collection to determine the specimen movement that will be compensated. They showed fivefold speed improvement in data collection as opposed to previous EMT implementations. A fundamental challenge of this approach stems from the inability to accurately set the sample eucentric height; resulting in potential errors in the calibration curves. As a consequence, additional tracking and focusing procedures must generally be employed during the data collection particularly at high magnifications. Ziese *et al.* (2002) also mentioned that it might be possible to model and thus predict the overall image movement after a few measurements of image shift and defocus change. This comment inspired us to develop the predictive strategy that fulfills the need for accurate image tracking and focusing without requiring additional images. This method, based upon the dynamic prediction of image movement in the spatial domain using previously acquired tomographic images is both fast and very robust. Another contributing factor to the robustness of this approach is that instead of collection from say $-70°$ to $+70°$, the data collection is performed in two loops that each proceed from zero tilt to the corresponding end angle. By starting at low tilts, the severity of any eucentric error is thus significantly reduced. Only five extra images are required at low magnification to eliminate stage tilting backlash, recenter the target, and track the specimen movement at the second tilt angle in each loop. Since the image at low magnification presents a much larger field of view containing more contrast for measurement, much lower exposures are typically employed to further reduce the extra dose. There is no need to record any additional images for tracking and focusing throughout the data collection. Furthermore, this method exempts users from defining tracking and focusing areas (Rath *et al.*, 1997) or extensive precalibration of stage movements (Ziese *et al.*, 2002) and thus results in enhanced productivity and simplicity. More recently, Mastronarde (2005) extended the predictive approach to compustages that may severely deviate from geometric rotation, allowing the predictive method to fall back to the traditional approach of tracking and focusing when the statistical error of prediction becomes sufficiently large.

Given the extremely low signal-to-noise ratio (SNR) of cryo-EMT, users generally collect numerous EMT data sets and then perform time-consuming gold-bead based or iterative reconstruction on each in the hope of finding a data set of the desired quality. To remedy this problem, we (Zheng *et al.*, 2007a,b) developed a real-time reconstruction scheme that performs the weighted back-projection on a small Linux cluster in parallel with the data collection. The 3D volume at moderate resolution is then made available at the end of each data collection, giving users immediate feedback on sample preparation and experimental settings. Efforts were also devoted to combine the multiscale grid scanning scheme implemented in Leginon, a single particle data collection software package (Carragher *et al.*, 2000; Potter *et al.*, 1999; Suloway *et al.*, 2005), with the predictive EMT data collection scheme in UCSF Tomography. This resulted in a fully automated serial acquisition scheme of numerous EMT data sets (Suloway *et al.*, 2009). Inspired by the successful multiscale mapping of Leginon, we developed a dual-axis target-mapping scheme and integrated it into the UCSF Tomography software system for automated sequential acquisition of dual-axis EMT data collection (Zheng *et al.*, 2009). This made UCSF Tomography fully automated for the entire process from target finding to data collection and to reconstruction of 3D volumes at moderate resolution with almost no user intervention.

The six-year service of UCSF Tomography both internally and at other sites has proven that this is an efficient and robust data collection system. While the functions provided in UCSF Tomography were reported in several papers during the past five years, presenting them altogether in this chapter gives users a complete view of this powerful tool and is thus deemed instrumental to facilitate its practical use.

2. PREDICTIVE EMT DATA COLLECTION

2.1. Geometric model

EMT data collection stands in the middle of the pipeline from target finding to final reconstruction. Therefore, not only does the successful data collection ensure that the effort spent on finding targets not futile, but it also almost guarantees successful EMT reconstruction. In this sense, the predictive data collection strategy serves as the foundation of UCSF Tomography. The predictive approach is premised on the ability to model the consequences of specimen tilt as a rigid body rotation around a single tilt axis (Ziese *et al.*, 2002). Figure 12.1 sketches this model, where t denotes the tilt axis and n stands for the axis perpendicular to the tilt axis. The sample point P is the characteristic point at which the optical axis intersects the specimen. At zero degree tilt, the location of P is expressed by (n_0, z_0) where n_0

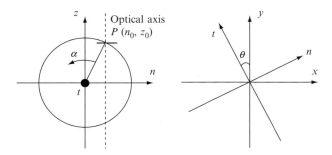

Figure 12.1   Geometric model of specimen rotation around tilt axis (Zheng *et al.*, 2004).

represents the offset between the tilt axis and the optical axis, and z_0, known as noneucentricity, denotes the distance between the sample point and the tilt axis in z direction. θ is the angle between the y-axis and the tilt axis. Assume that a specimen starts to tilt at $\alpha = 0°$, where α is the stage tilt angle, the translation of the specimen at any tilt angle relative to the starting angle can be written in the $n–z$ plane as follows:

$$\Delta n = n_0 * \cos\alpha + z_0 * \sin\alpha - n_0 \tag{12.1}$$

$$\Delta z = z_0 * \cos\alpha - n_0 * \sin\alpha - z_0 \tag{12.2}$$

Since the image movement is measured in the CCD plane, that is, the $x–y$ plane, the x, y coordinates as functions of tilt angle can be derived by projecting Δn onto the x and y axes as given in Eqs. (12.3) and (12.4):

$$\Delta x = n_0 * \cos\alpha * \cos\theta + z_0 * \sin\alpha * \cos\theta - n_0 * \cos\theta \tag{12.3}$$

$$\Delta y = n_0 * \cos\alpha * \sin\theta + z_0 * \sin\alpha * \sin\theta - n_0 * \sin\theta \tag{12.4}$$

Equations (12.2)–(12.4) thus fully describe the spatial movement of the sample provided that the unknown geometric parameters (θ, n_0, z_0) can be determined by some means or other. Under this assumption, the sample displacement can thus be predicted instead of being measured for each tilt angle. The optical system is then adjusted accordingly to compensate the predicted movement prior to taking a tilt image at each angular step. Because of the explicit dependency on the geometric parameters in the prediction, the values of θ and n_0 need to be determined before data collection, whereas z_0 can be estimated and refined dynamically during data collection.

Since the stage x-axis is also the α-tilt of the Tecnai Compustage, the orientation of the tilt axis, θ, can be readily measured by two images of which the second one is taken at the location shifted along the stage x-axis relative to the first one. The two images are cross-correlated with each other to yield a displacement vector whose angle relative the CCD y-axis is then calculated. This angle provides an estimate of θ whose accuracy (typically $\pm 1.0°$) is sufficient for the prediction of the specimen movement.

The physical orientation and the location of the tilt axis are expected to remain fixed unless the goniometer is given a service adjustment. The position of the optical axis should also remain relatively unchanged between two consecutive alignments. It is therefore assumed that n_0, the offset between the tilt axis and the optical axis, can be precalibrated and treated as a known constant to predict the specimen movement during data collection. To determine n_0, focus is measured within a range of α tilt angles using the beam-tilt method (Koster et al., 1992). Least squares curve fitting is then applied to the measured focus values based upon Eq. (12.2). In our implementation, the focus measurement is performed every $10°$ within the angular range of $\pm 50°$. In addition to its use in the prediction of specimen movement, the value of n_0 provides a good diagnosis of the goniometer alignment. Goniometer service is recommended when n_0 exceeds 2.0 μm.

2.2. Dynamic determination of z_0

z_0 can be minimized by setting the eucentric height. However, mechanical imperfections in goniometers make it almost impossible to tune z_0 to zero even assisted by automated routines. As a result, the uncorrected residue of z_0 can still be severe enough to shift the object from the field of view at high magnifications and lead to a significant focus change. In addition, not only does z_0 depends on the actual sample being used but it also varies from point to point on the grid owing to nonuniform thickness and flatness of the specimen. These uncertainties thus necessitate dynamic determination of z_0 during the course of data collection for each selected sample location. To do so, note that the usage of Eq. (12.3) is twofold, meaning that it can also be used to solve for z_0 from the right-hand side given a set of data points of $(\alpha, \Delta x)$. Since n_0 and θ have been precalibrated, z_0 is the only unknown constant for a specific sample point. Mathematically, one data point of $(\alpha, \Delta x)$ is sufficient to solve for z_0. In reality, it is better to use least squares fitting from multiple data points to minimize the consequences of errors in measuring the x translation. Equation (12.5) is thus derived from Eq. (12.3) and used to dynamically estimate and refine z_0 during the course of data collection. This dynamic determination of z_0 based upon the acquired tomographic images is the cornerstone of this proposed data collection scheme.

$$z_0 = \frac{n_0 * \cos \theta * \sum_i (1 - \cos \alpha_i) * \sin \alpha_i + \sum_i \Delta x_i * \sin \alpha_i}{\cos \theta * \sum_i \sin^2 \alpha_i} \qquad (12.5)$$

As mentioned earlier, a single data point of $(\alpha, \Delta x)$ is sufficient to solve for z_0. Therefore, the prediction scheme can be invoked as early as when two tomographic images have been collected. The required data point can be obtained by cross-correlation of these two images and plugged into Eq. (12.5) to generate the first estimate of z_0. This estimated z_0 and the precalibrated n_0 and θ are then sufficient to predict x, y, and z shifts for the third image using Eqs. (12.2)–(12.4). When the stage is tilted to the third tilt angle, the beam and image shift coils are adjusted according to the predicted Δx and Δy while Δz is compensated by changing the objective lens current. Upon completion of these adjustments, the third image is acquired and then correlated against the previous one to give rise the second data point of $(\alpha, \Delta x)$. As the first estimate of z_0 is based upon a single data point and thus may not accurately describe the geometry, this newly acquired data can be combined with the previous one and used to refine the estimate of z_0 based upon Eq. (12.5). The refined z_0 is then used to predict x, y, and z for the fourth image. This predict-refine procedure is repeatedly preformed throughout the data collection. It is worthwhile to point out that the global geometric rotation is only the first order approximation to stage tilting. Many stages exhibit more or less systematic localized behaviors that deviate from the global geometric rotation model. This kind of aberrant behaviors inspired us to update z_0 with only the most recent three points instead of using all historical data. This strategy allows the prediction scheme to quickly adapt to changes in the stage behavior by absorbing the local deviation in stage tilting into the z_0 update; thus minimizing the adverse effect of stage imperfections on prediction accuracy.

2.3. Experimental verification

High resolution EMT data collection generally takes place at magnifications higher than $50,000\times$. Under these conditions, many practical factors including mechanical imperfections and nonuniform sample thickness and flatness have prohibitive effects on setting the precise eucentric height. It is therefore critical that an automation strategy works not only for accurately set eucentric height but can also tolerate reasonable amounts of eucentric error. To verify the robustness of this approach, measurements of specimen shifts as functions of stage tilt angle were repeated for the same sampling point at various stage z positions above and below the eucentric height and compared with predicted values.

To minimize the deleterious effects of eucentric errors, the data collection is divided into two loops, and in each loop the stage tilts only in one direction from the starting angle to the corresponding end angle. After the maximum angle of the first loop has been reached, the stage returns to its starting position, and the data collection resumes in the second loop by tilting the stage in the opposite direction. When the stage returns to the initial angle, mechanical imperfections and sample drift typically shift the sample point from its original position even if the microscope has been fully restored to its original state. To automatically correct for this shift, two low magnification ($\sim 14,000\times$ nominal) images are taken at the initial angle with the second one acquired when the stage returns to the initial position. The measured shift by cross-correlation is then compensated by adjusting the beam and image shift coils followed by the second loop of data collection. Note that using low magnification not only reduces the extra dose but also allows detection of large returning shifts. While this approach has the disadvantage of needing two extra low magnification images, the benefits of beginning at low tilt and employing a dynamic updating of z_0 leads to a dramatic enhancement in robustness.

A demonstration of the predictive data collection scheme was performed using the standard FEI holder on a FEI T20 transmission electron microscope equipped with a side-entry compustage. Image shift and focus change as functions of the tilt angle were measured simultaneously and are compared with the predicted values in Figs. 12.2–12.4. Figures 12.2 and 12.3 show x–y translations as functions of the tilt angle at three different noneucentric settings, corresponding to z height errors of 0, -1, $-2\mu m$. The tilt images were acquired every $2°$ starting from $0°$ at a CCD magnification of $62,560\times$. Since the prediction procedure requires at least one data point of $(\alpha, \Delta x)$ in order to get the first estimate of z_0 and the first data point can only be obtained when the second tilt image has been acquired, the prediction actually starts at the third tilt image acquired at $4°$. Even at the nominal setting of $z = 0$ μm, as shown in Figs. 12.2A and 12.3A, the sample still exhibits a shift in nearly 2000 pixels (~ 480 nm) in the x direction at this high magnification. As was pointed out in the previous section, it is impractically challenging even for a highly experienced and exacting user to tune z_0 to zero. Very small residual values of z_0 can result in severe image shifts even at relatively low tilts. This severity can be further amplified at higher magnifications. Furthermore, under cryo-conditions the eucentric height must be set at a remote location, again leading to increased sample noneucentricity. Therefore, the two-loop data collection scheme reaps the very significant advantage over the single-loop data collection in terms of reducing the adverse effect of noneucentricity. As can be seen from Figs. 12.2 and 12.3, the program precisely predicted the image shifts at three distinct z heights. At $z = -2\mu m$, up to ± 8000 pixel shifts have been predicted within ± 60 pixels corresponding to ± 14.4 nm at $62,560\times$. The error between the measured and predicted results within $\pm 70°$ is typically less than 60 pixels and rarely exceeds that value.

The early releases of UCSF Tomography did not compensate the displacement of the second tilt image. The sample eucentricity thus needed to be set with sufficient precision in order to minimize this displacement. As single particle tomography receives growing acceptance, there is often

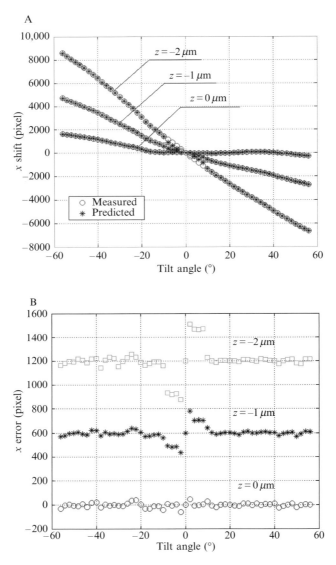

Figure 12.2 (A) Measured and predicted image shift versus tilt angle in the CCD x direction at three values of missetting the sample eucentricity at $62,560\times$ CCD magnification. $z = 0$ μm denotes that the sample point was manually made eucentric by an experienced microscope operator. (B) Errors between the measured and predicted values. Each pixel is 0.24 nm (Zheng *et al.*, 2004).

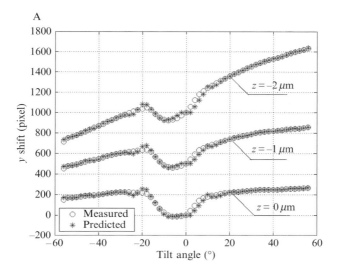

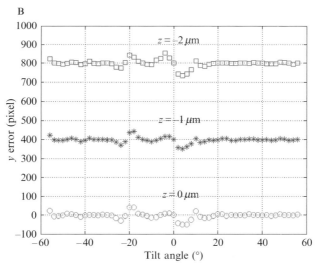

Figure 12.3 (A) Measured and predicted image shift versus tilt angle in the CCD y direction at three stage z heights at $62,560\times$ CCD magnification. Note that z is the relative stage z height or error in eucentricity. The curves are intentionally offset vertically by 400 pixels to improve clarity. (B) Errors between measured and predicted values. Each pixel is 0.24 nm (Zheng *et al.*, 2004).

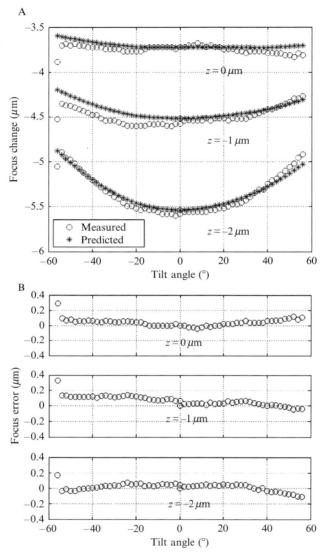

Figure 12.4 (A) Measured and predicted focus change versus tilt angle at three z heights at $62,560\times$ magnification. Note that z in the figure is the relative stage z height. (B) Errors between measured and predicted values (Zheng *et al.*, 2004).

the desire to collect tilt series at very coarse angular steps ($>5°$) and high magnifications ($>50,000\times$). Even slight eucentric errors can be disastrous under these conditions. The recent releases of UCSF Tomography have solved this problem by first tracking the specimen shift at the second tilt angle at low magnification. Since two auxiliary images have been taken at starting angles and low magnification for recentering the specimen, the specimen shift can be measured with another low magnification image taken at the second tilt angle and correlated with the one at the initial angle. The measured shift is then corrected via adjusting the beam and image shift coils prior to taking the second tomographic image. In the meantime, this measured shift also serves as the first data point to jump-start the prediction scheme at the third tilt angle. This greatly extends applicability and robustness.

Figure 12.4A shows the measured and predicted focus changes as a function of tilt angle at three z height settings, while the errors between the predicted and measured values are given in Fig. 12.4B. As in Figs. 12.2 and 12.3, $z = 0$ μm represents a manually set eucentric height. The maximum difference between the measured and predicted values is less than 0.15 μm except the left end point at $-56°$ where a remarkable focus jump (~ 0.2 μm) was measured in this particular test case. Several lines of evidence suggest that this error is actually due to an error in the measurement of the focus and not the prediction. First, using the same holder with other samples, such a big jump is not seen. Second, from Eq. (12.2) a 0.2 μm focus change due to a $2°$ stage tilting implies a grossly unrealistic 6 μm shift of z_0. Yet the simultaneously measured x and y shift curves show no discontinuities (Figs. 12.2A and 12.3A). Thus, we can safely exclude severe noneucentricity and instability of the sample within the holder as causes; leading us to believe that the focus measurement at that angle may be in error. It is important to note that our approach corrects focus changes throughout data collection solely based upon predicted rather than the measured focus values. As a matter of fact, no focus measurements are performed during the entire data collection. The anomalous focus measurement is thus not a concern in the prediction. We also noticed that Fig. 12.4A shows a slight discontinuity (0.001, 0.062, and 0.05 μm at $z = 0$, -1, and -2 μm, respectively) at $\alpha = 0°$. Although these discontinuities are small compared to the prediction errors, this is a consequence of dividing the data collection into two loops. When the stage returns to the starting angle from the positive end angle, mechanical imperfections make it impossible to exactly set the stage to the original x, y, and z position. As mentioned before, this is why we record images at lower magnification to allow the sample to be precisely recentered. In an effort to minimize dose, we do not measure the focus when the stage returns to the initial tilt angle. As a consequence, there is a slight change in z height resulting in the observed focus discontinuity.

2.4. System implementation

Our automated data collection program, UCSF Tomography, was developed based on this predictive approach for FEI Tecnai computer-controlled electron microscopes and has been constantly improved since its first release.

The current release has been extensively tested on a broad range of FEI instruments including T12, T20, F20 with conventional side-entry room temperature or cold stages as well as the Polara 300 kV F30 FEG with liquid He stage and a GATAN postcolumn energy filter. The control of the microscope is achieved via a Windows 2000 (or XP). A great amount of effort has been devoted to supporting cameras from various vendors including GATAN (USA), TVIPS (GERMANY), and FEI (USA).

A scripting server provided by FEI allows for control of stage movement and the optical system in many scripting languages. Our data collection program is built on top of this server and can be roughly sketched as three functional layers. The top layer is the user interface (UI) implemented in Microsoft Visual C++. Figure 12.5 shows the data collection portion of the UI. This portion of the UI provides three image windows. The left one shows the image acquired at the previous tilt angle, the middle for the image taken at the current tilt angle, and the third window displays the corresponding cross-correlation image. A log field continuously displays information as the data collection progresses. Users can interrupt the execution at any time if necessary. The data collection workflow is fulfilled at the bottom layer and is implemented in Microsoft JScript. We chose this programming language because of its simplicity in linking with the Tecnai

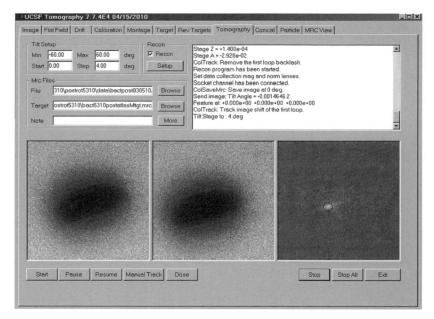

Figure 12.5   User interface for the UCSF Tomography data collection.

scripting server for microscope control. Since the UI and the data collection are implemented in different programming languages, Microsoft COM technology is employed in the middle layer as a communication channel.

3. REAL-TIME TOMOGRAPHIC RECONSTRUCTION

While the speed and ease-of-use of collecting tilt series had been significantly improved through the implementation of the predictive strategy, reconstruction of 3D volumes still throttled the EMT pipeline. As a result of lacking the immediate feedback from 3D reconstruction, cryo-EMT data collection was quite inefficient because the extremely low SNR makes sample assessment from 2D images very difficult. Thus, numerous EMT data sets had to be acquired and time-consuming gold-bead or iterative reconstruction were then performed on each in the hope of finding a data set of the desired quality. It was thus highly desirable to have the corresponding 3D volume available at the end of each data collection as a guide for subsequent data collection, allowing the user to make better decisions about targets for data collection and even data collection schemes. This requires that reconstruction be performed within the same time frame as data collection, that is, real-time reconstruction. While the goal was not to replace iterative methods for the refinement and optimization of the final tomogram, real-time reconstruction offers several distinct advancements toward simplifying the entire process. Because the quality of the output tomograms is quite high, they serve as a screen for selecting the best data sets to take through an iterative refinement process, leading to a significant improvement in the efficiency of using system resources. They also provide useful information on specimen thickness and the angular offset of the specimen, two parameters important for iterative reconstruction schemes. Most importantly, this scheme is enormously beneficial to those experiments where desired target is rare or difficult to distinguish.

3.1. General strategy

Real-time reconstruction imposes rigorous time constraints due to the relatively rapid data collection process. Several means were employed to ensure that the intensive computation can be completed within the time frame of the data collection. First of all, weighted back-projection (e.g., Frank *et al.*, 1987; McEwen *et al.*, 1986), a single pass method, was chosen to perform the 3D reconstruction. While this method is relatively fast compared to iterative approaches, for moderate sized volumes (512^2 by 100–300) the computation is still too intense to be conducted on the PC that is already heavily engaged in data collection. A Linux cluster formed by a head node and at least four dual-processor compute nodes connected via gigabit Ethernet provided sufficient computational power to perform the reconstruction concurrently with the data collection system. In addition to the back-projection calculation, data alignment represents another time-consuming process. In a real-time approach, one must work with the data as collected; thus only nearest-neighbor type image alignment approaches can be used. While these calculations could be done on the cluster, it is possible to achieve significant time savings by reusing the nearest-neighbor alignment data generated during data collection. Both the tilt image and the alignment data are sent to the cluster for immediate reconstruction. To ensure the load evenly balanced, the head node is used to handle data communications, binning the full resolution images to lower resolution, mass normalization, and applying alignment parameters. The corrected projection data are broadcast to all of the compute nodes which then calculate a separate z-volume region by weighted back-projection. The head node then assembles the full reconstruction upon completion of processing. Owing to technical advances, we are now converting our codes to allow the required real-time reconstructions to be done on one or more attached graphic cards (GPUs).

3.2. Parallel real-time reconstruction

Once a 2D projection is acquired at the new tilt angle, this image together with the corresponding tilt angle (α), angle of tilt axis (θ), and the image translation shift $(\Delta x, \Delta y)$, are sent to the cluster where the head node performs the required binning and mass normalization. When a subset of the projections has been received, the reconstruction program fits the intensities for those projections to an exponential absorption model. Then, as each projection becomes available, the reconstruction program bilinearly interpolates it to the aligned coordinate system where the tilt axis is vertical, scales its intensities, and sends it to other processors for back-projection. The organization of the back-projection calculation on the other processors takes advantage of the parallel projection geometry with the tilt axis oriented along the vertical axis. With that geometry, the center of the reconstructed voxel, (i, j, k), projects through the tilt angle, a, to the position, (i_0, j_0), as given in Eqs. (12.6) and (12.7).

$$i' = i * \cos \alpha + k * \sin \alpha \tag{12.6}$$

$$j' = j \tag{12.7}$$

Because a horizontal line (constant j') in a projection backprojects to a plane of constant j in the reconstruction, we have each processor work with a subset of the lines in the projection to reconstruct a subset of the planes.

A processor initializes its subset of the reconstruction to zero. When a processor receives a line from a projection, the processor weights the intensity values to compensate for the unequal sampling of the Fourier components and optionally applies a low pass filter to the values. To update the reconstruction for the newly received line, the processor considers each voxel and computes its projected position according to Eqs. (12.6) and (12.7). If the position falls within the measurement bounds, the processor adds the intensity interpolated at the projected position to the voxel. The interpolation weight for a pixel in the projection is the fraction of the projected voxel contained within that pixel. After all the projections have been received, the value of each reconstruction voxel is normalized by the number of contributing projections.

3.3. System design and implementation

In order to establish fast and robust communication between the data collection system and its counterpart, a TCP/IP socket channel over a Gigabit network was implemented. The data collection system is synchronized with the reconstruction program only when a new tilt image is acquired and ready to be sent through the socket channel, while asynchronous operation can be employed most of the time on both sides. Once the reconstruction program receives the tilt image and the alignment data, necessary reconstruction image computation can proceed independently. In the meantime, the data collection system can continue its work without waiting for the completion of processing the current image on the reconstruction side. For this reason, data communication was implemented using a multithreading technique by which sending data through the socket channel is executed in a dedicated thread, namely the data transmission thread (DTT). The DTT is, for most of the time, in sleeping state, yielding the full power of CPU to threads that perform more CPU demanding jobs. It wakes up only when a new tilt image has been acquired and is ready for transmission. The entire data collection can be viewed as a loop of cycles. Each cycle corresponds to the collection of a tilt image at individual angular steps and starts with the acquisition of a tilt image, followed by tilting the goniometer stage to the next angular step. Since the goniometer stage needs some time to stabilize after each tilt, tilting the stage immediately after each image acquisition allows the stage to relax while the subsequent computation is performed without interruption. The newly acquired image is then cross-correlated with the previous one. This yields the specimen shifts that are needed for both prediction and reconstruction. The tilt image along with the computed alignment data is saved into an MRC-format file. In the meantime, the DTT is notified and becomes active. The socket channel is examined within the DTT to ensure the other end is ready to receive. As soon as the data have been sent into the channel, the DTT goes to sleep again, which marks the end of the current cycle in the loop. Since the sample stage has already been set to the next angular position and has had time to stabilize, the next cycle begins by acquiring the new tilt image.

The reconstruction software was developed in the MPICH 1.2.5 implementation of the MPI standard as the interface for a parallel computing environment. To simplify cluster setup and management we use Rocks (Papadopoulos *et al.*, 2001). One thread (namely, the head node) of execution communicates with the data collection system. The head node bins the images, computes the mass–normalization parameters for intensity scaling, interpolates the projections to an aligned grid, converts the projected intensities, distributes the aligned and converted images to the other threads of execution, and assembles the full reconstruction at the end. To reduce the amount of communication and eliminate duplicate computations, each of the other threads is dedicated to reconstructing a subset of the xz planes through the reconstruction volume. Each such thread keeps its part of the reconstruction in memory so the amount of memory available for each thread and the size of the reconstruction constrain the minimum number of threads and how those threads are mapped to processors in the cluster.

3.4. Experimental verification

It is commonly appreciated that both thick (>0.1 μm) and cryo low–dose samples are more difficult for EMT than thin and high–dose room temperature specimens. Setting the eucentric height for a thick sample is challenging and any uncorrected eucentric error will cause large image shifts when the stage is tilted. Correlating two thick images with large shifts can be problematic both because the common area decreases and also because stretching is less able to correct for tilt-related differences. Cryo-EMT data collection is especially difficult due to its very low SNR. In order to stringently test the real-time reconstruction, we chose Epon embedded HeLa cells (for details see Belmont *et al.*, 1987, 1989) that were sectioned to a nominal thickness of 0.3 μm as a representative thick sample and tobacco mosaic virus (TMV) as a representative cryo-sample. The experiments were conducted at liquid nitrogen temperature on an FEI G2 Polara 300 kV F30 equipped with a postcolumn GATAN energy filter and a 4096^2 pixel lens coupled, cooled CCD camera (Ultracam, GATAN). The tilt series were collected with a 20 eV energy loss window centered at 0 eV, a CCD magnification of 88,000×, and on chip binning of 2× for final image size of 2048^2 pixels. The total dose received by the TMV specimen was less than 50 e/Å^2. Both data sets were acquired at -2μm defocus. The real–time reconstruction was performed at $512^2 \times 256$ pixels binned fourfold from the original dimensions. For comparison, 10 rounds of iterative alignment (Keszthelyi, in preparation) and 100 rounds of TAPIR reconstruction were also performed on images of the same size. The well-defined TMV structure also helps evaluate the resolution of the real–time reconstruction (Champness *et al.*, 1976). Figure 12.6 presents the 2D projection at zero tilt, the central image from both real–time reconstruction and the TAPIR reconstruction. While the iterative TAPIR reconstruction yields a higher resolution reconstruction owing to its ability to correct for residual alignment errors and its ability to minimize the consequences of the missing wedge of tilt data, it comes at a cost of nearly 3 h of computation on a 10 node (20 processor) cluster. It is clear that the real–time reconstructed volume provides significantly improved visualization compared to the 2D projection at $\alpha = 0°$. This information is clearly sufficient for the purpose of surveying sample quality and even for some final applications.

Figure 12.7 presents the original $0°$ tilt projection TMV image and the central z slice from the reconstructed volume. To clarify arrows are added to link the matching filaments in Fig. 12.7A and B. The comparison

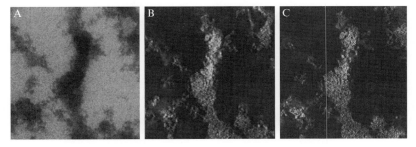

Figure 12.6 Projected and reconstructed images at magnification of $62{,}560\times$ on a HeLa cell chromosome. (A) Tilt image at $\alpha = 0°$, (B) central xy section of a stack from real–time reconstructed volume, and (C) central xy section of a stack after iterative alignment and TAPIR reconstructed volume (Zheng *et al.*, 2007a,b).

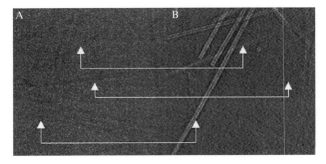

Figure 12.7 Projected and real–time reconstructed images of TMV at CCD magnification of $88{,}000\times$ (A) Tilt image at $\alpha = 0°$, (B) central z slice of the reconstructed image (Zheng *et al.*, 2007a,b).

between Fig. 12.7A and B clearly shows that the real-time reconstruction significantly improves the TMV visualization. While the TMV central cavity (4 nm in diameter) is almost invisible in the zero-tilt image presented in Fig. 12.8A, it is quite prominent in Fig. 12.7B. With such improvement in visualization, surveying cryo-specimens become quite practical.

It should be noted that while simply adding more processors would allow larger reconstructions to be calculated in real-time, residual alignment errors probably limit the advantages. By comparing to the fiducial marker alignment (Zheng *et al.*, 2007a,b), the real-time alignment errors at the reconstruction scale are found in both *x*- and *y*-axis less than 1 pixel except for a few points at angles beyond 55°. The growing alignment error at high tilts hints that stretching alone may not be sufficient to correct for tilt-related differences, which are more severe for thick specimens. Further efforts should be made on preventing the deterioration of alignment accuracy at high tilt angles.

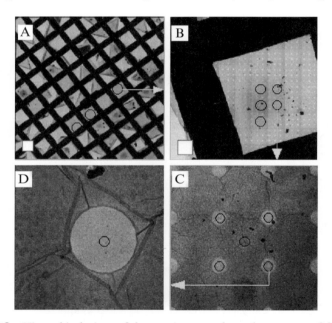

Figure 12.8 Hierarchical view of the specimen to show the process of finding the structures of interest. All images were acquired on FEI Tecnai Polara TF30 electron microscope equipped with a GATAN 4 k UltraCam CCD camera. The square frames in the lower left corner of (A) and (B) represent the corresponding CCD imaging area, respectively. (A) Atlas map constructed at 480× nominal magnification. (B) Atlas map constructed at 3000× nominal magnification (M mode). (C) A tile image that forms the atlas map shown in (B). (D) An image taken at 14,500× nominal magnification. The arrows show where the location of the pointed map or images is originated in the precedent map or image (Zheng *et al.*, 2009).

4. Dual-Axis EMT Data Collection

Dual-axis EMT reduces the missing wedge–induced resolution loss by taking two complementary tilt data sets of the same target along two orthogonal axes. The potential of this powerful approach was long hampered by the practical challenges in finding the original targets that are dramatically displaced due to the inherent noneucentric specimen rotation. For the FEI Tecnai Polara microscope, the specimen rotation is achieved by means of the so-called "flip–flop" rotation cartridge (GATAN, USA) that allows the specimen to be rotated by roughly 90° between either a flipped or the flopped position (for a detailed description see Iancu *et al.*, 2005). JEOL also introduced a similar rotation cartridge for its microscopes. With these cartridges, the rotations are rather crude and rotation is accompanied by displacements up to 300 μm (Zheng *et al.*, 2009). Not only is the manual search for the original targets very inefficient and tedious but the added dose during the manual quest is uncontrollable. We developed a hierarchical alignment scheme that is both efficient and robust at the cost of only a very small additional dose. This scheme allows to collect tomographic data from an arbitrary number of target sites in one grid orientation and then to find and collect the orthogonal data sets with little or no user intervention. Inspired by the successful multiscale mapping in Leginon (Carragher *et al.*, 2000; Potter *et al.*, 1999; Suloway *et al.*, 2005), our alignment is performed in three levels to gradually pinpoint the original targets. At the lowest level, the grid lattice is used to determine the rotation angle and translational shift resulting from specimen rotation via auto- and cross-correlative analysis of a pair of atlas maps constructed before and after specimen rotation. The target locations are further refined at the next level using a pair of smaller atlas maps. The final refinement of target positions is done by aligning the target contained image tiles. The batch processing nature of this hierarchical alignment allows multiple targets to be initially selected in a group and then sequentially acquired. Upon completion of the data collection on all the targets along the first axis and after specimen rotation, the hierarchical alignment is performed to relocate the original targets. The data collection is then resumed on these targets for the second axis. Therefore, only one specimen rotation is needed for collecting multiple dual-axis tomographic data sets.

4.1. Finding targets for sequential EMT data collection

Dual-axis EMT data collection proceeds with four major operations involved, namely, (1) finding targets, (2) sequential data collection before specimen rotation, (3) rotational alignment following specimen rotation, and (4) sequential data collection for the second axis. Single-axis EMT data collection involves only the first and second operations. While the rotational alignment is specific for dual-axis EMT data collection, from operational perspective there is no difference between the two sequential data collections taking place before and after specimen rotation.

The goal of target finding is to generate a stack of images of selected targets (objects to be reconstructed) along with the corresponding stage coordinates as the input to the sequential data collection scheme. EMT data collection can then be repeatedly performed without user intervention on these targets. First, a large scale atlas map ($\sim 1000^2$ μm^2) needs to be built that provides an aerial view of the grid and also serves for the subsequent rotational alignment given the large displacement induced by specimen rotation. Users have two options of how to build this large map depending on the system configuration. For microscopes without energy filters, it is recommended that this map be constructed at the lowest magnification (1700×) in M-mode since both the objective lens and aperture remain on and in place. With our cameras, under these conditions it takes 52 min to build an atlas of 800^2 μm^2 and is probably worth the time since this avoids the need to restabilize the microscope after turning on the objective lens and provides a much higher magnification view for target selection. However, due to a higher minimum M-mode magnification (3000×) and slower camera readout, it would take an unacceptable amount of time to build maps of the same size on our Tecnai Polara equipped with a lens-coupled GATAN UltraCam 4 k CCD camera and a GATAN energy filter. As an alternative, maps can be constructed at 480× magnification without the objective aperture inserted, as is shown in Fig. 12.8A. This magnification is one of the few that do not need tuning of the EFTEM cross-over alignment. A map covering about 1000^2 μm^2 is then composed of 121 images and takes only 15 min to build with 0.1 s exposures. This map is referred to as LM-mode map for simplicity.

Upon completion of building the LM-mode map, the objective lens is turned on by switching to the lowest magnification in M-mode and the objective aperture is reinserted. The overhead time of subsequent operations is used and in general sufficient to stabilize the objective lens. Guided by the LM-mode map, a smaller map of a promising area is built and shown in Fig. 12.8B. By the same convention, this map is named the M-mode map. In practice, if several promising areas are present but scattered in the LM-mode map, we build multiple M-mode maps, one for each identified area of the grid. This alternative strategy is most suitable for dual-axis tomographic data collection.

The next step is to select potential targets. Potential subareas of appropriate ice thickness can be selected by clicking on the atlas map. As shown in Fig. 12.8C, each click results in a zoomed-in view of that subarea by loading the nearest tile image that was saved during the map construction. In the meantime, a red circle is marked on the map to label the visited subarea.

Figure 12.8B, for example, indicates that a total of five subareas were visited. Candidate holes are then picked manually and marked on the rendered tile image shown in Fig. 12.8C where four holes were selected. A tile image with selected holes is referred to as target tile to distinguish those without any selection. The corresponding stage coordinate of the target tile along with the relative shift of each candidate hole is stored for later collection of images at intermediate magnifications, one per each selected hole. We also designate the first selection made on each target tile to be the site used only for tuning eucentric height, defocus, and energy filter zero-loss alignment, if installed. The red circle near the center in Fig. 12.8C is for this purpose. Candidate holes are always grouped together with a tuning site per target tile. This step is repeated until all subareas of interest on the map are examined. At the end, a list of stage coordinates is generated and each is associated with a sublist of relative positions of the candidate holes.

As can be seen in Fig. 12.8C, 3000× magnification is generally too low to determine the presence of structures of interest within each hole. An image at an intermediate magnification is thus preferred (one per each selected hole) to provide a closer look with a very small amount of dose. Figure 12.8D presents such an image (reference image hereafter) acquired at 14,000× magnification with 4× on-chip binning and 0.1 s exposure. Acquisition of the reference images are automated by first recentering the subarea of each target tile on the CCD. This can be implemented by taking a new image at the same stage location and magnification where the target tile is acquired. The stage positioning error is measured by correlative analysis between the target tile and this newly acquired image. This error along with the relative shift of each candidate hole is compensated using the image/beam shift coils. A reference image is then taken at the intermediate magnification for each centered hole. When all the candidate holes in this target tile have been imaged, the automation procedure then begins to process the next target tile and repeats until finished. A stack of reference images is generated and saved into an MRC file at the end of this step. Users then review the acquired reference images to screen for structures of interest (targets). Any reference image with found targets is marked with red circles to label their locations and referred to as target image. Only target images will be revisited during the following data collection while all other reference images are skipped.

4.2. Sequential EMT data collection

The complete selection of targets makes it possible to sequentially collect the EMT data set at each target without any user intervention. As mentioned earlier, targets are grouped with a tuning site per target tile that covers only an area of 14^2 μm^2 on the Polara electron microscope. It is reasonable to assume that the eucentric variation within a target tile is negligible. Setting eucentric height is therefore performed only once per target tile. The automation proceeds in such a way that the stage is first driven to the tuning site to automatically set eucentric height and true focus. If an energy filter is installed, this tuning site is also visited for zero-loss alignment prior to moving to a new target. The following task centers the next target on CCD within the current group. Since each target is associated with a target image in which the target is identified with the relative location to the image center, recentering this target simply involves the correction of the stage positioning error that can be measured via cross-correlation with a new image acquired at the same stage coordinate and magnification. The sum of the stage positioning error and the target relative shift is compensated by deflecting the image/beam shift coils. The user-specified defocus is then set by invoking the autofocus routine at the target location but using an intermediate magnification and a short exposure to minimize the dose received (Zheng *et al.*, 2007a,b). This target is now centered on the CCD at the desired eucentric height and properly defocused with the energy filter slit aligned to the zero-loss peak and ready for tomographic data collection (Zheng *et al.*, 2004). Upon completion of the data collection on current target, a zero-loss alignment is performed again by shifting the stage to the tuning site. When finished, we then repeat the previous steps from centering the next target image all the way to acquiring the tomographic tilt series. When all targets linked to the current target tile are collected, the next target group is loaded into computer memory followed by setting eucentric height at its associated tuning site. The targets within this group are then centered, focused, and collected in the same fashion. This whole process continues until all targets have been collected.

4.3. Rotational alignment

Upon completion of the data collection for the first axis followed by specimen rotation, rotational alignment needs to be performed to relocate the original targets. A second LM-mode map is constructed and correlated with its counterpart made before specimen rotation to yield an initial alignment. Changes in the grid lattice between these two maps are used to measure the grid orientation and image magnification changes by performing autocorrelation of each atlas map (for a detailed description see Zheng *et al.*, 2009).

Since the alignment parameters are drawn from the correlative analysis of two LM-mode atlas maps that cover about 1000^2 μm^2 grid area, the computed locations do not bear sufficient precision for centering the targets at the data collection magnification. Further refinement of target locations is necessary, starting with using M-mode maps. Figures 12.9A and B show two M-mode maps made before and after rotation, respectively. Again, rotational cross-correlation is employed for each pair of the M-mode maps to

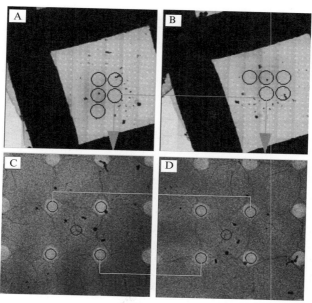

Figure 12.9 Rotational alignment of a pair of M-mode maps and relocation of target holes. (A) M-mode atlas map before specimen rotation. (B) M-mode atlas map after specimen rotation. (C) Target holes in a tile image before specimen rotation. (D) Target holes relocated after specimen rotation (Zheng *et al.*, 2009).

measure the residual translational shift. Using these values, all the targets shown in red circles in Fig. 12.9A are transformed to the postrotation map as shown by the red circles in Fig. 12.9B. As an example, the light blue line linking a pair of circles in Fig. 12.9A and B illustrates the location of a target before and after rotation.

The final round of alignment is at the level of the target tiles and provides precise matching of hole centers. Figure 12.9D shows the newly acquired image corresponding to the target tile displayed in Fig. 12.9C. While the target holes are manually picked by users from the original prerotation maps (Fig. 12.9C), the red circles in Fig. 12.9D are all computational results of the alignment process. Upon successful alignment of the current target tile, each associated target image acquired at the intermediate magnification is also digitally rotated to generate a new target image with an updated stage coordinate. Therefore, there is no need to collect any new target images at the intermediate magnification. This procedure is automatically repeated until all target tiles and the associated target images have been processed. The conclusion of the hierarchical alignment gives a list of target images that are digitally rotated from those acquired before specimen rotation.

The associated stage positions are mapped onto the postrotation coordinate system. The sequential data collection can then be resumed and runs in the same fashion as the initial single-axis data collection.

4.4. Experimental verification

20S Proteasomes were vitrified in liquid ethane onto Cu 200 mesh Quantifoil 1/4 holey carbon films (Quantifoil, Jena, Germany) using FEI Mark I Vitrobot. The experiment was performed on our FEI Polara equipped with the GATAN "flip–flop" rotation stage at LN2 temperature and with an 86,000× CCD magnification. The energy filter slit was centered on the zero-loss peak with 20 eV slit width. Each data set was collected in the angular range of ±60° and at every 4°. It took about 17 min to take a data set that includes centering the target, running autoeucentricity and autofocus routines, and performing GIF zero-loss alignment. Thirteen targets were selected allowing time to collect them all, rotate the specimen, and resume data collection for the second axis within the same day. However, it should be emphasized that there are no constraints on the number of targets that can be selected. Upon completion of acquiring the tomographic data on all 13 targets, the specimen was manually rotated and the hierarchical alignment scheme was then invoked with all 13 targets successfully found and 12 of 13 data sets were acquired without any problem. The only failed data set was due to a glitch in the zero-loss alignment that failed to center the slit on the zero-loss peak and not any problems with the hierarchical alignment. Figure 12.10 presents one pair of dual-axis data sets where the top row displays three tilted images before the specimen rotation and the corresponding images after rotation are shown in the bottom row. Given the small size of 20S Proteasomes (150 Å), the 300 keV acceleration voltage, and the low dose per image (2.44 e/Å^2) visualization in a single frame is difficult. As a result, we show this pair of dual-axis data because the presence of ice crystals aids in demonstrating the robustness of the hierarchical alignment. Clearly, the ice crystals are fully covered throughout the data collection both before and after specimen rotation. The prerotation data set was collected at the stage xyz coordinate (−95.20, 152.96, 78.73) microns while the postrotation data was at (−287.96, −82.90, 79.08). Therefore, the specimen rotation has displaced the target by 304 μm from its original location. Our scheme was still able to discover its location with sufficient precision to allow tomographic data collection at magnifications as high as 80,000×.

Zheng et al. (2009) also analyzed the auxiliary dose distributed at each step prior to sequential data collection for this setup. For single-axis EMT data collection, the total auxiliary dose is about 0.404 e/Å^2 while dual-axis EMT data collection tallies 0.729 e/Å^2 extra dose.

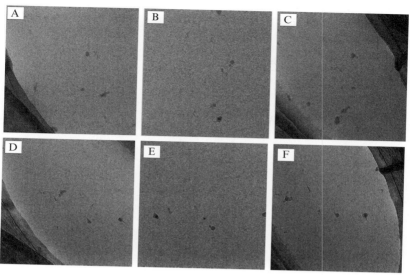

Figure 12.10 A pair of tomographic tilt data sets collected before and after specimen rotation. The top row displays three tilted images before specimen rotation and the bottom rows shows the counterparts at the same angles after specimen rotation. (A) $\alpha = -60°$, (B) $\alpha = 0°$, (C) $\alpha = +60°$, (D) $\alpha = -60°$, (E) $\alpha = 0°$, and (F) $\alpha = +60°$ (Zheng *et al.*, 2009).

 5. SUPPORTING AUTOMATION PROCEDURES

It is imperative to set the specimen to a reasonably accurate eucentric height and desired defocus prior to each EMT data collection. To exempt users from the burden of finding an auxiliary site for setup, low-dose automation procedures were developed in UCSF Tomography that allow these operations to be performed on site with only a very small amount of dose. The dramatic dose reductions were achieved via using much lower magnifications than used for collection of the actual tilt series. Since images acquired at low magnifications typically contain more features and a large defocus setting is used to further enhance the contrast, weaker beam setting and shorter exposure times can be used to minimize dose without reducing measurement precision.

5.1. Autofocus procedure

Targets can be automatically focused by comparing images recorded with different beam tilts. Koster *et al.* (1992) derived the mathematical expression for beam-tilt induced image shift $\vec{s}$:

$$\vec{s} = (f[F] + a[A] + b[B]) \cdot \vec{t} \tag{12.8}$$

where $\vec{t}$ stands for the beam–tilt vectors, f is the defocus and $[F]$, a 2D matrix, represents the effect of magnification and spherical aberration. a and b are the values of astigmatism in the two axes. $[A]$ and $[B]$ are the corresponding 2D matrices that reflect the contribution of astigmatism to beam–tilt induced image shift. These matrices need to be precalibrated before Eq. (12.8) can be used for measuring focus. Zheng *et al.* (2007a) extended this method to take stage drift and microscope misalignment into account. The revised version allows focus to be measured at low magnification ($\sim 14{,}000\times$ nominal reading) with sufficient precision (± 0.1 μm). If a fixed beam–tilt vector is used for both precalibration and focus measurement, Eq. (12.8) can be simplified into a vector form:

$$\vec{s} = f\vec{F} + \vec{A} \tag{12.9}$$

$\vec{F}$ and $\vec{A}$ correspond to the contributions from defocus and astigmatism to the beam–tilt induced image shift $\vec{s}$, respectively. Equation (12.9) can be further modified to reflect the contribution of mechanical and specimen drift by addition of a new term $\vec{d}$:

$$\vec{s} = f\vec{F} + \vec{A} + \vec{d} \tag{12.10}$$

To minimize the effect of drift, three images are taken: the first is recorded when the beam is tilted to the negative angle, the second at the positive angle, and the third at the negative angle again by tilting the beam back from the positive angle. The second image is cross-correlated with the first and then the third in order to measure the image shifts along the direction from negative tilt to positive tilt, expressed as $\vec{s_1}$ and $\vec{s_2}$, respectively. Since the drift vector reverses its direction if it is measured in the reversed temporal order, the drift vector contained in $\vec{s_1}$ is opposite to that in $\vec{s_2}$. By assuming that the drift has a constant rate and direction during the short period when these three images are taken and the time interval between two neighboring beam-tilted images is the same, the drift in $\vec{s_1}$ can be treated as being equal but opposite to that in $\vec{s_2}$. Therefore the average of $\vec{s_1}$ and $\vec{s_2}$, as is shown in Eq. (12.11), should have minimal contribution from the drift.

$$\vec{s_m} = \frac{\vec{s_1} + \vec{s_2}}{2} = f\vec{F} + \vec{A} \tag{12.11}$$

The determination of $\vec{F}$ and $\vec{A}$ can be precalibrated together by measuring $\vec{s_m}$ at two defocus settings by applying the same amount of defocus (Δf) above and below true focus, as are given in Eqs. (12.12) and (12.13).

$$\overrightarrow{F} = \frac{\overrightarrow{s}_{m1} - \overrightarrow{s}_{m2}}{2\Delta f} \tag{12.12}$$

$$\overrightarrow{A} = \frac{\overrightarrow{s}_{m1} + \overrightarrow{s}_{m2}}{2} \tag{12.13}$$

With the precalibrated $\overrightarrow{F}$ and $\overrightarrow{A}$, the focus can thus be measured based upon Eq. (12.14) by taking three images as suggested above for drift minimization.

$$f = \frac{(\overrightarrow{s}_m - \overrightarrow{A}) \cdot \overrightarrow{F}}{\left|\overrightarrow{F}\right|^2} \tag{12.14}$$

Due to practical limitations in the accuracy of the microscope alignment, switching magnifications can alter the focus. A precalibration approach is used to correct for this focus misalignment on either a test sample or an unimportant sample area. First true focus is set at the magnification where the EMT tilt data will be taken, then switched to the focusing magnification to perform the calibration of $\overrightarrow{F}$ and $\overrightarrow{A}$. The switch from high to low magnification may induce a change in focus (f_a). Therefore, Eq. (12.13) can be modified to add this term:

$$\overrightarrow{D} = f_a\overrightarrow{F} + \overrightarrow{A} = \frac{\overrightarrow{s}_{m1} + \overrightarrow{s}_{m2}}{2} \tag{12.15}$$

The average of the two displacement vectors obtained at two defocus settings now represents the combined effect from astigmatism and focus misalignment. Therefore, the calibrated $\overrightarrow{D}$ is stored along with the calibration of $\overrightarrow{F}$ and later used to determine the defocus unbiased by astigmatism and focus misalignment during data collection.

5.2. Autoeucentricity procedure

Manual adjustment to eucentric height is typically performed by wobbling the stage within an angular range (typically $\pm 15°$) while tuning the z height until specimen movement is minimal. One automated version was developed based upon this strategy. The general idea is to measure the specimen movement between $0°$ and some tilt angles, for example $\pm 15°$, as opposed to visual observation in manual operation. Images taken at these angles are cross-correlated to determine the shift from which the noneucentricity is estimated based upon geometric rotation. The stage z height is then adjusted accordingly. The major drawback of this approach stems from its direct detection of the specimen shift that can be very large when the eucentric error is severe. Without any prior knowledge of how severe the eucentric error is, users are forced to perform this operation at very low magnifications within a small angular range to avoid the abortive scenario that the specimen is shifted out of the field view. Several rounds of refinement may be necessary at gradually increased magnifications or within larger angular ranges. Given the large variability of eucentric errors over the grid due to the nonuniform thickness and flatness, this approach is quite inefficient and less robust.

The second strategy is based upon the concept of microscope eucentric focus alignment and has been adopted in Leginon (Suloway *et al.*, 2005, 2009) and UCSF Tomography. It is reasonably assumed that each stage has an intrinsic eucentric height that is deemed to be fixed unless the stage has been given a major service. This height can thus be labeled with a specific objective lens current (eucentric focus). When the objective lens is set to this current, the specimen will exhibit some defocus if its z position is not aligned to the stage eucentric height. Therefore, the eucentric error can be detected by measuring this defocus, which is then corrected by adjusting the stage z height instead of the lens current. This approach is much more robust and efficient than its counter-part based upon wobbling the stage, allowing detection of a very broad range of eucentric errors ($\sim 100\ \mu m$) at intermediate magnifications ($\sim 14,000\times$ nominal). When the eucentric error is significant ($> 10\ \mu m$), the eucentric focus based routine typically needs to be repeated twice in order to correct for the stage backlash in z direction.

As mentioned earlier, specimens must be tuned to the eucentric height and desired defocus prior to each EMT data collection. These improved autoeucentricity and autofocus procedures make it practical to perform these operations at much lower magnifications under low-dose conditions. According to our estimate, an extra dose of at total of only 0.404 e/$\mathring{A}^2$ is needed prior to each single-axis EMT data collection (Zheng *et al.*, 2009).

6. SUMMARY

UCSF Tomography provides full automation along the EMT pipeline that involves target finding, sequential acquisition of numerous EMT tilt sets, and real-time reconstruction at moderate resolution both for single and dual axes. The highly robust and efficient predictive data collection scheme is especially suited for high performance cryo-EMT data collection. The multiscale target finding and rotational alignment scheme enables seamless connection between searching targets to sequential EMT data collection at numerous targets for both single and dual axes. Real-time reconstruction at moderate resolution was achieved using a small Linux cluster to perform weighted back-projection in parallel with data collection, thereby providing users with a reconstructed 3D volume at the end of each data collection. This gives users immediate feedback on a broad range of information from sample preparation to experimental settings and allows efficient screening for the best data sets for final reconstruction at full resolution. Better decisions or experimental adjustment can thus be made right away to influence the subsequent data collections. All these features have been integrated into UCSF Tomography, allowing users to perform routine both cryo and room temperature, single- or dual-axis EMT data collection that may span several days without manual intervention. The extensive acceptance both inside and outside of UCSF strongly attests to the robustness, efficiency, and ease of use enabled by UCSF Tomography. We are continuously committed to further improvements on this software system for the benefit of the structural biology community.

ACKNOWLEDGMENTS

This work was supported by funds from the Howard Hughes Medical Institute and the W. M. Keck Advanced Microscopy Laboratory at UCSF.

REFERENCES

Baumeister, W. (2002). Electron tomography: Toward visualizing the molecular organization of the cytoplasm. *Curr. Opin. Struct. Biol.* **12,** 679–684.

Belmont, A. S., Sedat, J. W., and Agard, D. A. (1987). A threedimensional approach to mitotic chromosome structure: evidence for complex hierarchical organization. *J. Cell Biol.* **105,** 77–92.

Belmont, A. S., Braunfeld, M. B., Sedat, J. W., and Agard, D. A. (1989). Large-scale chromatin structural domains within mitotic and interphase chromosomes in vivo and in vitro. *Chromosoma* **98,** 129–143.

Braunfeld, M. B., Koster, A. J., Sedat, J. W., and Agard, D. A. (1994). Cryo automated electron tomography—Towards high resolution reconstructions of plastic embedded structures. *J. Microsc.* **172**(2), 75–84.

Carragher, B., Kisseberth, N., Kriegman, D., Milligan, R. A., Potter, C. S., Pulokas, J., and Reilein, A. (2000). Leginon: An automated system for acquisition of images from vitreous ice specimens. *J. Struct. Biol.* **132,** 33–45.

Champness, J. N., Bloomer, A. C., Bricogne, G., Butler, P. G., and Klug, A. (1976). The structure of the protein disk of tobacco mosaic virus to 5A resolution. *Nature* **259**(5538), 20–24, 1–8.

DeRosier, D. J., and Klug, A. (1968). Reconstruction of three dimensional structures from electron micrographs. *Nature* **217,** 130–134.

Dierksen, K., Typke, D., Hegerl, R., Koster, A. J., and Baumeister, W. (1992). Toward automatic electron tomography. *Ultramicroscopy* **40**(1), 71–87.

Dierksen, K., Typke, D., Hegerl, R., and Baumeister, W. (1993). Toward automatic electron tomography II. Implementation of autofocus and low-dose procedures. *Ultramicroscopy* **49**(1), 109–120.

Downing, K. H., Koster, A. J., and Typke, D. (1992). Overview of computeraided electron microscopy. *Ultramicroscopy* **46**, 189–198.

Frank, J., McEwen, B. F., Radermacher, M., Turner, J. N., and Rieder, C. L. (1987). Three-dimensional tomographic reconstruction in high-voltage electron microscopy. *J. Electron Microsc. Tech.* **6**, 193–205.

Grimm, R., Barmann, M., Hackl, W., Typke, D., Sackmann, E., and Baumeister, W. (1997). Energy filtered electron tomography of ice-embedded actin and vesicles. *Biophys. J.* **72**(1), 482–489.

Hart, R. (1968). Electron microscopy of unstained biological material: The polytropic montage. *Science* **159**, 1464–1467.

Iancu, C. V., Wright, E. R., Benjamin, J., Tivol, W. F., Dias, D. P., Murphy, G. E., Morrison, R. C., Heymann, J. B., and Jensen, G. J. (2005). A "flip–flop" rotation stage for routine dual-axis electron cryotomography. *J. Struct. Biol.* **151**, 288–297.

Keszthelyi, B., in preparation.

Koning, R. I., and Koster, A. J. (2009). Cryo-electron tomography in biology and medicine. *Ann. Anat.* **191**, 427–445.

Koster, A. J., Chen, H., Sedat, J. W., and Agard, D. A. (1992). Automated microscopy for electron tomography. *Ultramicroscopy* **46**, 207–227.

Koster, A. J., Braunfeld, M. B., Fung, J. C., Abbery, C. K., Han, K. F., Liu, W., Chen, H., Sedat, J. W., and Agard, D. A. (1993). Toward automatic three dimensional imaging of large biological structures using intermediate voltage electron microscopy. *MSA Bull.* **23**(2), 176–188.

Koster, A. J., Grimm, R., Typke, D., Hegerl, R., Stoschek, A., Walz, J., and Baumeister, W. (1997). Perspectives of molecular and cellular electron tomography. *J. Struct. Biol.* **120**, 276–308.

Leis, A., Rockel, B., Andrees, L., and Baumeister, W. (2009). Visualizing cells at the nanoscale. *Trends Biochem. Sci.* **34**(2), 60–70.

Mastronarde, D. N. (2005). Automated electron microscope tomography using robust prediction of specimen movements. *J. Struct. Biol.* **152**, 35–51.

McEwen, B. F., Radermacher, M., Rieder, C. L., and Frank, J. (1986). Tomographic three-dimensional reconstruction of cilia ultrastructure from thick sections. *Proc. Natl. Acad. Sci. USA* **83**, 9040–9044.

Papadopoulos, P. M., Katz, M. J., and Bruno, G. (2001). NPACI rocks: Tools and techniques for easily deploying manageable linux clusters. Cluster 2001: IEEE International Conference on Cluster Computing.

Potter, C. S., Chu, H., Frey, B., Green, C., Kisseberth, N., Madden, T. J., Miller, K. L., Nahrstedt, K., Pulokas, J., Reilein, A., Tcheng, D., Weber, D., *et al.* (1999). Leginon: A system for fully automated acquisition of 1000 electron micrographs a day. *Ultramicroscopy* **77**, 153–161.

Rath, B. K., Marko, M., Radermacher, M., and Frank, J. (1997). Low-dose automated electron tomography: A recent implementation. *J. Struct. Biol.* **120**, 210–218.

Steven, A. C., and Aebi, U. (2003). The next ice age: Cryo-electron tomography of intact cells. *Trends Cell Biol.* **13**(3), 107–110.

Suloway, C., Pulokas, J., Fellmann, D., Cheng, A., Guerra, F., Quispe, J., Stagg, S., Potter, C. S., and Carragher, B. (2005). Automated molecular microscopy: The new Leginon system. *J. Struct. Biol.* **151**, 41–60.

Suloway, C., Shi, J., Cheng, A., Pulokas, J., Carragher, B., Potter, C. S., Zheng, S. Q., and Agard, D. A. (2009). Fully automated, sequential tilt-series acquisition with Leginon. *J. Struct. Biol.* **167**, 11–18.

Zheng, Q. S., Braunfeld, M. B., Sedat, J. W., and Agard, D. A. (2004). An improved strategy for automated electron microscopy tomography. *J. Struct. Biol.* **147,** 91–101.

Zheng, S. Q., Kollman, J. M., Braunfeld, M. B., Sedat, J. W., and Agard, D. A. (2007a). Automated acquisition of electron microscopic random conical tilt data sets. *J. Struct. Biol.* **157,** 148–155.

Zheng, S. Q., Keszthelyi, B., Branlund, E., Lyle, J. M., Braunfeld, M. B., Sedat, J. W., and Agard, D. A. (2007b). UCSF tomography: An integrated software suite for real-time electron microscopic tomographic data collection, alignment, and reconstruction. *J. Struct. Biol.* **157,** 138–147.

Zheng, S. Q., Matsuda, A., Braunfeld, M. B., Sedat, J. W., and Agard, D. A. (2009). Dual-axis target mapping and automated sequential acquisition of dual-axis EM tomography data. *J. Struct. Biol.* **168,** 323–331.

Ziese, U., Janssen, A. H., Murk, J.-L., Geerts, W. J. C., Van Der Krift, T., Verkleij, A. J., and Koster, A. J. (2002). Automated high-throughput electron tomography by pre-calibration of image shifts. *J. Microsc.* **205**(2), 187–200.

CHAPTER THIRTEEN

Correlated Light and Electron Cryo-Microscopy

Ariane Briegel,*,† Songye Chen,* Abraham J. Koster,‡
Jürgen M. Plitzko,§ Cindi L. Schwartz,¶ and Grant J. Jensen*,†

Contents

Abstract

Light and electron cryo-microscopy have each proven to be powerful tools to study biological structures in a near-native state. Light microscopy provides important localization information, while electron microscopy provides the resolution necessary to resolve fine structural details. Imaging the same sample by both light and electron cryo-microscopy is a powerful new approach that combines the strengths of both techniques to provide novel insights into cellular ultrastructure. In this chapter, the methods and instrumentation currently used to correlate light and electron cryo-microscopy are described in detail.

1. Introduction

Electron cryo–microscopy (ECM), including electron cryo-tomography (ECT), is a uniquely powerful tool for investigating cellular ultrastructure to high ("macromolecular") resolution with minimal preparative artifacts.

* Division of Biology, California Institute of Technology, California Boulevard, Pasadena, California, USA
† Howard Hughes Medical Institute, California Institute of Technology, California Boulevard, Pasadena, California, USA
‡ Department of Molecular Cell Biology, Faculty of Biology and Institute of Biomembranes, Utrecht, The Netherlands
§ Max-Planck-Institut für Biochemie, Abteilung Molekulare Strukturbiologie, Martinsried, Germany
¶ Boulder Laboratory for 3-D Electron Microscopy of Cells, Department of MCD Biology, University of Colorado, Boulder, Colorado, USA

Methods in Enzymology, Volume 481
ISSN 0076-6879, DOI: 10.1016/S0076-6879(10)81013-4

The identification of cellular components in cryo-micrographs and cryo-tomograms can be a challenge, however. Sometimes the structures of interest are large enough and have such characteristic appearances that they can be recognized immediately in the image. Synapses, microtubules, bacterial micro-compartments, and magnetosomes are examples (Iancu *et al.*, 2009; Komeili *et al.*, 2006; Koning *et al.*, 2008; Lucic *et al.*, 2005; Scheffel *et al.*, 2006). Similarly, large macromolecular complexes can also sometimes be identified by their shape through "template matching" approaches (Chapter 11, Vol. 483). In this procedure, a three-dimensional image (usually a tomogram) is searched by cross-correlation for features that resemble an object with known structure (the template). Template matching was recently used, for instance, to map the locations and orientations of several large macromolecular complexes in the pathogenic bacteria *Leptospira interrogans* and *Mycoplasma pneumoniae* (Beck *et al.*, 2009; Kühner *et al.*, 2009). Unfortunately, because of resolution and signal-to-noise limitations as well as molecular crowding in the cytoplasm, template matching has only been possible for just a few of the largest complexes (typically larger than 500 kDa).

In addition to pattern or shape recognition, another way structures of interest have been identified in ECM images is through comparison of images of mutant strains wherein the abundance, stability, or localization of the structure is perturbed. A cytoskeletal filament near the constriction site of dividing *Caulobacter crescentus* cells was identified as FtsZ this way for example. While filaments similar in shape, orientation, and length were more numerous in cells overexpressing FtsZ, they were entirely absent in cells depleted of FtsZ. Furthermore, when mutations were introduced into FtsZ that were expected to stabilize filaments, even more numerous and longer filaments were observed (Li *et al.*, 2007). While comparisons of mutants can help identify certain structures, the results are unfortunately indirect, mutations can intro-duce artifacts, and the work can be time consuming.

In order to circumvent these problems, substantial efforts have been invested into developing an electron dense tag that could be used to mark features of interest directly in ECM images (Chapter 9). Metallothionein tags have now been used, for instance, to label proteins by nucleating gold clusters inside *Escherichia coli* cells grown in the presence of gold-salts (Diestra *et al.*, 2009), but these methods are still in an early stage of development.

Correlated light and electron microscopy now appears to be a powerful alternative. A wonderful arsenal of specific and/or genetically fusible fluor-ophores has been developed that can be tracked dynamically by light microscopy (LM). Techniques to locate these same tags in conventional (chemically fixed, plastic-embedded) EM preparations have already been developed, including immunogold labeling and the use of a tetracysteine biarsenical system to locally photoconvert diaminobenzadine in the prox-imity of the fluorophore into osmophilic polymers (for a more detailed review on these methods, see e.g., Giepmans, 2008). In order to correlate fluorescent light microscopy (FLM) images with EM images of cells in their native (frozen-hydrated) states, the samples must be frozen. Thanks to the recent development of a set of cryo-LM stages, FLM images can now be recorded of samples after they are frozen. Combining FLM and ECM utilizes the strength of both types of microscopy: the ability to easily and specifically tag a protein of interest and then follow it dynamically at low magnification by FLM is combined with the high resolution and near-native sample preservation of ECT. Several approaches to correlated FLM/ECM are described and compared in detail here.

2. Correlative FLM/ECM with Freezing after FLM Imaging

Correlated FLM/ECM can be done by freezing the sample either before or after FLM. The main advantage of doing the FLM at room temperature is that high-powered, oil-immersion lenses with large numerical apertures (NAs) can be used, maximizing resolution. This can be especially important when working with small cells such as bacteria, where the dimensions of the structures of interest are often comparable to or smaller than the resolution limit of the LM (which is typically about 200 nm). FLM at room temperature can also be done with any typical FLM instrument and without any specialized cryo-stage (CS). The experiment has three basic steps:

(a) cells on an EM grid are imaged in the FLM at room temperature
(b) grids are plunge-frozen
(c) individual cells are imaged in the cryo-EM

(a) *Cells on an EM grid are imaged in the FLM at room temperature.* Theoretically, living cells could be observed moving and growing on an EM grid in an LM, and then at some critical moment, be plunge-frozen and observed by EM. Such an experiment would be limited by the speed at which ultrastructures changed within the cell and the time required to remove the sample from the LM and plunge-freeze. In practice, the only examples to date of correlated FLM/ECM where the FLM was done at room temperature have been of fixed cells. Fixation protected the cells from damage that otherwise occurred when they were immobilized on EM grids (Briegel *et al.*, 2008). While this fixation does introduce minor artifacts through cross-linking of cellular components, many structural relationships are well preserved.

Bacterial cells can be fixed as follows: a 2 ml aliquot of the bacterial cell culture is gently pelleted (5 min at 1500rcf), resuspended in a fixing solution (0.1% glutaraldehyde, 3% formaldehyde, and 20% sodium phosphate at pH 7 (for 15 min at room temperature or alternatively 45 min on ice)), washed three times in washing buffer (140 mM NaCl, 3 mM potassium KCl, 8 mM Na$_2$HPO$_4$, and 0.05% Tween), and finally resuspended in the imaging buffer (50 mM glucose, 10 mM EDTA, and 20 mM Tris–HCl pH 7.6) (see Briegel *et al.*, 2008). Alternatively, 800 μl of cell culture can be added to 200 μl 5× concentrated fixing solution (12.5% paraformaldehyde in 150 mM Na–phosphate buffer, pH 7.5) and incubated for 15 min at room temperature. After gentle centrifugation (5 min at 1500rcf), the pellet is washed twice in Na–phosphate buffer and resuspended in ∼40 μl fresh buffer.

Following fixation, the cells can be immobilized onto specialized "Finder" TEM grids, which are marked with symbols, letters, or numbers to facilitate refinding targets in both the FLM and EM. We find the "London," or type "H2" finder-grids the most convenient to use (see Fig. 13.1). EM finder-grids can be purchased precoated with a holey-carbon support film (we commonly use R2/2 Quantifoil grids (Quantifoil Micro Tools GmbH, Jena, Germany)), and are first glow discharged for several minutes (note that different labs use very different glow discharging times ranging from only a few seconds to several minutes). A 5 μl droplet of 0.5 mg/ml sterile filtered poly-L-lysine (Sigma P1524) is added to each grid and subsequently dried in a 60 °C oven. A 4 μl droplet of fixed cell solution is then added to a poly-L-lysine treated grid, gently blotted with filter paper and rinsed in buffer.

The grid is then immediately placed onto a droplet of buffer on a glass slide and is covered directly with a standard cover slip. More sensitive objects such as mammalian cells may need to be protected from direct contact with the cover slip. A spacer such as a rubber gasket can be used to provide adequate space between the sample and the cover slip. The sample can now be imaged in any FLM, using any desired objective lens. If oil immersion is used, avoid contact of the immersion oil with the edges of the cover slip to prevent contamination of the sample. Overview maps of large areas of the grid are typically taken to pinpoint areas of interest on the finder grid, sometimes followed by high-magnification images of specific cells.

(b) *Grids are plunge-frozen.* After FLM imaging, the grid needs to be plunge-frozen. First, excess buffer is injected between the cover slip and the glass slide to "float" the cover slip off from the top of the grid. The grid is then quickly picked up with tweezers and placed in the plunge-freezer at 100% humidity to prevent drying of the sample (note: not all plunge-freezers allow humidity-control). A 4 μl droplet of BSA-treated 10 nm gold solution is added and the grid is plunge-frozen in liquid ethane or an ethane/propane mixture (Chapter 3). The grid can then be stored (essentially indefinitely) at liquid nitrogen temperature, shipped to an EM facility, or imaged immediately in an electron microscope (EM).

(c) *Individual cells are imaged in the cryo-EM.* Refinding the same cells imaged by FLM in the EM is the most challenging step. Even though the

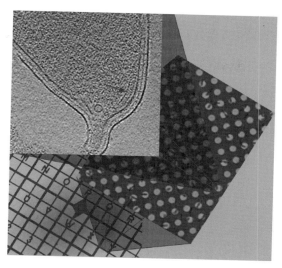

Figure 13.1   *C. crescentus* cells with fluorescently labeled chemoreceptors (mCherry-tagged MCPA) are seen immobilized on a poly-L-lysine-coated holey-carbon support film (Quantifoil R2/2). The letters and numbers on the finder grid help locate targets in the light and electron microscopes (bottom left). FLM images (center right) are recorded of favorable regions of the grid using standard FLM instruments. Oil-immersion lenses with high-NAs can be used, since the sample is imaged at room temperature in the FLM. Here individual cells appear as dark, curved rods with bright red spots at their tips spanning the circular holes. After the sample is imaged in the FLM, the grid is retrieved from underneath the cover slip, plunge-frozen, and inserted into a cryo-EM. Tomograms of the same individual cells imaged previously by FLM are recorded, providing views of structures in nearly native states with macromolecular resolution (upper left corner). The positions of the fluorescent spots are seen to correlate perfectly with the locations of large protein arrays in the cryo-tomograms (red arrow), confirming that these structures are in fact the chemoreceptor arrays. (See Color Insert.)

cells have been immobilized on the grids, some shift their position or detach completely when the grid is removed from the LM and plunge-frozen. Finding the correct targets can be tedious and is not always possible. Software packages such as TOM (Nickell *et al.*, 2005), SerialEM (Mastronarde, 2005), or Leginon (Suloway *et al.*, 2005, 2009) can be used to generate whole grid maps that help, but care must be taken to verify that the same cells are in the same places before conclusions can be drawn. Typically only a half or less of the areas imaged in the FLM are found frozen in suitably thick vitreous ice in unperturbed locations. Once such cells are found, ECM images are recorded (usually full ECT tilt-series) as described elsewhere in this volume (Chapter 12).

While this correlation approach facilitates FLM investigation without restrictions regarding the objective lenses and maximizes the achievable resolution, it only works for structures that are unperturbed by light fixation or are relatively stable (i.e., do not change in the seconds to minutes that expire between the time FLM images are recorded and the grid is plunge-frozen). Once objects are identified, their ultrastructure can be analyzed in cryo-tomograms of unfixed (and therefore more native) samples. This strategy was followed for example to first confirm the identity of chemoreceptor arrays in the bacterium *C. crescentus* by correlated FLM/ECM, but then the architecture of these large polar arrays was analyzed in cryo-tomograms of unfixed cells (Briegel *et al.*, 2008).

3. Correlative FLM and ECM, with Freezing before FLM Imaging

Due to the recent development of nitrogen-cooled cryo–LM stages, previously vitrified cells grown on EM grids (or vitreous sections) can now be directly imaged in FLM. The main advantages of this approach are that (a) the correlation between FLM and ECT is direct and without any dislocations or disruptions caused by lifting cover slips, adding fiducial markers, blotting and/or plunge-freezing; (b) no extra sample preparation steps such as light fixation are needed, making the strategy more general and reducing potential artifacts; (c) the method works for cryo-sections; and (d) the frozen–hydrated grids can be screened in the FLM to identify favorable regions and search for cells or cellular structures tagged with fluorophores. As in ECM, however, the samples need to be kept below approximately $-140\ ^\circ$C to prevent crystallization of the amorphous ice, so special instrumentation is required.

The experiment has three basic steps:

(a) fluorescently labeled cells are either (1) vitrified onto commercially available EM finder-grids using standard plunge freezing methods (Chapter 3) or (2) high-pressure-frozen and then cryo-sectioned (Chapter 8), in which case the cryo-sections are collected onto the EM finder-grids;
(b) FLM images are recorded while the grid is located in a cryo–LM stage;
(c) the grid is transferred into the cryo-EM, cells or cryo-sections are relocated and imaged by ECM.

Although commercial low-temperature LM stages have been available for some time, they had to be adapted to hold the standard EM grids and keep them below $-140\ ^\circ$C. At least three different cryo-LM stages have been developed for this purpose, either "from scratch" with novel designs (Sartori *et al.*, 2007; Schwartz *et al.*, 2007), or modified from a commercial low-temperature stage (van Driel *et al.*, 2009). These stages have been made to work with most conventional fluorescent LMs and standard LM techniques such as epi-fluorescence, large area scanning, and z-stack microscopy. Because of the low-temperature of the sample, however, oil- or water-immersion objective lenses with high–NA cannot be used. The resolution of cryo-FLM is therefore limited by the NA of the dry objective lens. This diffraction-based limit is given by $r = (0.61\lambda) /NA$, where r is the resolution limit and λ is the wavelength of light used. Thus for NA = 0.75 (the highest NA of the objective lenses that have been used with these CSs; Rigort et al., 2010; van Driel et al., 2009), the theoretical resolution limit is estimated to be $\sim$400 nm, and experimentally further aggravated with increasing ice thickness. This resolution was verified by direct measurement of the point spread function (PSF) of the fluorescent signals from subresolution ($\sim$200 nm) yellow/green fluorescent beads (van Driel et al., 2009).

At the beginning of these studies, it was unclear whether standard fluorophores would fluoresce at such low-temperatures. Wonderfully, not only did the fluorophores still fluoresce, but bleaching was slowed down significantly, enabling the integration of fluorescence signal over a longer period of time than would be possible at room temperature (Schwartz et al., 2007).

Figure 13.2 shows the schematic of the CS described in Sartori et al. (2007) and the picture of the LM setup with the CS. This CS is designed for inverted light microscopes, but the frozen-hydrated EM grid can also be illuminated by transmitted light. So in addition to the fluorescence images, phase-contrast images of the sample can be obtained. The vitrified sample grid is loaded in the precooled stage and no other loading station is needed. An insert within the sealed liquid nitrogen (LN_2) reservoir is designed to place a grid storage box for grid transferring. The grid is positioned in the center of the brass-metal holder and held down by a 1 mm-thick brass clamp ring. The grid is kept within the metal block in a cold N_2 atmosphere, which in turn is kept at LN_2 temperature through direct contact with the LN_2 reservoir. There are two outlets in the metal holder for the LN_2 reservoir, one for the LN_2 refill and the other for cold N_2 vapor release. The grid temperature is kept stable for about 30 min with a full LN_2 reservoir, and up to 2 h with manual refills.

The CS is effectively insulated from the ambient atmosphere by an isolation box, which is made of a thermoplastic layer similar to conventional Styrofoam and adapted to the motorized LM stage. The whole box is then covered by a lid with a central opening for the condenser insertion. As shown in the enlarged inset in Fig. 13.2, the grid and the brass clamp ring are further insulated by two glass cover slips of 0.17 mm thickness. This way a "dry-air" slab is created as an isolation layer between the grid and the air objective lens by a plastic ring sealed with another glass cover slip (0.17 mm in thickness). The outside surface of the lower cover slip is constantly flushed with dry N_2 gas at room temperature during the

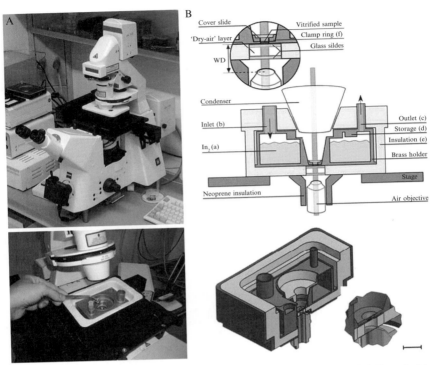

Figure 13.2   The cryo-stage and LM set up from Sartori *et al.* (2007). (A) Cryo-holder mounted on an LM, general view (*above*) and a close-up showing the interior of the cryo-holder. (B) Schematic of the whole cryo-holder assembly. Section of the cryo-holder in operational mode on the stage of the optical microscope (*above*) and a view in perspective. Insets show magnified views of the central part (*scale bar 24 μm*). Reproduced from Histochemistry and Cell Biology with permission (Lucic *et al.*, 2008).

experiments to prevent condensation and frost formation. A long working distance (LWD) 40× objective with 7.7–8.3 mm WD and 0.55NA was used in Sartori *et al.* (2007).

   This CS was later improved and commercialized by FEI (Eindhoven, The Netherlands) (Rigort *et al.*, 2010). Figure 13.3 shows a schematic of this second-generation cryo-correlative stage (Cryostage²). The major improvements compared to the previous version are: (1) There is only one glass cover slip between the grid and the objective lens, so a higher NA LWD objective lens can be used, for example, 63×, NA 0.75 and WD 1.8–2.4 mm from Zeiss, to obtain higher resolution. The optical aberrations generated by the presence of the second glass cover slip in the optical path are also eliminated. (2) It employs an automated LN_2 pumping system (Norhof LN2 Microdosing system, Series #900, Maarsen, The Netherlands), which monitors and adjusts the LN_2 level and the temperature to allow

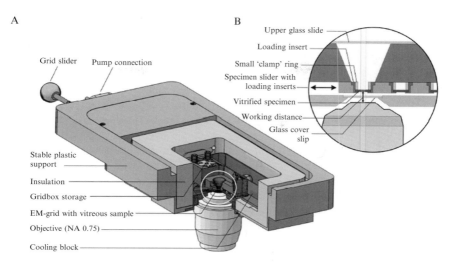

Figure 13.3 Schematic illustration of the cryo-correlative stage (Cryostage^2) for an inverted light microscope. (A) Cut-away perspective view of the Cryostage^2 holder with the 63× long distance objective (NA 0.75), depicting housing and insulation requirements. (B) Magnified view of the central part (white circle in A), pointing at the imaging position and specimen slider with four loading positions for vitrified EM grids. Bar: 40 mm. Inset (B) not drawn to scale. Reproduced from *Journal of Structural Biology* with permission (Rigort *et al.*, 2010).

longer time experiments. (3) It holds up to four frozen-hydrated grids, and each can be slid into the center window for viewing. (4) It can be used with so called Autogrids (FEI), also called C-clip rings (see Fig. 13.4) to facilitate transfer and manipulation of the fragile EM grids. These Autogrids used with the latest generation of electron microscopes guarantee steadier specimen support with their rigid reinforcement ring and stabilize the grid during multiple transfer and handling steps (Rigort *et al.*, 2010).

Transferring the grids from the grid box inside the stage minimizes ice-contamination. As shown in the inset (B) of Fig. 13.3, the grids are stored beneath the main cooling block when not being transferred or imaged to further prevent ice-contamination. During imaging, the frozen-hydrated sample is protected by the cover glass slide and the dry N_2 gas purging the stage.

The correlation between cryo-FLM and EM images is done through reference points. Correlation is first roughly determined by visual inspection. For better correlation, or more precise determination of the position on the grid, coordinate transformation-based software was developed as a part of the TOM package. For each image tile, a phase-contrast image and a fluorescence-signal image are recorded. The phase-contrast image is used to determine the orientation of the map with respect to the TEM map.

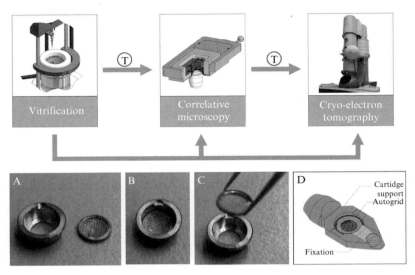

Figure 13.4 Schematic illustration of the necessary transfer steps (T) from plunge-freezing via FLM to ECT. To improve the success rate and to minimize damage or loss of the fragile specimen, reinforced autogrids can be used, which are inserted and placed in the "cups" of the correlative cryo-stage (A–C). These autogrids can then be placed in modified Polara cartridges, containing a simple spring load mechanism (D), or directly transferred to the latest generation of electron microscopes (Titan Krios, FEI Company, Eindhoven, NL).

The fluorescence-signal map shows the signal of the labeled specimen features that are of interest for TEM imaging and will be superimposed (see following). After acquisition of the LM map, the specimen is inserted into the TEM and another map is acquired. The LM map needs to be registered with the TEM map and then transformed into the same coordinate system to enable an overlay. Maps can be scaled, rotated, shifted, and mirrored with respect to each other. Furthermore, the contrast of a frozen-hydrated specimen in a light microscope is formed by a phase shift of the photons interacting with the sample (phase contrast), whereas the contrast formation mechanism of an electron microscope in low-magnification mode (with objective lens switched-off) is based on absorption contrast. Therefore, the maps will show similar but not identical features, for example, cells in the TEM will be black, whereas in the light microscope they will be transparent and only visible by phase contrast of their outer membrane (Lucic *et al.*, 2007). Given reference points, such as the symbols of the finder grid or the corners between grid bars, the software calculates the transformation between the EM coordinates and the LM image coordinates. The transform is then used to either find the position on the LM image that corresponded to the EM coordinates of a feature of interest identified in EM, or to find EM coordinates of a feature identified on the LM image (Fig. 13.5).

A

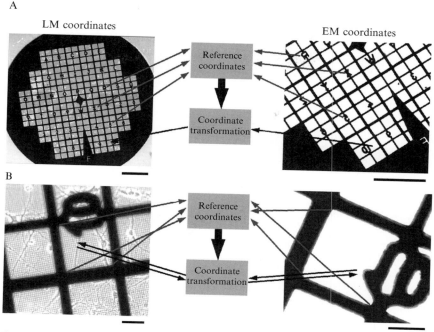

Figure 13.5 Schematic representation of the correlation method. (A) Positions of grid symbols in both LM (left) and EM (right) are used as reference coordinates (gray arrows) that are required to establish a coordinate transformation. The application of this transformation to the EM position of the region of interest facilitates the identification of the field where the predetermined region is located on the LM map (black arrows). Scale bars 500 μm. (B) Positions of grid corners in both LM (left) and EM (right) are used as reference coordinates (gray arrows) that are required to establish a precise coordinate transformation. This transformation is used either to find the position on the LM image that corresponds to the feature of interest identified in EM, or to find the EM position of the feature identified on the LM image (black arrows). Scale bars 50 μm. Reproduced from Journal of Structural Biology with permission (Lucic et al., 2007).

Simple cross–correlation approaches to register the maps are unreliable and computationally intensive, since all rotations, mirroring and scaling parameters need to be sampled in order to find the correct orientation of the map. Instead, feature-based methods such as the scale-invariant feature transform (SIFT) algorithm (Lowe, 2004) or its successor, speeded up robust features (SURF) (Bay et al., 2008), are. The advantage of SIFT is that key points can be detected and matched regardless of differences between maps in SNR, scaling, shift, rotation, and contrast. The fluorescence maps have a lower point resolution due to the lower magnification; however, and key points cannot be identified with pixel-accuracy due to the different contrast formation mechanisms between both types of maps. Furthermore, spatial aberrations of the LM objective will cause distortions in the LM map. Therefore, the overlay can only show approximately where features of interest are located on the TEM grid. Typical errors of superposition are in the range of 1–2 μm. Nevertheless, this technique greatly facilitates the search for tomography positions on the grid when searching for rare events in the specimen. For example, to identify cells with a certain phenotype that have been labeled with a fluorescent dye, the overlay can provide a rough orientation aid for the user to locate these cells among others, which do not exhibit this phenotype. SIFT has been implemented in the new TOM toolbox called TOM^2 (to be released).

Further modifications were made to the Cryostage^2 at Caltech as shown in Fig. 13.6. The grid holder was modified to hold up to 2 FEI TF30/Polara cartridges. This modification allows frozen-grids to be loaded into cartridges and then imaged in both the FLM and EM without further manipulation. The advantages are that (1) the grid has exhibited higher mechanical stability in the FLM, probably because the clamp mechanism in the Polara cartridges is more robust; (2) less ice–contamination and damage occurs during transfer between the FLM and EM, since the handling is simplified; (3) the coordinate transformation between FLM and EM images is always the same, so that targets can be easily relocated in the EM without any help from software.

A protocol to use such a modified FEI CS is as follows:

1. Load the frozen–hydrated grids (with whole cells or cryo–sections) into cartridges in the Polara cryo–transfer station. Note the cartridges must be custom–modified to fit in the cryo–LM stage. The grid is held down by a copper clip ring.
2. Keep the cartridges in liquid nitrogen during storage and transfer.
3. Place the CS on the light microscope. Connect the stage to the liquid nitrogen pump and cool down the stage to liquid nitrogen temperature ($-190\ ^{\circ}$C).

Figure 13.6 Schematics of the Caltech-modified stage. (A) Picture of the stage with a modified Polara cartridge mounted. (B) Schematic drawing of the modified specimen slider, showing the housing for two Polara cartridges.

4. Load the cartridges into the CS. To prevent ice-contamination, when the cartridges (and hence the grids) are not being imaged, store them beneath the main cooling block.
5. Move one cartridge/grid into the viewing window and focus the objective lens onto the grid.
6. Find samples and take images. The grid is usually scanned and images montaged to give an overview.
7. Transfer the cartridges into the cryo-EM.
8. At low magnification, scan the grid and relocate targets. Record ECM images (usually full ECT tilt-series).

The other two CSs are made for upright light microscopes (Figs. 13.7 and 13.10). Both stages have heaters that can be used to maintain a constant desired temperature.

In the stage designed by Schwartz et al. (2007), the grid is cooled by cold N_2 gas evaporated from LN_2 and kept at $-150\ ^\circ C$. The temperature of the grid is monitored and used to regulate the flow of cold N_2 gas by adjusting the heater in the Dewar. The grid is clamped between the silver plate and a small attached cover plate with a v-shaped recess to allow imaging of the central area of the grid, and the loading is difficult. The depression that contains the specimen must be covered with a Styrofoam sleeve that slides up and down on the objective, to minimize ice-contamination on both the grid and the lens surface during operation. A Zeiss objective lens (40×, NA 0.63, WD < 1 mm) and a Nikon ELWD lens (40×, NA 0.66, WD ~ 5 mm) were both used (Schwartz et al., 2007). Because N_2 gas rather than LN_2 is used to cool the sample, there is very little mechanical vibration, and long exposure times (~ 10 s) can be used without drift or focus change.

Instec. Inc. (Boulder, CO, USA) has revamped and commercialized this CS for upright microscopes (model # CLM77K). The CLM77K is designed to fit on any modern upright microscope and a CS for use with inverted microscopes (CLM77Ki) is in final stages of design. Among its many improvements, the CLM77K has ports to allow imaging with both phase and fluorescence, allows the viewing of multiple grids, and the cooling system keeps grids well below devitrification temperatures ($\ll -140\ ^\circ C$) (Fig. 13.8).

The CLM77K stage holds up to nine grids in a shuttle device called the grid holder tongue (GHT), which is preloaded in a separate Styrofoam box under LN_2 and then placed into the stage. The GHT has a thermocouple attached that allows for monitoring the GHT temperature at all times. It sits directly on a cold plate, cooled by LN_2 that is pulled by a pump through channels bored within the plate. There are two ports for nitrogen gas. One port is used for flowing N_2 over the viewing windows to keep them frost-free. The other port flows N_2 into the sample chamber to keep the chamber purged and free of water vapor. The N_2 flow within the CS is controlled by

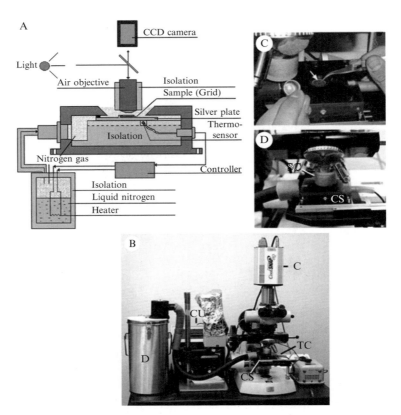

Figure 13.7 The cryo-stage schematic and the LM set up from Schwartz *et al.* (2007). (A) Schematic diagram showing the three parts of the cryo-LM stage: The cryo-stage itself with the silver plate that holds the specimen grid, the Dewar that contains liquid nitrogen (*dark blue*) and a heater that blows cold nitrogen gas (*speckled blue*) over the grid, and the temperature controller (*red*) that reads the temperature from the thermo-sensor and controls the heater within the Dewar. (B)–(D) Pictures of the cryo-stage attached to a Zeiss Universal microscope. The Dewar (*D*) is on the left and is connected to the cryo-stage (CS) via thermally insulated tubing. The grid area is protected by a Styrofoam cylinder (SC) during imaging. The temperature control (TC) probe is attached to the right side of the stage and feeds back to the controller unit (CU). The digital camera (C) is used to collect the images. Reproduced from Journal of Microscopy with permission (Schwartz *et al.*, 2007). (See Color Insert.)

small hex-screws. There are quartz viewing windows that allow phase imaging in conjunction with the fluorescent imaging. The working distance is about 5 mm, requiring LWD lenses. Examples of lenses used are: Nikon ELWD 20× (WD = 11 mm, NA = 0.4), ELWD 50× (WD = 8.7 mm, NA = 0.55), and SLWD 100× (WD = 6.5 mm, NA = 0.7).

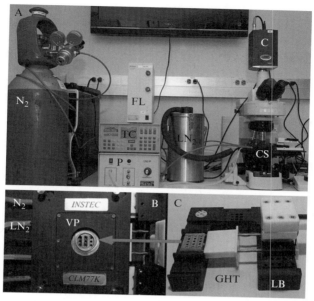

Figure 13.8 The Instec CLM77K cryo–stage. (A) The cryo-stage (CS) is placed into an existing upright LM stage, in this case, a Nikon 80i with a fluorescent light (FL) source. The temperature controller (TC) is plugged directly into the cryo-stage and the pump (P) pulls LN$_2$ through the cryo-stage to keep it cold. A digital camera (C) is used to record images that are later used to correlate with EM maps. (B) The CLM77K cryo-stage has a large quartz viewport (VP) that accommodates all nine grid positions of the grid holder tongue (GHT). Dry N$_2$ gas flows across the upper and lower viewports to keep them frost-free. (C) The GHT is shown here mounted in its loading block (LB). The block would be immersed in LN$_2$, the grids removed from their storage box, and placed in either of the nine-grid positions. Photographs taken by Ian Rees.

A protocol for using the Instec CLM77K CS:

1. Place the CLM77K CS on your microscope stage with the ports facing either to the right or left, whichever is most convenient for the operator. Connect the service cable to the mK1000 temperature controller. Be sure the stage is at an XYZ location so that all of the objective lenses can be moved on the turret and focusing is possible. Sometimes the LN$_2$ hoses can freeze up, limiting movement of the stage.

2. Remove excess moisture from the CS by heating to 110 °C for 5–10 min. In addition, attach a dry N$_2$ gas line to the LN$_2$ inlet port to purge the inside of the LN$_2$-cooling channel of the cold plate for at least 1 min.

3. Fill the LN$_2$ Dewar and attach the LN$_2$ inlet and outlet tubing as well as the dry N$_2$ gas tubing. Cool down the CLM77K CS to − 193 °C using the mK1000 temperature controller. Adjust the flow of the dry N$_2$ gas so

you can hear a slight hissing when your ear is very near the stage. You will have to adjust the dry N_2 gas flow depending on ambient humidity to keep the viewing windows frost-free.

4. When the CS is cooled, prepare the GHT, the GHT platform, and the sample covers by placing them in LN_2. A shallow Styrofoam box works best. Using good cryo-technique (precooling tools, limiting breath contamination, keeping grids submerged or right at the LN_2 level, etc.) load grids into the GHT, placing the sample covers on as you go. You can also slide the shutter forward to cover previously loaded grids. Completely close the shutter when you are finished loading grids.

5. With the loaded GHT in LN_2 and near the CLM77K CS, quickly transfer the GHT into the CS (use less than 5 s to prevent possible devitrification). Push the GHT into the CS until it is fully engaged and a "click" is heard.

6. Using appropriate objectives, find your regions of interest using either phase imaging or fluorescence. A motorized XYZ stage is highly recommended. If your LM software is capable, you can record calibrated XY stage positions, which can be used later for correlation with EM maps in SerialEM.

7. To remove the grids, essentially, you work in backward order by closing the GHT shutter, removing the GHT from the CS and immersing in LN_2, removing the sample covers, and placing grids back into their storage boxes.

8. When finished, warm the CLM77K CS to room temperature by stopping the LN_2 pump. When warmed to room temperature, heat the stage to 110 °C and purge the LN_2 channel with dry N_2 gas as above.

9. After loading the grid into the electron microscope, in Boulder we use SerialEM (Mastronarde, 2005) to make maps that we can correlate with imported cryo-LM images. SerialEM is used to map the entire grid at a low magnification ($\sim 150\times$). We can use that information to make a rough correlation to particular grid squares we imaged in the light microscope. Then, at higher magnification ($\sim 3000\times$), we map the grid square containing our area of interest we found in the light microscope (Fig. 13.9A). We import the light microscope images as maps (Fig. 13.9B and C) and then mark points in both the EM and LM maps to make them in register. Once they are in register, we can click on any area of the LM image and go to it's corresponding position in the EM. In Fig. 13.9, we mapped a vitreous section of PtK2 cells stained with the vital dye Hoescht 33342, which targets DNA.

Van Driel *et al.* chose to modify a commercially available CS Linkam THMS 600 (Fig. 13.10). The sample and the silver block support are cooled by LN_2 pumped from a Dewar into the silver block. The evaporated N_2 is recycled and used to purge the specimen chamber. The temperature of the

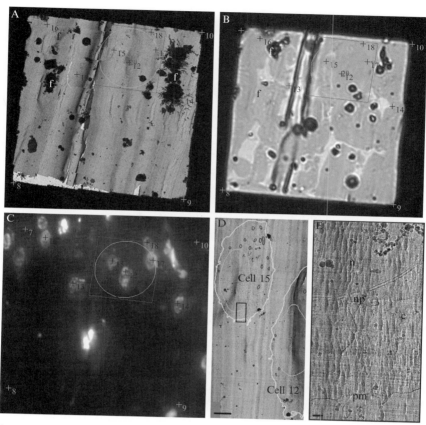

Figure 13.9 SerialEM can be used to correlate electron microscopy (EM) maps to light microscopy (LM) maps. (A) EM map of a grid square of interest. (B) Brightfield LM map showing location of grid square for registration purposes (yellow numbers at each corner). Frost (white f's where frost is present, black f's where frost is absent) can move around during transfer from the LM to the EM and consequently, frost present in the LM may be absent in the EM or vice versa. (C) Fluorescence LM map showing locations of Hoescht 33342 positive nuclei (red numbers). (D) High-resolution EM map of two cells found using the fluorescent map in C (area marked by green circle and blue square). Cells are outlined with the plasma membrane in green, nuclei in blue, and mitochondria in red. (E) Close-up of the black-boxed area in D. The nucleus (n) and cytoplasm (c) are clearly evident with a nuclear pore (np) found in the nuclear envelope. The plasma membrane (pm) is present. Scale bars: D = 2 μm, E = 300 nm. (See Color Insert.)

silver block is monitored with a platinum resistor sensor and regulated by a controller. The original stage holds a specimen on a round glass cover slip that is clamped in a stainless steel specimen-loading cartridge, which enters the stage via a side-entry port. The glass cover slip was replaced by a 0.8 mm

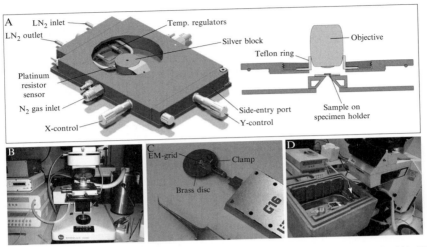

Figure 13.10 The cryo-fluorescence LM setup from van Driel *et al.* (2009). (A) 3D rendering of the freezing stage without a lid, and schematic cross-section through the Teflon ring lid and silver block. (B) Photograph of the setup with accessory devices mounted on a Leitz fluorescence microscope. The nitrogen Dewar on the right of the microscope is connected to the stage via an insulated tube. The temperature controller and nitrogen pump are shown on the left hand side, bottom unit and second from the bottom unit, respectively. (C) Modification of the specimen cartridge to enable loading of EM grids. The forceps serves to open and close the clamp. (D) Styrofoam box construction used to minimize frost-buildup during specimen transfer to and from the freezing stage. A small window (arrow) cut in one of the sides of the Styrofoam box slides over the side-entry port of the freezing stage. A stainless steel basin placed on the bottom of the Styrofoam box is used to attach a Teflon holder in which the specimen cartridge can be secured during loading and unloading of an EM grid. Notice that the freezing stage is mounted on a different microscope (Leica DMR) than the one shown in (B). Reproduced from European Journal of Cell Biology with permission (van Driel *et al.*, 2009).

thick brass disc with a circular depression of 3 mm diameter in the center, and the grid was held in place by a pivoting clamp mechanism similar to a room temperature EM specimen holder. A separate Styrofoam box by the side-entry port of the stage was used for loading the grid under LN_2 and reducing ice-contamination during transferring. The insulating glass window on the original stage was also replaced by a Teflon objective collar. A resolution of 400 nm was obtained with a Leica 100×, NA 0.75, WD 4.7 mm objective lens without cover slip correction. Coordinate translation software routine was written in Matlab to correlate the FLM and EM images.

Since the description of the system in van Driel *et al.* (2009), the 0.8 mm thick brass disk was replaced by a disc of sapphire, which is both transparent and has a good thermal conductivity. The disc is covered with a thin stainless steel disc that contains five circular holes of 3 mm diameter to hold five EM grids. In addition, the clamping mechanism was modified to hold the five grids simultaneously. With these adaptations, in addition to the fluorescence imaging, bright field imaging can also be used.

A protocol to use such a modified Linkam CS is as follows:

1. Prepare the freezing stage by activating the liquid nitrogen pump, purging the specimen chamber with dry nitrogen gas, and cooling down the silver block to liquid nitrogen temperature ($-196\ °C$).
2. Place the stainless steel basin for loading and unloading grids in the Styrofoam box, and attach the box to the freezing stage.
3. Fill the Styrofoam box with LN_2 and put the Perspex cover on top of it to prevent ice-contamination.
4. Put the specimen cartridge, without EM grids, into the stainless steel basin.
5. Remove the Perspex cover. Load the frozen-hydrated samples on EM finder-grids into the specimen cartridge, and secure the grids with the clamping mechanism.
6. Within the Styrofoam box, open the side-entry port of the freezing stage, and transfer the cartridge inside the freezing stage.
7. Focus the objective lens and manipulate the xy-controls of the freezing stage to move the grids into the viewing window.
8. Take image stitches of areas of interest, and make sure the stitches also contain a character of the finder grid for subsequent localization in EM.
9. Transfer the cartridge back from the freezing stage into the LN_2-filled steel basin in the Styrofoam box. Unload the grids and store them in an EM storage box.
10. From the EM storage box, load a grid into a cooled EM cryo-holder under LN_2.
11. Insert the cryo-holder into the cryo-EM.
12. For retrieving of the fluorescent area in cryo-EM, upload a fluorescence-image stitch into the Matlab Script LCOTRES (Fig. 13.11).
13. At low magnification, use the xy-controls of the goniometer to locate a recognizable feature and center this on the fluorescent screen. This could be a corner of a character on a finder grid. Click on the corresponding point in the uploaded fluorescence image. Repeat these steps for at least another three points. At this stage, the coordinates of the fluorescent image are mapped to the xy-goniometer controls of the cryo-EM.
14. Now, click a feature of interest in the fluorescence-image stitch, and confirm that you would like to have the stage move to that position. The program will control the goniometer to the correct xy-coordinates. This can be done at any magnification, and has an accuracy of about 1 μm.
15. Record cryo-EM image. If useful, collect a tilt series.

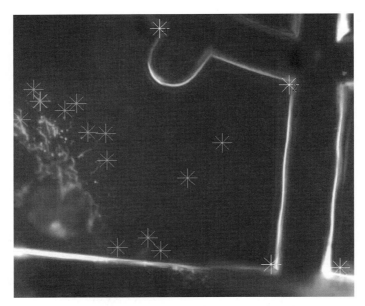

Figure 13.11   Leiden COordinate TRanslation Software (LCOTRAS). Part of an uploaded fluorescence image stitch showing fluorescently labeled cells, and part of a character of a finder grid. The green points indicate the positions that could be uniquely identified at very low magnification on the TEM. Subsequently, by clicking on features of interest, as indicated by the red points, the EM stage is moved to the corresponding area on the EM grid. The calibration will yield an accuracy of about 1 μm. When imaging cells, the selected point of interest might appear too thick for cryo-EM. Therefore, by clicking on other areas shown in the FM image, quick navigation over the EM grid is possible. (See Color Insert.)

As examples of the types of samples and questions that can be addressed using the above techniques, in α-tubulin-GFP—expressing PtK cells, microtubules were identified by fluorescence in the cryo-LM, then correlated to the high-resolution cryo-EM images and ECT slices (Schwartz *et al.*, 2007). Cryo-tomograms of cultured neurons labeled with FM dye showed neuronal processes in correlation with the fluorescence images obtained with cryo-LM (Lucic *et al.*, 2007). The tomogram in Fig. 13.12 shows fine structures of membrane invaginations and an extracellular vesicle, for example, which suggest endocytotic activity. This is further supported by the exact correlation to a fluorescence spot, obtained from cryo-LM using FM dye to label endocytotic processes. Mammalian cell mitochondrial ultrastructures were studied in intact cells (van Driel *et al.*, 2009) as well as in cryo-sections (Gruska *et al.*, 2008). The correlated FLM/ECM method allowed for selecting mitochondria in thin areas of the cultured human

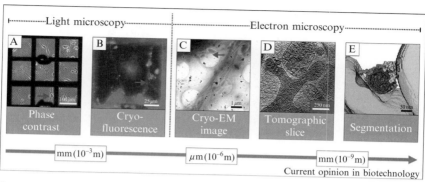

Figure 13.12 Correlative microscopy of mammalian cells. Illustrative example for correlating a cellular region from the scale of the light microscope to the high magnification and relatively small field of view of the electron microscope. The red arrows indicate areas enlarged in subsequent panels. (A) Live cell, phase-contrast image and (B) cryo-fluorescence image of neurons labeled with FM1-43 grown on EM "Finder"-grids. (C) Corresponding cryo-EM image of the region exhibiting various neuronal processes surrounding the area where the tomogram was recorded. (D) Tomographic slice showing two neuronal processes, and an extracellular vesicle connected to one of the processes and to a protrusion from the other. (E) Surface rendering showing the extracellular vesicle (blue), two neighboring neuronal processes (gray), connections between them (yellow and orange), and some vesicle-bound molecular complexes (external: green; internal: red). The vesicle is shown in a cut-away view to expose the complexes. Figure reproduced from Current opinion in Biotechnology with permission (Plitzko *et al.*, 2009). (See Color Insert.)

umbilical vein endothelial cells and the generation of cryo-tomograms of whole human mitochondria for the first time (van Driel *et al.*, 2009). It was shown that the distributions of mitochondria within the cytoplasm of stained HL-1 cells could be observed using vitreous sections of 50, 100, or 150 nm, and later relocated in the cryo-EM (Gruska *et al.*, 2008). With higher contrast and lateral resolution from thin sections, *in situ* ECT of mitochondria in P19 stem cells and HL-1 cardiomyocytes revealed ATP synthases and discrete inorganic deposits (Gruska *et al.*, 2008).

In an attempt to recover the high resolution of an oil-immersion light microscope, but on a frozen-hydrated sample, Le Gros *et al.* built a cryogenic immersion light microscope for correlative X-ray tomography (Le Gros *et al.*, 2009) (Fig. 13.13), which could also potentially be used for correlated FLM/ECM. This microscope uses a custom low-temperature, high–NA (1.3) objective lens, which is a hybrid construction from a Fratelli Koristka lens 30× NA 1.0 and a Spencer Lens Company 82× NA 1.33 oil lens, and liquid propane as immersion medium. The images obtained by this microscope on cells frozen in glass capillaries or grown on cover slips are comparable to conventional room temperature FLM images.

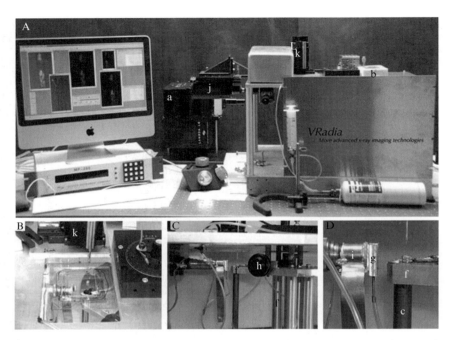

Figure 13.13 Photographs of cryogenic immersion light microscope. (A) Photograph of the prototype cryo-microscope in operation. (B) Top view of the specimen imaging compartment and specimen rotation instrumentation. (C) Partially disassembled microscope, showing details of the sample access port with the rotating platform positioned for use of the propane plunge freezer. (D) Close-up of the low-temperature objective and cryogenic immersion fluid container with glass capillary positioned above the container. The components labeled are: (a) illumination for bright field imaging; (b) exit port for cryo-transfer; (c) reservoir for propane plunge freezer; (f) rotating cryo-platform with conduction cooling rods immersed in liquid nitrogen; (g) liquid propane immersion fluid container; (h) condenser lens; (j) CCD detector; (k) dual-purpose sample stage/propane plunge freezer. Reproduced from Journal of Microscopy with permission (Le Gros *et al.*, 2009).

Finally, Fig. 13.14 shows the integration of a laser scanning fluorescence microscope in a standard FEI Tecnai 12 TEM (Agronskaia *et al.*, 2008). The sample stage of the TEM was modified to allow tilt angles from $+90°$ to $-70°$. The laser scanning fluorescence microscope module was mounted on the side port. For vacuum compatibility, limited space and less electron optical aberrations at positions close to the electron optics, a single element glass molded aspherical objective lens with NA 0.55 and WD 3.1 mm was used with a laser scanning geometry. Diffraction limited resolution of ~ 550 nm was obtained. For FLM imaging, the stage was tilted to $90°$ with the electron beam off, the LM module moved toward the sample and the focusing was accomplished

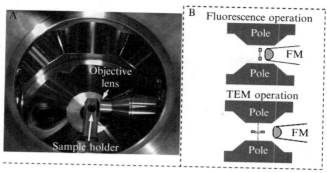

Figure 13.14 (A) Photograph showing the tip of the fluorescence microscope and the specimen stage inside the open TEM column. (B) Schematic drawing of the setup in FM and TEM operation modes: for FM operation the FM module is positioned between the TEM pole pieces and the TEM stage is tilted to 90°°; for TEM operation the FM module is retracted and the stage is tilted to 0°°. Reproduced from Journal of Structural Biology with permission (Agronskaia *et al.*, 2008).

by adjusting the Z-axis of the stage. For EM imaging, the LM module was retracted and the stage was rotated back to the 0° position. At the time of writing, although not published yet, preliminary data suggest that the iLEM is suitable for cryo-applications as well.

The instrumentation for correlated FLM/ECM is expected to continue to improve as more labs get involved. New protocols involving advanced LM and EM techniques are also expected. For example, cryo-FLM has now been used to guide focused-ion-beam/cryo-SEM-based micromachining in preparation for ECT (Rigort *et al.*, 2010). "Super-resolution" fluorescence microscopy methods such as PALM/STORM and STED are revolutionizing FLM (for a review, see Hell, 2008), and may be amenable to cryo-applications as well. On the other hand, it is important to note that any method based on tags and markers has liabilities because the tag itself can perturb the native function and/or localization of structures inside the cell. As a sobering example, Werner *et al.* (2009) recently performed a comprehensive screen of the localization of every open reading frame in *C. crescentus*. Of the 185 proteins that exhibited specific localization patterns with a GFP tag on the C-terminus, only 58 (less than 1/3) exhibited the same pattern when the tag was shifted to the N-terminus. To overcome potential artifacts, ideally investigators could use correlated FLM/ECM to identify structures in ECM images, but then reimage untagged (wildtype) strains by ECM only and draw their conclusions from that most-native data. With such controls, correlated FLM/ECM is likely to become a powerful tool that bridges the resolution gap between light and electron microscopies.

ACKNOWLEDGMENTS

We thank Dr. Alasdair McDowall for help with correlated FLM/ECM at Caltech. Work at Caltech was supported by the HHMI, NIH grants R01 AI067548 and P01 GM066521 to GJJ, and gifts from the Gordon and Betty Moore Foundation. Work at Boulder was supported by NIH Research Resource Grant #RR00592 to A. Hoenger. Work at the Max-Planck-Institute of Biochemistry in Martinsried was supported by the European Commission's 7th Framework Programme (grant agreement HEALTH-F4-2008-201648/ PROSPECTS).

REFERENCES

Agronskaia, A. V., Valentijn, J. A., van Driel, L. F., Schneijdenberg, C. T. W. M., Humbel, B. M., van Bergen en Henegouwen, P. M. P., Verkleij, A. J., Koster, A. J., and Gerritsen, H. C. (2008). Integrated fluorescence and transmission electron microscopy. *J. Struct. Biol.* **164**(2), 183–189.

Bay, H., Ess, A., Tuytelaars, T., and Van Gool, L. (2008). SURF: Speeded up robust features. *CVIU* **10**(3), 346–359.

Beck, M., Malmström, J. A., Lange, V., Schmidt, A., Deutsch, E. W., and Aebersold, R. (2009). Visual proteomics of the human pathogen *Leptospira interrogans*. *Nat. Methods* **6**(11), 817–823.

Briegel, A., Ding, H. J., Li, Z., Werner, J., Gitai, Z., Dias, D. P., Jensen, R. B., and Jensen, G. J. (2008). Location and architecture of the *Caulobacter crescentus* chemoreceptor array. *Mol. Microbiol.* **69**(1), 30–41.

Diestra, E., Fontana, J., Guichard, P., Marco, S., and Risco, C. (2009). Visualization of proteins in intact cells with a clonable tag for electron microscopy. *J. Struct. Biol.* **165**, 157–168.

Giepmans, B. N. (2008). Bridging fluorescence microscopy and electron microscopy. *Histochem. Cell Biol.* **130**, 211–217.

Gruska, M., Medalia, O., Baumeister, W., and Leis, A. (2008). Electron tomography of vitreous sections from cultured mammalian cells. *J. Struct. Biol.* **161**(3), 384–392.

Hell, S. W. (2008). Microscopy and its focal switch. *Nat. Methods* **6**(1), 24–32.

Iancu, C. V., Morris, D. M., Dou, Z., Heinhorst, S., Cannon, G. C., and Jensen, G. J. (2009). Organization, structure, and assembly of α-carboxysomes determined by electron cryotomography. *J. Mol. Biol.* **396**(1), 105–117.

Komeili, A., Li, Z., Newman, D. K., and Jensen, G. J. (2006). Magnetosomes are cell membrane invaginations organized by the actin-like protein MamK. *Science* **311**, 242–245.

Koning, R. I., Zovko, S., Barcena, M., Oostergetel, G. T., Koerten, H. K., Koster, A. J., and Mommaas, A. M. (2008). Cryo electron tomography of vitrified fibroblasts: Microtubule plus ends *in situ*. *J. Struct. Biol.* **161**, 459–468.

Kühner, S., van Noort, V., Betts, M. J., Leo-Macias, A., Batisse, C., PRode, M., Yamada, T., Maier, T., Bader, S., Beltran-Alvarez, P., Castano-Diez, D., Chen, W.-H., *et al.* (2009). Proteome organization in a genome-reduced bacterium. *Science* **27**(5957), 1235–1240.

Le Gros, M. A., McDermott, G., Uchida, M., Knoechel, C. G., and Larabell, C. A. (2009). High-aperture cryogenic light microscopy. *J. Microsc.* **235**, 1–8.

Li, Z., Trimble, M. J., Brun, Y. V., and Jensen, G. J. (2007). The structure of FtsZ filaments *in vivo* suggests a force-generating role in cell division. *EMBO J.* **26**, 4694–4708.

Lowe, D. G. (2004). Distinctive image features from scale-invariant keypoints. *Int. J. Comput. Vis.* **60**(2), 91–110.

Lucic, V., Yang, T., Schweikert, G., Förster, F., and Baumeister, W. (2005). Morphological characterization of molecular complexes present in the synaptic cleft. *Structure* **13**(3), 423–434.

Lucic, V., Kossel, A. H., Yang, T., Bonhoeffer, T., Baumeister, W., and Sartori, A. (2007). Multiscale imaging of neurons grown in culture: From light microscopy to cryo-electron tomography. *J. Struct. Biol.* **160**(2), 146–156.

Lucic, V., Leis, A., and Baumeister, W. (2008). Cryo-electron tomography of cells: Connecting structure and function. *Histochem. Cell Biol.* **130**, 185–196.

Mastronarde, D. N. (2005). Automated electron microscope tomography using robust prediction of specimen movements. *J. Struct. Biol.* **152**, 36–51.

Nickell, S., Förster, F., Linaroudis, A. A., Del Net, W., Beck, F., Hegerl, R., Baumeister, W., and Plitzko, J. M. (2005). TOM software toolbox: Aquisition and analysis for electron tomography. *J. Struct. Biol.* **149**(3), 227–234.

Plitzko, J. M., Rigort, A., and Leis, A. (2009). Correlative cryo-light microscopy and cryo-electron tomography: From cellular territories to molecular landscapes. *Curr. Opin. Biotechnol.* **20**(1), 83–89.

Rigort, A., Bäuerlein, F. J. B., Leis, A., Gruska, M., Hoffmann, C., Böhm, U., Eibauer, M., Gnaegi, H., Baumeister, W., and Plitzko, J. M. (2010). Micromachining tools and correlative approaches for cellular cryo-electron tomography. *J. Struct. Biol.* doi:10.1016/j.jsb.2010.02.011.

Sartori, A., Gatz, R., Beck, F., Rigort, A., Baumeister, W., and Plitzko, J. M. (2007). Correlative microscopy: Bridging the gap between fluorescence light microscopy and cryo-electron tomography. *J. Struct. Biol.* **160**(2), 135–145.

Scheffel, A., Gruska, M., Faivre, D., Linaroudis, A. A., Plitzko, J. M., and Schüler, D. (2006). An acidic protein aligns magnetosomes along a filamentous structure in magnetotactic bacteria. *Nature* **440**, 110–114.

Schwartz, C. L., Sarbash, V. I., Ataullakhanov, F. I., McIntosh, J. R., and Nicastro, D. (2007). Cryo-fluorescence microscopy facilitates correlations between light and cryo-electron microscopy and reduces the rate of photobleaching. *J. Microsc.* **227**, 98–109.

Suloway, C., Pulokas, J., Fellmann, D., Cheng, A., Guerra, F., Quispe, J., Stagg, S., Potter, C. S., and Carragher, B. (2005). Automated molecular microscopy: The new Leginon system. *J. Struct. Biol.* **151**, 41–60.

Suloway, C., Shi, J., Cheng, A., Pulokas, J., Carragher, B., Potter, C. S., Zheng, S. Q., Agard, D. A., and Jensen, G. J. (2009). Fully automated, sequential tilt-series acquisition with Leginon. *J. Struct. Biol.* **167**, 11–18.

van Driel, L. F., Valentijn, J. A., Valentijn, K. M., Koning, R. I., and Koster, A. J. (2009). Tools for correlative cryo-fluorescence microscopy and cryo-electron tomography applied to whole mitochondria in human endothelial cells. *Eur. J. Cell Biol.* **88**, 669–684.

Werner, J. N., Chen, E. Y., Guberman, J. M., Zippilli, A. R., Irgon, J. J., and Gitai, Z. (2009). Quantitative genome-scale analysis of protein localization in an asymmetric bacterium. *Proc. Natl. Acad. Sci. USA* **106**(19), 7858–7863.

CHAPTER FOURTEEN

PHASE PLATES FOR TRANSMISSION ELECTRON MICROSCOPY

Radostin Danev *and* Kuniaki Nagayama

Contents

Okazaki Institute for Integrative Bioscience, National Institutes of Natural Sciences, Okazaki, Japan

Methods in Enzymology, Volume 481
ISSN 0076-6879, DOI: 10.1016/S0076-6879(10)81014-6

Abstract

Phase plates are a new technique in the field of cryo-electron microscopy. They provide improved contrast and signal-to-noise ratio in images of radiation sensitive specimens. Thin film phase plates are being tested in biological applications and have demonstrated benefits for single particle analysis and cryo-tomography. There are still unsolved problems, such as reliability of manufacturing and deterioration of performance with time. Several other types of phase plates are currently under development and may become available for cryo-microscopy in near future. Presented is a short overview of the current state of the field as well as ideas for the future directions. Also included is a detailed description of the instrumentation requirements and the experimental procedures for phase plate application.

1. INTRODUCTION

The resolution limit of the transmission electron microscope (TEM) was gradually improved by developments and is currently on a subangstrom level while the phase contrast method has remained the same throughout the years. Defocus phase contrast (DPC) is currently the standard technique for low dose phase contrast imaging of cryo-specimens. It utilizes inherent (i.e., spherical aberration) or induced (i.e., defocus, hence the name of the technique) aberrations to produce intensity variations in the image corresponding to phase variations in the object wave. This approach can image adequately only parts of the object's Fourier spectrum leaving other areas blank. The loss of phase information is particularly noticeable in the low frequency region, corresponding to specimen features larger than a few nanometer. These properties give DPC images a pronounced high–pass filter appearance and an overall low contrast.

Since the early years of TEM, there have been proposals for phase contrast devices similar to those successfully used in light microscopy, namely "phase plates." The theoretical principles of phase plates are very similar between the light and the electron microscopes. In practice however, the implementation of devices for electron optics is much more challenging. The features of the device must be roughly three orders of magnitude smaller which poses manufacturing difficulties. Furthermore, the different physics of electron–matter and photon–matter interactions constitute more constraints on the phase plates for electrons.

Phase plates for TEM can be divided into two groups by the symmetry of the modulation pattern they produce. In the first group are the Zernike type phase plates which generate circularly symmetric modulation pattern. The images produced using such phase plates exhibit isotropic contrast features around objects. The theory of Zernike type phase plates has been studied in detail, see for example (Beleggia, 2008; Danev and Nagayama, 2001b). The second group of phase plates is the Hilbert type. These devices modulate the diffracted wave asymmetrically giving rise to anisotropic contrast in the images. The contrast features are similar to those produced by differential interference contrast in light microscopy. The Hilbert type of phase plates for TEM was developed in the last few years. More details about the theory of such devices can be found in (Barton et al., 2008; Danev and Nagayama, 2004).

Figure 14.1A illustrates the optical arrangement of a TEM equipped with a Zernike phase plate. The phase plate is located on a diffraction plane after the objective lens. Figure 14.1B shows curves illustrating the shape of the phase contrast transfer function (CTF) with and without a phase plate. The main effect of the phase plate is changing the sine-type CTF of the conventional TEM to a cosine-type. The curves are complementary with one having a zero where the other has a maximum. The phase plate CTF starts at a point labeled k_{CO}, which is the so-called "cut-on" frequency (Danev et al., 2009). It is determined by the size of the phase plate central hole and can be calculated by the formula:

$$k = \frac{r}{\lambda f} \tag{14.1}$$

where k is the modulus of the spatial frequency, r is the real space distance from the center of the diffraction plane, λ is the electron wavelength, and f is

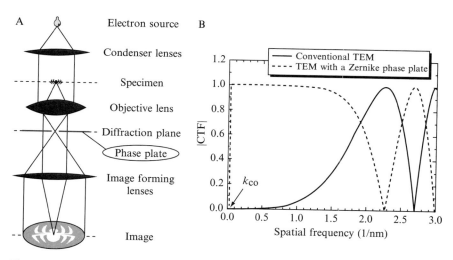

Figure 14.1 (A) Optical layout of a transmission electron microscope equipped with a Zernike phase plate. (B) Moduli of phase contrast transfer functions without (solid line) and with (dashed line) a Zernike phase plate. k_{CO} is the cut-on frequency of the phase plate; Parameters: defocus 0, spherical aberration 5 mm, acceleration voltage 300 kV.

the focal length of the objective lens. The beneficial effect of the phase plate is the almost flat CTF region between the cut-on frequency and ~ 1.5 nm^{-1} (0.67 nm periodicity). In that region, the CTF of the conventional TEM has low magnitude which means that object information at those frequencies will be weakly represented in the image. Adjusting the defocus can improve the transfer for parts of the low frequency region but at the expense of reduced performance at higher frequencies and rapidly oscillating CTF modulation.

To illustrate the effect of the phase plate in practice, Figure 14.2A, B shows images of ice-embedded T4 bacteriophage acquired by DPC and Zernike phase contrast, respectively. The images are shown using the same intensity scale. The higher contrast of the Zernike image is due to improved transfer for the low spatial frequencies. In addition, fine, fiber-like structures are easier to recognize because of the uniform transfer for a wide portion of the spectrum. Figure 14.2C, D shows the amplitudes of the Fourier transforms of the images in Fig. 14.2A, B. The DPC spectrum (Fig. 14.2C) shows moderate presence of low frequency information around the center

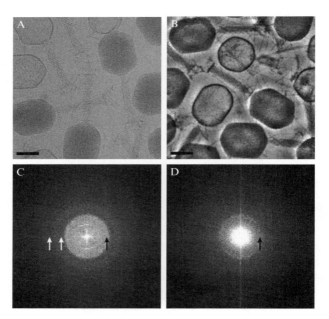

Figure 14.2 Images of ice-embedded T4 bacteriophage. (A) Defocus phase contrast image, defocus 1.6 μm. (B) Zernike phase contrast image close to focus. (C) and (D) moduli of the Fourier transforms of the images in (A) and (B) respectively. White arrows in (C) indicate the first two zeros of the contrast transfer function. Black arrows in (C) and (D) indicate the ring corresponding to the 2.3 nm periodicity of the DNA packed in the phase capsids. Experimental conditions: acceleration voltage 200 kV, electron dose 20 e$^-$/Å^2. Scale bars: 50 nm.

of the spectrum and ring-shaped areas of reduced amplitude due to CTF zeros (white arrows). The ZPC-TEM spectrum (Fig. 14.2D) does not exhibit CTF zeros and has very strong presence of low frequency information. The bright ring at about 1/3 of the Nyquist frequency in both spectra (black arrows) corresponds to the 2.3 nm periodicity of the DNA packed in the T4 phage capsids.

In more quantitative terms, a recent simulation study by (Malac et al., 2008) used Gaussian phantom objects to estimate the required dose for reliable detection. A Zernike phase plate provided better than threefold decrease in the required dose for objects larger than 3 Å. In another study, Chang et al. (2010) used simulated TEM images of ice-embedded macromolecules with molecular weights in the range 100–500 kDa to study the minimum number of particles for a single particle reconstruction with a preset target resolution. For 100 kDa molecules, the phase plate provided roughly 10 times reduction in the number of required particles. For a 500 kDa particle, the improvement was between 1.4 and 2 times depending on the resolution. These results are in agreement with a recent experimental work using a thin film Zernike phase plate and GroEL (800 kDa) as a test specimen (Danev and Nagayama, 2008) which demonstrated roughly 1.3 times reduction in the number of required particles.

2. TYPES OF PHASE PLATES AND HISTORICAL NOTES

This chapter discusses the types of phase plate devices according to their design and the physics principles they utilize. For each device, there are notes about the history and the present state of its development.

2.1. Thin film phase plates

The thin film phase plate is the oldest type of phase plate tested in practice. The initial idea to use material film as a phase shifting media was proposed by (Boersch, 1947). The Zernike version consists of a thin film with thickness adjusted for $-\pi/2$ phase shift and a small hole in the center. The first experiments used phase plates made of collodion films (Agar et al., 1949; Kanaya et al., 1958). These attempts were hampered by the low performance of the TEMs at the time and the rudimentary techniques used to prepare the phase plates. Later efforts (Badde and Reimer, 1970; Johnson and Parsons, 1973; Willasch, 1975) used phase plates made from amorphous carbon but with relatively large central hole (>6 μm) which limited the contrast improvement for low spatial frequencies. Most of the early attempts at using thin film phase plates were aimed at improving the resolution and reducing the effect of spherical aberration. The specimens were either carbon film or negatively stained materials and the electron dose was not a major concern. The improvements that were observed were not impressive enough to justify the additional effort and experimental limitations of the phase plates. That led to a pause in thin film phase plate research until (Nagayama, 1999) devised a scheme for object wave reconstruction, called "complex reconstruction." The new scheme required Zernike phase contrast data and initiated a new set of trials on the development of thin film phase plates for TEM (Danev and Nagayama, 2001a,b). The initial results employing a carbon film phase plates with ∼1 μm central hole showed remarkable improvement in contrast and details for negatively stained specimens (Danev and Nagayama, 2001b). Since then there have been numerous applications, most of which cryo-EM, demonstrating benefits for polymers (Tosaka et al., 2005), ECO/F-BAR-domain dimer-covered liposomes (Shimada et al., 2007), ice-embedded lipid/DNA complexes (Furuhata et al., 2008), ice-embedded lipid nanotubes (Yui et al., 2008), ice-embedded Influenza A viruses (Yamaguchi et al., 2008), single particle analysis of ice-embedded GroEL (Danev and Nagayama, 2008), vitrified cells (Fukuda et al., 2009), and cryo-electron tomography of ice-embedded T4 phages (Danev et al., 2010).

The strong topographic contrast produced when the Zernike phase plate was slightly misaligned promoted the development of the Hilbert phase plate (Danev and Nagayama, 2004; Danev et al., 2002). It consists of a thin film edge positioned very close to but not intercepting the zero order diffraction beam. The thickness of the film is adjusted for $-\pi$ phase shift, which is twice thicker than the Zernike phase plate. Hilbert phase contrast images exhibit strong topographic contrast similar to that of differential interference contrast in light microscopy. In an ideal case, the envelope of the CTF is the same as that of the Zernike phase contrast. By numerical phase shifting of both halves of the Fourier space, it is possible to demodulate a Hilbert image and convert it to a Zernike type image (Barton et al., 2008; Danev and Nagayama, 2004, 2006). Several published works use results acquired using a Hilbert phase plate, as applied to ice-embedded cyanobacteria (Kaneko et al., 2005), vitrified mammalian cells (Setou et al., 2006), and electron tomography of resin embedded cells (Barton et al., 2008).

The present state of thin film phase plate research was discussed in detail recently by (Danev et al., 2009). Amorphous carbon is still the material of choice for phase plate preparation. Its positive qualities include: mechanical strength, chemical inertness, easy preparation and transfer, good electrical conductivity, amorphous structure, and low electron scattering. Its single negative aspect is aging—the performance of carbon film phase plates deteriorates within a few days to a week necessitating frequent changes of the phase plates. The cause of aging is presently unknown but the hypothesis is that it is due to physical or chemical changes on the surface of the film leading to reduced conductivity and susceptibility to beam induced electrostatic charging. Better materials have not been found yet mainly due to lack of research in that direction. The preparation techniques are another area in need of more investigation. At present, the phase plates are manufactured by manual transfer of the evaporated carbon film to the aperture through floating on water surface. This process is the main source of contaminants and cause of variability in performance. Precise control of the conditions during the transfer is practically impossible due to the miniature size of the grids and capillary effects. A promising direction for future research are the nanofabrication technologies, such as MEMS (Microelectromechanical systems), which in combination with appropriate materials could improve the consistency of quality and possibly extend the lifetime of the phase plates.

Figure 14.3 shows control images of phase plates after producing the holes in a single beam focused ion beam machine. The images are noisy due to the low ion dose used for imaging in order to avoid damage to the carbon film. Figure 14.3A shows an image of a Zernike phase plate on a 100 μm aperture. The white arrow indicates the 0.5 μm central hole and the black arrows point to marker holes close to the aperture edge which are used as an aid during phase plate alignment in the TEM. Figure 14.3B shows an image of a Hilbert phase plate supported on a 100 μm aperture. The long edges of the rectangular cutout in the center are used for Hilbert phase contrast imaging. Manufacturing a long straight edge on a carbon film is practically challenging due to its tendency to curl. Smaller rectangular cutouts prevent curling and provide two or more edges that can be used as a phase plate.

Multihole apertures are used to improve the chances of producing a good quality phase plate. Figure 14.4 shows a light microscopy image of a custom made 2 mm molybdenum multihole aperture disk covered with amorphous carbon film. A phase plate can be prepared in each of the twenty-five 100 μm diameter holes.

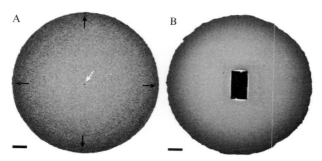

Figure 14.3 Low dose focused ion beam images of thin film phase plates made from amorphous carbon. (A) Zernike phase plate, white arrow indicates the central hole, black arrows indicate marker holes close to the aperture edge. (B) Hilbert phase plate. The square opening was made by cutting along the perimeter. Some film fragments remain on the shorter edges due to curling. Scale bars: 10 μm.

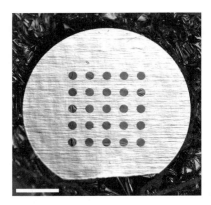

Figure 14.4 A light microscopic image of a custom multihole 2 mm aperture disk made of molybdenum. The disk was thoroughly cleaned by sonication in organic solvents and sputter-coated with gold before an amorphous carbon film was transferred on it by floating on water surface. Scale bar: 500 μm.

In summary, the advantages of the thin film phase plates are: easy manufacturing (with the caveats discussed above), low cut-on frequency, and no hard obstructions in the beam path. Their disadvantages include: fixed phase shift, inconsistent quality of manufacturing, deterioration of the performance with time (aging), and a small loss (∼ 13%) of information due to scattering by the material of the film.

2.2. Electrostatic phase plates

The use of an electrostatic field in the construction of a phase plate was first suggested by Boersch (1947). He proposed to place a miniature Einzel lens on the path of the zero order diffraction beam. The lens consists of three consecutive ring-shaped electrodes with the outer two at ground potential and the middle one acting as a control electrode.

Figure 14.5A shows a drawing illustrating the principle of a Boersch phase plate. The device lies in the diffraction plane and is centered on the zero order diffraction beam. Electrons passing through the Einzel lens experience phase shift with magnitude determined by the excitation voltage of the central electrode. The electrostatic field is confined inside the lens by the outer grounded electrode, so electrons scattered by the specimen and passing outside of the device are not phase shifted. Although this design was proposed more than 60 years ago, the fabrication challenges were overcome only recently thereby allowing practical research and experiments to begin. Figure 14.5B illustrates a typical implementation of a Boersch phase plate. The Einzel lens is in the center of an aperture supported by one to three support bars. At present, there are several groups working on Boersch type

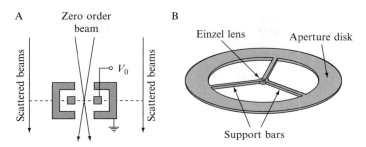

Figure 14.5 (A) A schematic of an Einzel lens positioned on the zero order diffraction beam and functioning as an electrostatic phase plate. The phase shift can be adjusted by varying the voltage on the inner ring-shaped electrode. The outer electrode is at ground potential and acts as an electrostatic shied. (B) An example sketch of a Boersch electrostatic phase plate device.

devices (Alloyeau *et al.*, 2010; Majorovits *et al.*, 2007; Shiue *et al.*, 2009). Most devices are based on the design proposed by Matsumoto and Tonomura (1996) consisting of inner and outer electrodes separated by an insulating layer. An exception is the work of Cambie *et al.* (2007) where they propose and test a "drift tube" device in which the electrodes are separated by vacuum. This design avoids the possible problems of leak currents and electrostatic charging of the insulating layer at the expense of slightly larger dimensions and weaker confinement of the electrostatic field.

The main advantage of the Boersch type of phase plates is real-time control of the amount of phase shift by varying the excitation voltage of the Einzel lens. This opens exciting possibilities for wavefront reconstruction (Gram *et al.*, 2010; Van Dyck, 2010). In addition, unlike thin film phase plates there is no loss of electrons at higher spatial frequencies due to scattering by a material film. Their main disadvantage is obstruction of parts of the diffraction plane by the mechanical components of the device, especially the area around the zero order beam where the Einzel lens is. The material parts on the beam path are also prone to beam induced contamination and electrostatic charging, which puts stringent requirements on the cleanness of the device and necessitates measures to avoid deposition of contamination while inside the microscope. At present, this type of phase plate is still at an early experimental stage but the proof of concept results are very encouraging and actual applications are expected very soon.

2.3. Magnetic phase plates

The ability of magnetic potential to shift the phase of electron waves was first demonstrated by Tonomura *et al.* (1986). They used trapped magnetic flux in a ring covered by superconducting material preventing the leakage of field. The experiment was aimed at measuring the Aharonov–Bohm (AB)

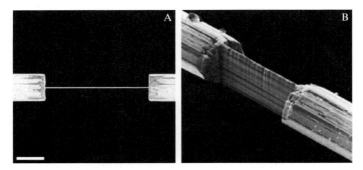

Figure 14.6 Focused ion beam images of the ion beam trimmed central portion of a platinum wire used in a linear magnet phase plate. Scale bar: 10 μm.

effect—a quantum mechanical phenomenon in which charged particles interact with the electromagnetic potential in a field-free space. The group proposed to use the AB effect in phase plates but there were no experiment trials in that direction until very recently.

In order to avoid the challenges of manufacturing and supporting of a small magnetic ring, Nagayama (2008) proposed and is currently investigating a device utilizing a linear magnet. The magnet was made by evaporation of cobalt on a platinum wire which was trimmed to $\sim$500 nm width from original thickness of 10 μm by a focused ion beam machine. Figure 14.6A shows a top view and Fig. 14.6B shows an angled view of the thinned part of the platinum wire. The wire is supported on an aperture disk (not shown). The magnetic flux through the strip induces a phase shift difference between electron waves passing on both sides of the wire. Positioning the zero order beam very close to one side of the strip, without touching it, will create a phase shift difference between the two halves of the diffraction plane. This asymmetric phase modulation makes this device a Hilbert type phase plate.

The magnetic design for a phase plate is still in very early stages of development. Phase shift effects have been confirmed (unpublished data) but there are still problems with magnetic domain formation along the linear magnet and electrostatic charging due to impurities on the surface of the wire. If these problems are solved, the linear magnet phase plate will have an advantage compared to thin film Hilbert phase plates because except for the narrow area blocked by the wire there is no loss of electrons due to scattering by a material film.

2.4. Photonic phase plates

A very recently proposed idea is to use high intensity focused laser light to introduce phase shift to the electron wave (Müller *et al.*, 2010). Figure 14.7A shows the optical arrangement for such a device. A simple

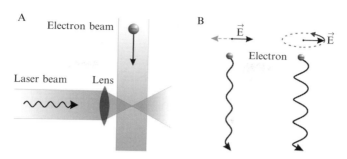

Figure 14.7 (A) Schematic illustration of the optical arrangement for a photonic phase plate. (B) Classical physics illustration of the effect of oscillating electrostatic field on a moving electron. In a linearly polarized field (left) the electron wiggles side-to-side. In a circularly polarized field (right) the electron moves along a spiral.

semiclassical explanation can be given for the principle of operation of the device. The intense oscillating electrostatic field at the laser focus causes the electrons to wiggle along their path. This elongates their optical paths as well as causes a small increase in their kinetic energy which in turn shortens their wavelength. The combination of the two effects leads to phase retardation of the electron wave. Figure 14.7B shows two simplified scenarios— the left side illustrates an electron in linearly polarized electrostatic field; the right side an electron in circularly polarized field. The case of circularly polarized field generates more phase shift for the same field intensity. The biggest practical challenge of the photonic phase plate is the required laser power, which is in the order of several kilowatt. The power requirement increases rapidly with the increase of acceleration voltage. Having a continuous wave laser with power of more than about a hundred watts will not be practical for phase plate applications due to physical size, electric power requirements, safety of operation, etc. A possible solution proposed by Müller *et al.* (2010) is to increase the intensity at the laser focal spot by an optical resonance cavity. Another possibility would be to build a device which decelerates the electrons locally (in-column decelerator) and apply the laser in that area. Such devices already exist in the form of an electrostatic mirror (see Section 2.5). Placing the laser focus close to the zero energy plane of the mirror will reduce the laser power requirements significantly.

Photonic phase plates are still at the concept stage. Theoretical evaluations by Müller *et al.* (2010) show that building a test device is feasible based on the current level of laser technology and optics. A photonic phase plate will have a big advantage over other types of phase plates by not placing any material objects on the beam path thereby solving the most difficult problems of the current experimental phase plates: electrostatic charging and electron scattering. In addition, it will provide real-time control of the phase shift amount, similar to that of electrostatic phase plates.

2.5. Electrostatic mirror pixel-wise phase shifter

Another novel idea for a TEM phase plate is to use an electrostatic mirror at an appropriate place along the beam path (Okamoto, 2010). Figure 14.8 illustrates one possible configuration for such a device. The electron beam coming from a TEM column is deflected by the magnetic beam separator and directed toward the electrostatic mirror. Before the mirror, there may be lens(es) for focusing. Inside the mirror, the electrons are gradually decelerated until they stop at the zero energy plane. After momentarily stopping the electrons are accelerated backward by the mirror and deflected toward the remaining part of a TEM column by the beam separator. At a close distance behind the zero energy plane is located a pixel-wise phase shifting device. It comprises an array of individual pointed electrodes arranged in a raster-like 2D pattern. The voltage of each electrode/pixel is controlled independently. The 2D array of pixels can be excited to a predefined voltage pattern producing a desired electrostatic potential distribution at the zero energy plane, located just in front of the array. Consequently, the electron wavefront can be modulation by a freely controllable 2D phase shift pattern which can be varied in real-time. Such a device will allow the formation of any kind or shape of phase plate, including Zernike and Hilbert. In addition, proper phase shift patterns could compensate for spherical and higher order aberrations. Instead of the pixel-wise modulator, a focused laser can be applied at the zero energy plane. This will produce a photonic phase plate as described in Section 2.4.

Electrostatic mirror devices are well studied theoretically and prototypes exist for low electron energies (~ 15 keV) (Okamoto, 2010). Adapting the

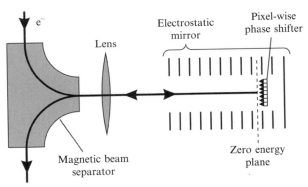

Figure 14.8 A diagram of a pixel-wise phase shifting device inside an electrostatic mirror. Electrons are decelerated inside the mirror, stop at the zero energy plane, and then are accelerated back toward the beam separator. The pixel-wise phase shifter creates a preprogrammed electrostatic potential pattern and is located very close behind the zero energy plane, so that electrons cannot reach it and interact with its material.

design for the application described here will raise many engineering challenges related to the higher beam energy and the requirement to have a suitably magnified diffraction plane at the zero energy plane. The research has already started but there is no timeframe for the first experimental tests.

2.6. Anamorphotic phase plates

The anamorphotic phase plate for TEM was recently proposed by Schröder *et al.* (2007). Figure 14.9A illustrates the construction of the device. Embedded electrodes are exposed on the inner wall of a narrow slit-shaped aperture. The number and arrangement of the electrodes can vary depending on the desired behavior of the device and the manufacturing capabilities. The anamorphotic phase plate requires an optical plane exhibiting anisotropic magnification where the diffraction pattern is highly compressed in one direction. Such optical planes can be produced by specialized multipole lens optics or by specially adapted aberration correctors (Schröder *et al.*, 2007). Figure 14.9B shows the optical arrangement of the anamorphotic diffraction pattern inside the phase plate aperture. The narrow slit helps to localize the electrostatic field generated by the electrodes. The aspect ratio of the slit determines its ability to confine the field. Higher aspect ratios allow better localization but are more demanding on the performance of the optics producing the anamorphotic diffraction plane. By varying the voltages on the electrodes, various amounts of phase shift and types of phase contrast can be produced. The phase shift profile is controlled only in one dimension, across the long side of the slit, resulting in strip-wise modulation in the Fourier space. In order to avoid anisotropic artifacts and to improve the overall performance, two anamorphotic phase plates can be applied at right angles. This however necessitates the availability of two anamorphotic optical planes.

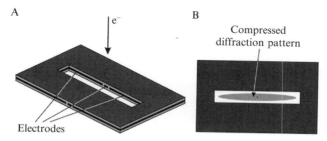

Figure 14.9 (A) A sketch of an anamorphotic phase plate. Embedded electrodes are exposed on the inner wall of a narrow slit-shaper aperture. Voltage applied to the electrodes creates localized electrostatic potential variations along the slit. (B) A compressed diffraction pattern centered inside the anamorphotic phase plate.

The research on the anamorphotic phase plate is ongoing. At present, it is at the simulation stage but a prototype may be available for testing in near future. If realized, this type of phase plate will provide advantages similar to the expected advantages of the photonic and electrostatic mirror designs, that is, no materials in the electron beam path.

3. Microscope Requirements and Modifications

3.1. Objective lens

There are several requirements which must be satisfied to make a TEM suitable for phase plate applications. The most important one is to have an aperture holder or the possibility to install one exactly on a diffraction plane. In many modern TEMs, especially those equipped with high resolution (short focal length) objective lens polepieces, the objective lens aperture is not located on the diffraction plane due to space restrictions. Such microscopes require extensive modifications, such as polepiece replacement and/or the installation of a transfer lens system (see Section 3.7), to make them capable of phase plate applications. From Eq. (14.1), the effective size of the phase plate in Fourier space is inversely proportional to the focal length of the objective lens, that is, longer focal length objective lenses produce larger diffraction patterns. In terms of phase plate design, this means that longer focal lens objectives can use physically larger phase plates. This relaxes the miniaturization requirements and can help overcome limitations of the phase plate manufacturing process. Typical objective focal lengths of microscopes currently used with thin film phase plates are ~ 5 mm (Danev et al., 2009, 2010).

3.2. Phase plate holder

The second requirement is to have an aperture holder capable of supporting phase plates. The holder should be compatible with the phase plate device and provide the necessary wire pass-through for heater, thermocouple, phase plate electrode voltages, etc. For phase plates having a material in the beam path (thin film, electrostatic, magnetic), heating is essential for preventing beam induced contamination. The holder should be able to heat and maintain the phase plate at a preset temperature, preferably by a closed loop temperature control system which improves temperature stability and minimizes thermal drift. A typical temperature range is of up to $\sim 300\ ^\circ C$ (Danev et al., 2009), but the optimum temperature for a phase plate depends on many factors, such as the cleanness of the TEM vacuum, phase plate materials, proximity to the specimen, etc. If the TEM is used for cryo-observations and does not have a transfer lens system (Section 3.7), then it requires proper thermal shielding between the heated phase plate holder and the specimen cryo-holder. A good practical solution is to have a cold trap surrounding the specimen and the phase plate holders with an additional cold fin in between the two holders. Currently only one manufacturer (JEOL, Tokyo) offers a phase plate holder as an optional accessory for their microscopes. Hopefully, in near future other TEM manufacturers or third party TEM accessory makers will offer competitive products.

3.3. Phase plate airlock

An airlock system for the phase plate holder is highly desirable. At present there are no commercially available phase plate airlock systems. We have heard of several ongoing projects in that direction, both by manufacturers and researchers, but no information has been published yet. An airlock can provide a lot of flexibility for research and in the case of thin film phase plates, alleviate the aging problem by allowing frequent, even daily exchanges. Without an airlock, exchanging the phase plate requires leaking the TEM column to atmospheric pressure which in most cases means at least a day of maintenance time and could interfere with the usage schedule of the microscope.

3.4. Illumination system

The requirement for a phase plate to be positioned exactly on a diffraction plane imposes some restrictions on the illumination system of the TEM (see Fig. 14.11 and the explanation in Section 4.1) (Danev et al., 2009). The position of the diffraction plane depends on the convergence angle of the beam at the specimen plane. In a properly designed and constructed system, the phase plate will be on the diffraction plane ("on-plane" condition) when the specimen illumination is parallel. For many existing microscopes that is not the case, that is, illumination is not parallel when the phase plate is "on-plane." In any case, in order to be able to achieve the on-plane condition for various beam intensities and sizes, the microscope must have a minimum of three condenser lenses thus providing three degrees of freedom for the illumination: crossover size (or "spot size," usually controlled by the first condenser lens), beam intensity at the specimen (usually controlled by the second condenser lens), and beam convergence angle at the specimen (usually controlled by a third condenser or a minicondenser lens). The condenser lens aperture is used to limit the illuminated area on the specimen. Microscopes equipped with only two condenser lenses or with three but without a minicondenser lens will have a limited range of beam intensities for which an "on-plane" condition can be achieved. Usually the limit is on the high side meaning that at higher magnifications the beam intensity will be too low and require impractically long exposure times. At present at least one manufacturer (FEI, Eindhoven) offers a system in their high-end TEMs which through software control of two or more condenser lenses can maintain a constant convergence angle while varying the beam size or intensity on the specimen. Such or similar system would be highly beneficial for phase plate applications because it allows specimen illumination adjustments while keeping the phase plate "on-plane."

3.5. Field emission gun

The field emission gun (FEG) has become a standard part of high resolution cryo-microscopes. For phase plate applications, a FEG is highly recommended. FEGs are much brighter and produce smaller crossovers compared to other types of electron guns. The smaller zero beam crossover makes the centering of a phase plate much easier and permits the use of phase plates with smaller central openings which reduces the cut-on frequency and improves contrast.

3.6. CCD camera

A CCD camera is essential for phase plate applications. With a phase plate, the image does not show any noticeable change in contrast when varying the focus. The control software of most modern CCD cameras can display an animated pseudo-real-time Fourier transform of the image. Focusing and quality evaluation of the phase plates is easy to perform by observing the positions of the CTF zeros in the Fourier image.

3.7. Transfer lenses and aberration correctors

Using a phase plate at the back-focal plane of the objective lens has a lot of restrictions mainly due to space constraints. A better practical approach is to use a lens system which creates a second diffraction image on a plane below the objective lens. Such a lens system is usually called a "transfer lens" and may consist of one or two lenses or be a part of an aberration corrector.

Figure 14.10 shows an illustration of the principle of a transfer lens. One or two lenses are added after the objective lens creating a second diffraction image far from the objective lens where a phase plate can be applied. The transfer lens approach has numerous benefits. It alleviates the restrictions on the design of the objective lens polepiece and provides a dedicated instrumentation area for the phase plate. This opens the door to many possibilities, such as larger phase plate devices (i.e., photonic phase plates), no restrictions on phase plate heating, cold traps, *in situ* plasma cleaning, phase plate airlocks, etc. Another very important advantage of a transfer lens system is the ability to magnify the diffraction image, that is, to effectively increase the focal length of the objective lens. This can be especially beneficial with Boersch or similar types of phase plates for which

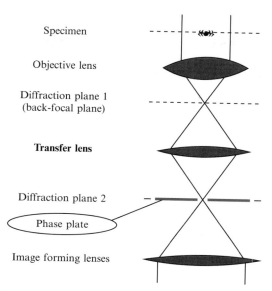

Specimen

Objective lens

Diffraction plane 1
(back-focal plane)

Transfer lens

Diffraction plane 2

Phase plate

Image forming lenses

Figure 14.10 An optical arrangement of a microscope equipped with a transfer lens. A second diffraction plane is created away from the objective lens where a phase plate can be installed with more instrumental flexibility and freedom of design.

manufacturing constraints limit the ability to shrink the device further. Magnifying the diffraction plane reduces the cut-on frequency and minimizes beam obstructions. Furthermore, the transfer lens provides flexibility for achieving an "on-plane" condition. Instead of adjusting the illumination system, the excitation of the transfer lens can be changed which shifts the position of the second diffraction plane. Transfer lens systems in the form of an additional intermediate lens were first constructed and used by Faget *et al.* (1962) and Johnson and Parsons (1973). A more recent implementation (Hosokawa *et al.*, 2005) uses a lens doublet which provides two degrees of freedom and can control independently the magnification and the position of the secondary diffraction plane. At present, there are no officially available transfer lens equipped instruments; however, TEM manufacturers have prototypes and are working on further development. In future, phase plates will probably become a part of aberration correctors and use diffraction planes produced in such optical systems (Schröder *et al.*, 2007).

4. PHASE PLATE OPERATION PROCEDURES

Below we discuss the practical procedures associated with the use of thin film phase plates. Most of the described steps can be as well applied to other kinds of phase plates.

4.1. Setting the phase plate on a diffraction plane

In most TEMs, the intensity of the beam on the specimen is adjusted by varying one of the condenser lenses ("Brightness" or "beam size" control knob). This adjustment also changes the distance between the beam cross-over and the specimen (Fig. 14.11). Movement of the beam crossover causes movement of the diffraction plane. At one particular setting of the condenser lens, the diffraction plane is on the phase plate (Fig. 14.11A) producing the "on-plane" condition. In that condition, the phase plate hole projection is infinitely large and the edge of the phase plate hole is not visible in the image (Fig. 14.11C). At other condenser lens settings, the diffraction plane is either above (not shown in Fig. 14.11) or below (Fig. 14.11B) the phase plate, making the phase plate to be "off-plane" and producing a projection of the phase plate hole in the image (Fig. 14.11D). Varying the condenser lens

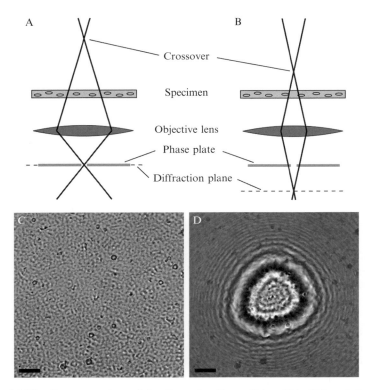

Figure 14.11 Demonstration of the "on-plane" condition (A), (C) and the "off-plane" condition (B), (D) for a phase plate. (A) and (B) When the condenser lens is varied, the crossover above the specimen moves causing movement of the diffraction plane. (C) In on-plane condition, the phase plate hole edge is at infinity and not visible in the image. (D) In off-plane condition, the phase plate hole projection is visible in the image. Scale bars: 200 nm.

around the on-plane setting "zooms" the phase plate hole in and out of view. In practice, the off-plane condition is used during phase plate alignment and the on-plane condition for image acquisition.

4.2. Phase plate centering

The first step of the phase plate centering is finding the phase plate hole. This step is performed in low magnification mode ("Search" mode), typically a defocused diffraction image (Fig. 14.12). Having a specimen in the view can make spotting the phase plate hole difficult. Try to find an area of the specimen where the support film is broken or where the specimen is thinner. Bring the phase plate into the beam path by selecting the desired phase plate aperture number on the phase plate holder. Finding the phase plate hole is much easier if the phase plate was produced with edge markers (Fig. 14.3A) and the marker holes were aligned with the axes of the aperture holder during installation. Use the aperture drives to find the outer edge of the phase plate aperture. Move along the edge until one of the markers comes into view. Use the aperture drive which acts perpendicular to the edge to move toward the center of the phase plate until the phase plate hole comes into view. Position the phase plate hole in the center of the screen. If the phase plate does not have edge markers, the centering process may take longer. Search around the central area of the phase plate aperture until the phase plate hole comes into view and center it.

The next step is fine centering. Switch the microscope to imaging mode. Expand the beam by overexciting the condenser lens and reduce magnification until the phase plate hole projection is visible on the screen. Move the phase plate toward the center of the screen using the aperture drives (if they are precise enough) and/or the beam tilt deflectors. Gradually

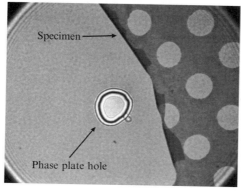

Figure 14.12 An example of the image on the screen with the microscope set to defocused diffraction mode used for initial centering of the phase plate.

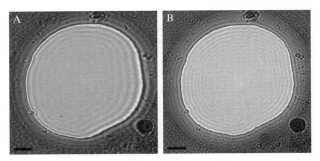

Figure 14.13 Image appearance with the phase plate "on-plane" but slightly misaligned (A) and properly centered (B). Scale bar: 200 nm.

reduce the size of the beam and (if necessary) increase the magnification while constantly keeping the phase plate hole centered using the aperture drives and/or beam tilt. The phase plate hole projection will grow in size until it is larger than the beam spot and the hole edge is no longer visible. The effect can be described as: the beam "zooming-in" into the phase plate hole. Once the phase plate hole is larger than the beam the phase plate can be considered to be on-plane.

While on-plane the phase plate centering can be fine-tuned by adjusting the beam tilt. Find a high contrast, preferably circular, feature on the specimen, such as a hole in the support film or a contaminant. Figure 14.13A shows an image of a hole in the specimen with the phase plate on-plane but slightly misaligned. The fringes on one side of the hole are more pronounced and with longer periodicity indicating that the phase plate central hole is not perfectly centered on the zero diffraction beam. By using the beam tilt controls try to make the fringes as symmetrical as possible around the object. Figure 14.13B shows an image after the phase plate was properly centered. Please note, the beam tilt deflectors should not move the beam on the specimen but only relative to the phase plate. If the beam moves relative to the specimen, check the beam deflector's tilt compensation.

4.3. Adjusting electron dose

As discussed in Section 3.4, the phase plate imposes some restrictions on the illumination settings. The electron dose for a given experiment must be adjusted with the phase plate in an "on-plane" condition. The procedure is time consuming but once performed, the condenser lens settings can be memorized and used in future observations. First, choose a magnification and electron dose for the experiment. Center and bring on-plane the phase plate at that magnification (see Sections 4.1 and 4.2). Check the beam intensity/electron dose by the microscope's beam current density meter or a CCD camera. If the electron dose needs adjustment, change the spot size or the minicondenser lens setting and make a note of the change. The phase plate will be out of alignment and probably off-plane after the change. Recenter and bring on-plane the phase plate, check the beam intensity again. If the measured value is closer to the desired value keep adjusting the spot size or the minicondenser lens in the same direction and realigning the phase plate until the target dose is reached. If the measured value is worse, change the direction of adjustment for the spot size or minicondenser lens and repeat the steps above. As a final step, adjust the beam size on the specimen by selecting an appropriate condenser lens aperture. Changing the condenser lens aperture may slightly affect the beam intensity. If the change is too large, repeat the steps above with the new condenser aperture. It may not be possible to achieve an exact match (typically it is possible to achieve less than ~20% difference) of the beam intensity to the planned value. In such cases, make small adjustments to the exposure time for fine tuning of the dose. Once the desired conditions are set, make notes of the condenser lenses values to simplify the setup of future experiments.

4.4. Focusing and phase plate condition evaluation

Unlike the conventional defocus method, a phase plate provides best performance when the TEM is set close to focus. With a phase plate, the images exhibit high contrast for a wide range of focus values with only subtle changes in fine detail, so focusing by observing the image is rather imprecise. Focusing is much easier to perform by observing the pseudo–real-time animated Fourier transform produced by the control software of a CCD camera. The cosine character of the CTF (Fig. 14.1B) means that in an area around the center of the Fourier image, the CTF intensity will be high followed by ring-shaped regions of high and low intensity corresponding to the maxima and minima of the oscillating part of the CTF.

Increasing the defocus shrinks the central area and brings the first CTF zero (white arrow, Fig. 14.14A) closer to the Fourier origin. The goal of the focusing procedure is to extend the central area of the CTF as far as possible thus achieving good information transfer for a wide range of spatial frequencies. In practice, the procedure is as follows: By observing the position of the first CTF minimum, adjust the focus so that it moves away from the center and toward higher frequencies (white arrows, Fig. 14.14A–C). At a particular focus, setting the central region will be widest (Fig. 14.14C) and further adjustment will start to shrink back the central area. That focus position is the optimal focus for the phase plate. Depending on the magnification and pixel size, the first zero may go beyond the Nyquist frequency of the CCD camera. In such cases the goal is exactly that—move the first zero so that it is beyond the Nyquist frequency.

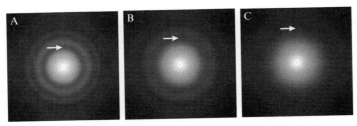

Figure 14.14 Focus series with a Zernike phase plate. The focus was gradually changed from (A) to (B) to (C) bringing the microscope closer to focus. White arrows indicate the first zero of the contrast transfer function, which moves further from the center as the microscope approaches focus. Nyquist frequency: 6.5 Å.

Using phase plates involves most of all evaluating their condition. Electrostatic charging, aging, and contaminants can cause deformation of the CTF and degrade performance. By observing the Fourier image, it is easy to assess the condition of a phase plate. The most common problem with phase plates is electrostatic charging and it can cause a wide range of effects. Easiest to notice is severe charging or anisotropic charging due to contaminants. In such cases, the CTF is heavily deformed and may exhibit very strange shapes. Moderate charging can cause astigmatism–like deformation or appearance of CTF zeros very close to the Fourier origin. Weak charging is not immediately noticeable and can have an effect similar to slight change of focus. In any case, the easiest way to evaluate the performance of a phase plate is by observing the CTF and measuring how far from the Fourier center can the first zero be brought by changing the focus. Really good phase plates in our microscope (JEOL JEM-2200FS) show a first zero at ~6 Å periodicity. Average quality phase plates typically achieve maximum first zero at ~10 Å. Depending on the application, a phase plate may still be usable even if its performance is not ideal. For example, in tomographic applications, the Nyquist of the CCD camera is typically at ~20 Å periodicity. If the first zero of the CTF can be brought beyond that point, the phase plate can be used for tomography.

4.5. Low dose observation with a phase plate

Low dose observations are usually based on several preset microscope states, typically named "Search," "Focus," and "Photo." "Search" mode can be used for finding the phase plate central hole and rough centering (see Section 4.2). "Photo" mode must be set to the desired magnification and beam intensity with the phase plate in "on-plane" condition as described in

Sections 4.2 and 4.3. "Focus" mode is usually set to observe an adjacent area of the specimen by offsets to the image shift and beam shift deflectors. Our experience shows that for best results with phase plates, the "Focus" mode should be set to exactly the same lens values as the "Photo" mode, except for the beam and image shift deflectors. This helps to avoid phase plate misalignment due to lens hysteresis. In high performance modern microscopes, it may be practical to use different magnification and illumination conditions for these two modes but the stability must be checked for each particular instrument. In practice, it is helpful to have constant offset values (rather than absolute values memorized for each mode) for the beam deflectors between the "Focus" and "Photo" modes. This ensures that if the phase plate is misaligned for some reason, such as thermal drift or specimen charging, recentering it in "Focus" mode will also center it for the "Photo" mode. We wrote simple extension software that augments the microscope's (JEOL JEM-2200FS) built-in low dose system. The software applies a predefined beam shift and beam tilt offsets when switching between the two low dose modes. So if, for example, the beam is tilted by 10 mrad in "Focus" mode and then the microscope is switched to "Photo" mode, the same amount of correction will be applied there. One can think of this setup as the beam deflector values of "Focus" and "Photo" modes being connected by a "rigid pole" (fixed offset values), so moving the beam in "Focus" mode automatically moves the beam in "Photo" mode. The offset values must be determined by comparing the values for an on-plane centered phase plate in each mode. First center and bring on-plane the phase plate in "Photo" mode. Switch to "Focus" mode, expand the beam, and reduce magnification if necessary, until the projection of the phase plate hole is visible. Center and bring on-plane the phase plate by using just beam tilt (not mechanical movement). Check the difference between the beam shift and beam tilt values for "Photo" and "Focus" mode and use the same offset during low dose operation.

Low dose data acquisition with a phase plate consists of the same steps as the conventional TEM, with a few additions. Find a suitable specimen area in "Search" mode. Switch to "Focus" mode and check the centering of the phase plate. If the phase plate is misaligned, recenter it using the beam tilt. Check and adjust the focus with a CCD camera. Blank the beam, switch to "Photo" mode, and acquire an image. After acquiring the image, lower the screen and check the centering of the phase plate. If the phase plate is not centered, center it using the beam tilt. If the misalignment was severe (phase plate hole edge visible in the image), recheck the deflector offsets between "Focus" and "Photo" modes as described in the previous paragraph. Adjustment to the offsets may be necessary after the first few images of an observing session, but are typically not necessary afterward.

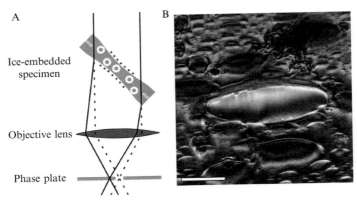

A — Ice-embedded specimen

Objective lens

Phase plate

B

Figure 14.15 Illustration of the effect of specimen charging on the phase plate alignment. (A) An electrostatically charged tilted specimen causes the beam to slightly tilt which shifts the diffraction pattern relative to the phase plate. (B) Image showing topographic-like contrast most probably due to specimen charging causing misalignment of the phase plate. Scale bar: 500 nm.

4.6. Specimen charging effects

In cryo-electron microscopy specimen, charging is a common occurrence. In most cases, it is not detectable in the image and experimenters are not aware of its presence. The phase plate however, makes specimen charging an obvious phenomenon. Figure 14.15A illustrates the effect. An asymmetric distribution of charges on the surface of a nonconducting specimen, such as when the specimen is tilted, can act as an electrostatic beam deflector and introduce beam tilt. A slight change of the beam tilt angle does not change noticeably the image. At the back-focal plane, however, the shift of the diffraction pattern relative to the phase plate effectively misaligns the phase plate producing topographic-like contrast effects in the image (Fig. 14.15B). In severe cases, specimen charging can cause the phase plate to go off-plane and produce visible phase plate hole projection in the image. The best practical advice for avoiding specimen charging is to always include as much as possible conductive specimen area (such as carbon film) in the illuminated spot. Presumably, secondary electrons emitted by the conductive regions greatly reduce the magnitude of charging of nearby insulating areas.

5. SUMMARY AND FUTURE PROSPECTS

Phase plate devices and techniques are still in their infancy. The development of phase plates will continue to be stimulated by the shortcomings of the DPC in cryo-microscopy applications. At present, only the thin film phase plate is being used for cryo-observations and has demonstrated practical benefits. It still has many problems, such as inconsistency of manufacturing, aging, and information loss. Mass production techniques and phase plate airlock systems could, in near future, alleviate the former two problems, but the information loss is an inherent feature for this type of phase plate. Several groups are advancing in the development of electrostatic phase plates. We expect that soon some of these devices will be tested in biological applications. In addition to the absence of information loss at high spatial frequencies, this type of phase plate has the advantage of real-time phase shift control, which can open the doors to new methods. Presently, phase plate experiments require manual operation which greatly reduces their attractiveness. Much needed developments are hardware and software for automated phase plate alignment, evaluation, focusing, and data collection. We are eagerly looking forward to first experimental trials with the novel phase plate designs which do not require placement of materials on the beam path—the photonic, anamorphotic, and electrostatic-mirror designs.

ACKNOWLEDGMENTS

We thank Robert Glaeser for critical reading of the manuscript. We gratefully acknowledge funding by CREST (Core Research for Evolutional Science and Technology), Japan Science and Technology Agency.

REFERENCES

Agar, A. W., Revell, R. S. M., and Scott, R. A. (1949). A preliminary report on attempts to realise a phase contrast electron microscope. *Proc. Eur. Reg. Conf. Electron Microsc. Delft.* **52**, 1–3.

Alloyeau, D., Hsieh, W. K., Anderson, E. H., Hilken, L., Benner, G., Meng, X., Chen, F. R., and Kisielowski, C. (2010). Imaging of soft and hard materials using a Boersch phase plate in a transmission electron microscope. *Ultramicroscopy* **110**, 563–570.

Badde, H. G., and Reimer, L. (1970). Der Einfluß einer streuenden Phasenplatte auf das elektronenmikroskopische Bild. *Z. Naturforsch.* **25a**, 760–765.

Barton, B., Joos, F., and Schröder, R. R. (2008). Improved specimen reconstruction by Hilbert phase contrast tomography. *J. Struct. Biol.* **164**, 210–220.

Beleggia, M. (2008). A formula for the image intensity of phase objects in Zernike mode. *Ultramicroscopy* **108**, 953–958.

Boersch, V. H. (1947). Über die Kontraste von Atomen in Electronenmikroskop. *Z. Naturforsch.* **2a**, 615–633.

Cambie, R., Downing, K. H., Typke, D., Glaeser, R. M., and Jin, J. (2007). Design of a microfabricated, two-electrode phase-contrast element suitable for electron microscopy. *Ultramicroscopy* **107**, 329–339.

Chang, W.-H., Chiu, M. T.-K., Chen, C.-Y., Yen, C.-F., Lin, Y.-C., Weng, Y.-P., Chang, J.-C., Wu, Y.-M., Cheng, H., Fu, J., and Tu, I.-P. (2010). Zernike phase plate cryoelectron microscopy facilitates single particle analysis of unstained asymmetric protein complexes. *Structure* **18**, 17–27.

Danev, R., and Nagayama, K. (2001a). (1) Complex observation in electron microscopy. II. Direct visualization of phases and amplitudes of exit wave functions. *J. Phys. Soc. Jpn.* **70**, 696–702.

Danev, R., and Nagayama, K. (2001b). (2) Transmission electron microscopy with Zernike phase plate. *Ultramicroscopy* **88**, 243–252.

Danev, R., and Nagayama, K. (2004). Complex observation in electron microscopy: IV Reconstruction of complex object wave from conventional and half plane phase plate image pair. *J. Phys. Soc. Jpn.* **73**, 2718–2724.

Danev, R., and Nagayama, K. (2006). Applicability of thin film phase plates in biological electron microscopy. *Biophysics* **2**, 35–43.

Danev, R., and Nagayama, K. (2008). Single particle analysis based on Zernike phase contrast transmission electron microscopy. *J. Struct. Biol.* **161**, 211–218.

Danev, R., Okawara, H., Usuda, N., Kametani, K., and Nagayama, K. (2002). A novel phase-contrast transmission electron microscopy producing high-contrast topographic images of weak objects. *J. Biol. Phys.* **28**, 627–635.

Danev, R., Glaeser, R., and Nagayama, K. (2009). Practical factors affecting the performance of a thin-film phase plate for transmission electron microscopy. *Ultramicroscopy* **109**, 312–325.

Danev, R., Kanamaru, S., Marko, M., and Nagayama, K. (2010). Zernike phase contrast cryo-electron tomography. *J. Struct. Biol.* **171**, 174–181.

Faget, J., Fagot, M. M., Ferré, J., and Fert, C. (1962). Microscopie electronique a contraste de phase. Academic Press, New York. Fifth International Congress for Electron Microscopy. A-7.

Fukuda, Y., Fukazawa, Y., Danev, R., Shigemoto, R., and Nagayama, K. (2009). Tuning of the Zernike phase-plate for visualization of detailed ultrastructure in complex biological specimens. *J. Struct. Biol.* **168**, 476–484.

Furuhata, M., Danev, R., Nagayama, K., Yamada, Y., Kawakami, H., Toma, K., Hattori, Y., and Maitani, Y. (2008). Decaarginine-PEG-artificial lipid/DNA complex for gene delivery: Nanostructure and transfection efficiency. *J. Nanosci. Nanotechnol.* **8**, 1–8.

Gram, B., Dries, M., Schultheiss, K., Blank, H., Rosenauer, A., Schröder, R. R., and Gerthsen, D. (2010). Object wave reconstruction by phase-plate transmission electron microscopy. *Ultramicroscopy* **110**, 807–814.

Hosokawa, F., Danev, R., Arai, Y., and Nagayama, K. (2005). Transfer doublet and an elaborated phase plate holder for 120 kV electron-phase microscope. *J. Electron Microsc.* **54**, 317–324.

Johnson, H. M., and Parsons, D. F. (1973). Enhanced contrast in electron microscopy of unstained biological material. III. In-focus phase contrast of large objects. *J. Microsc.* **98**, 1–17.

Kanaya, K., Kawakatsu, H., Ito, K., and Yotsumoto, H. (1958). Experiment on the electron phase microscope. *J. Appl. Phys.* **29**, 1046–1049.

Kaneko, Y., Danev, R., Nitta, K., and Nagayama, K. (2005). In vivo subcellular ultrastructures recognized with Hilbert differential contrast transmission electron microscopy. *J. Electron Microsc.* **54**, 79–84.

Majorovits, E., Barton, B., Schultheiß, K., Pérez-Willard, F., Gerthsen, D., and Schröder, R. R. (2007). Optimizing phase contrast in transmission electron microscopy with an electrostatic (Boersch) phase plate. *Ultramicroscopy* **107**, 213–226.

Malac, M., Beleggia, M., Egerton, R., and Zhu, Y. (2008). Imaging of radiation-sensitive samples in transmission electron microscopes equipped with Zernike phase plates. *Ultramicroscopy* **108**, 126–140.

Matsumoto, T., and Tonomura, A. (1996). The phase constancy of electron waves traveling through Boersch's electrostatic phase plate. *Ultramicroscopy* **63**, 5–10.

Müller, H., Jin, J., Danev, R., Spence, J., Padmore, H., and Glaeser, R. M. (2010). Design of an electron microscope phase plate using a focused continuous-wave laser. *New J. Phys.* **12**, 073011.

Nagayama, K. (1999). Complex observation in electron microscopy. I. Basic scheme to surpass the Scherzer limit. *J. Phys. Soc. Jpn.* **68**, 811–822.

Nagayama, K. (2008). Development of phase plates for electron microscopes and their biological application. *Eur. Biophys. J.* **37**, 345–358.

Okamoto, H. (2010). Adaptive quantum measurement for low-dose electron microscopy. *Phys. Rev. A* **81**, 043807.

Schröder, R. R., Barton, B., Rose, H., and Benner, G. (2007). Contrast enhancement by anamorphotic phase plates in an aberration corrected TEM. *Microsc. Microanal.* **13**, 136–137.

Setou, M., Danev, R., Atsuzawa, K., Yao, I., Fukuda, Y., Usuda, N., and Nagayama, K. (2006). Mammalian cell nano structures visualized by cryo Hilbert differential contrast transmission electron microscopy. *Med. Mol. Morphol.* **39**, 176–180.

Shimada, A., Niwa, H., Tsujita, K., Suetsugu, S., Nitta, K., Hanawa-Suetsugu, K., Akasaka, R., Nishino, Y., Toyama, M., Chen, L., Liu, Z.-J., Wang, B.-C., et al. (2007). Curved EFC/F-BAR-domain dimers are joined end to end into a filament for membrane invagination in endocytosis. *Cell* **129**, 761–772.

Shiue, J., Chang, C.-S., Huang, S.-H., Hsu, C.-H., Tsai, J.-S., Chang, W.-H., Wu, Y.-M., Lin, Y.-C., Kuo, P.-C., Huang, Y.-S., Hwu, Y., Kai, J.-J., et al. (2009). Phase TEM for biological imaging utilizing a Boersch electrostatic phase plate: Theory and practice. *J. Electron Microsc.* **58**, 137–145.

Tonomura, A., Osakabe, N., Matsuda, T., Kawasaki, T., and Endo, J. (1986). Evidence for Aharonov–Bohm effect with magnetic field completely shielded from electron wave. *Phys. Rev. Lett.* **56**, 792–795.

Tosaka, M., Danev, R., and Nagayama, K. (2005). Application of phase contrast transmission microscopic methods to polymer materials. *Macromolecules* **38**, 7884–7886.

Van Dyck, D. (2010). Wave reconstruction in TEM using a variable phase plate. *Ultramicroscopy* **110**, 571–572.

Willasch, D. (1975). High resolution electron microscopy with profiled phase plates. *Optik* **44**, 17–36.

Yamaguchi, M., Danev, R., Nishiyama, K., Sugawara, K., and Nagayama, K. (2008). Zernike phase contrast electron microscopy of ice-embedded influenza A virus. *J. Struct. Biol.* **162**, 271–276.

Yui, H., Minamikawa, H., Danev, R., Nagayama, K., Kamiya, S., and Shimizu, T. (2008). Growth process and molecular packing of a self-assembled lipid nanotube: Phase-contrast transmission electron microscopy and XRD analyses. *Langmuir* **24**, 709–713.

CHAPTER FIFTEEN

RADIATION DAMAGE IN ELECTRON CRYOMICROSCOPY

Lindsay A. Baker*,† *and* John L. Rubinstein*,†,‡

Contents

Abstract

In an electron microscope, the electron beam used to determine the structures of biological tissues, cells, and molecules destroys the specimen as the image is acquired. This destruction occurs before a statistically well-defined image can be obtained and is consequently the fundamental limit to resolution in biological electron cryomicroscopy (cryo-EM). Damage from the destructive interaction of electrons with frozen-hydrated specimens occurs in three stages: primary damage, as electrons ionize the sample, break bonds, and produce secondary electrons and free radicals; secondary damage, as the secondary electrons and free radicals migrate through the specimen and cause further chemical reactions; and tertiary damage, as hydrogen gas is evolved within the sample, causing gross morphological changes to the specimen. The deleterious effects of radiation are minimized in cryo-EM by limiting the exposure of the specimen to incident electrons and

* Molecular Structure and Function Program, The Hospital for Sick Children, Ontario, Canada
† Department of Biochemistry, The University of Toronto, Ontario, Canada
‡ Department of Medical Biophysics, The University of Toronto, Ontario, Canada

Methods in Enzymology, Volume 481
ISSN 0076-6879, DOI: 10.1016/S0076-6879(10)81015-8

cooling the sample to reduce secondary damage. This review emphasizes practical considerations for minimizing radiation damage, including measurement of electron exposure, estimation of absorbed doses of energy, selection of microscope voltage and specimen temperature, and selection of electron exposure to optimize images.

1. INTRODUCTION

In an electron microscope, electrons may interact with a specimen in one of two ways: electrons that scatter from the sample but retain their incident energy leave the structure unchanged; electrons that deposit some of their energy into the sample cause radiation damage and consequent structural changes. Contrast in bright field phase contrast electron microscopy (EM) comes primarily from the electrons that do not deposit energy into the specimen. If this type of scattering was the only type of interaction that occurred between the specimen and the electron beam, a modern electron microscope would be capable of atomic resolution tomography of a variety of biological specimens. Unfortunately, damaging interactions outnumber non-destructive interactions by a ratio of approximately 3:1 (Henderson, 1995), and destroy the specimen before statically well-defined images can be obtained. Therefore, electron beam damage is the fundamental limit to resolution in electron cryomicroscopy (cryo-EM) of frozen-hydrated specimens. This damage increases with the exposure of the sample to electrons. Its effects are visible first at high spatial frequencies in the image, but can subsequently be observed even at low resolution (Conway *et al.*, 1993; Glaeser, 1971).

Limiting electron exposure and using appropriate cryoprotection can effectively reduce radiation damage. However, low exposures result in undersampled, noisy images, and introduce complications into the process of building 3D models of biological specimens from cryo-EM images. With limited exposures, the ability of the electron detector to count every electron and produce similar signals from each electron (the detective quantum efficiency of the detector) becomes critical. Furthermore, during irradiation, specimens may charge and move as they degrade in the electron beam. Because of radiation damage to biological samples, those wishing to acquire images of radiation sensitive specimens in ice are forced to do so before the specimen has been able to equilibrate in the electron beam.

2. MEASURING ELECTRON EXPOSURE

While discussing EM, many authors have used the words *dose* and *exposure* interchangeably. However, in more precise terminology, dose refers to the energy absorbed by the specimen while exposure indicates the amount of radiation incident on the specimen. Clearly, there is a relationship between the exposure of electrons to which a sample is subjected and the dose of energy absorbed. An experimentalist can perform an approximate conversion between these two quantities if the linear energy transfer (LET) between the incident electrons and the sample is known.

In principle, a metal disk of known area inserted into the electron beam should produce a current as electrons strike it, with the magnitude of the current linearly related to the number of electrons hitting the disk. Faraday cups use this principle and are the most accurate way of measuring exposure in an electron microscope. From a simple disk, two sources of error, emission of low-energy secondary electron and backscattering of incident electrons, contribute to inaccuracies that would generally underestimate the measured exposure. Therefore, Faraday cups are designed to minimize these sources of error in two different ways. First, a small entrance aperture to the cup allows electrons into a larger chamber, dramatically reducing the probability of a backscattered or secondary electron successfully leaving the trap. Second, the inside of the cup is usually lined with a low atomic number material to reduce the amount of high-angle backscatter. The Faraday cup is usually introduced at the object plane of the microscope and therefore measures the flux of electrons through the specimen independent of the magnification and focus settings of the microscope.

The phosphor screen of an electron microscope makes a convenient measuring device for electron exposure. The current through the screen is proportional to the number of electrons striking the screen, even though the efficiency with which it detects electrons is decreased by the backscattering of electrons and secondary electron emission described above. Therefore, the efficiency of electron capture by the screen must be determined using a Faraday cup. Once the efficiency at a given voltage is known, a correction factor can be applied to current measurements. Because the phosphor screen is located at the image plane of the microscope, the magnification setting of the instrument must be taken into account when converting beam density at the screen to electron flux through the specimen. Also, if a specimen is present in the microscope when flux is measured, scattering of electrons by the specimen will reduce the apparent exposure. Therefore, measurement of the electron flux with the phosphor screen should be done without a sample in the instrument.

The darkening of photographic film is a third method for estimating electron exposures. As with a phosphor screen, when using film to estimate the flux of electrons through the specimen, one must account for the magnification of the microscope and avoid scattering of electrons by a specimen. Any type of photographic film will have a defined "speed". For electron image film, speed is the reciprocal of the exposure in $e^-/\mu m^2$ required to produce an optical density (OD) of 1.0 above the gross fog level of the film. OD is defined as

$$OD = -\log_{10}\left(\frac{I}{I_0}\right),\qquad\qquad(15.1)$$

where I is the intensity of light transmitted through the film after illumination with light of intensity I_0. With 100 keV electrons, the speed of the commonly used Kodak SO-163 film is 2.2 when developed in full strength Kodak D19 developer for 12 min. Therefore, an OD of 1.0 from SO-163 film indicates that the film has been exposed to ~ 0.46 e$^-/\mu$m^2. If one then considers the magnification of the specimen onto the image plane of the microscope, an image acquired at 50,000 k× magnification that produced this darkening of the film would suggest that the specimen had been exposed to ~ 11.4 e$^-/\text{Å}^2$. Unfortunately, the speed for SO-163 film is stated for 100 kV electrons only, which are not always used, and the apparent speed of the film is strongly dependent on the freshness of the developing solution, its temperature being the appropriate 20 °C (68 °F), and proper agitation. In our experience, darkening of photographic film is the least accurate way to measure electron exposures.

3. ENERGY ABSORBED BY SPECIMENS

The exposure of the sample to a flux of electrons is conveniently expressed in terms of electrons per Å^2 of specimen surface area (e$^-/\text{Å}^2$). The energy of the electrons is usually between 100 and 300 keV (1 eV $= 1.602 \times 10^{-19}$ J), although some microscopes employ accelerating voltages in the MV range. To convert from electron exposure to the dose of energy absorbed by the specimen, one must know the LET of incident electrons of a specific energy with a specific specimen. Glaeser has quoted values for the LET of electrons with protein based on the LETs of electrons with polyethylene, Lucite, and CO_2 (Glaeser et al., 2007). This analysis yielded LETs of 4.1 MeV cm^2/g at 100 kV, 2.8 at 200 kV, 2.3 at 300 kV, 2.1 at 400 kV, and 1.8 at 1 MV. The product of the electron exposure with the LET gives the energy deposited per gram of specimen (in eV), which may be converted to the SI unit for absorbed ionizing radiation, the Gray (Gy, with 1 Gy $= 1$ J/kg), as shown in Table 15.1.

4. RADIATION DAMAGE AND CHOICE OF ACCELERATING VOLTAGE

The probabilities of different types of interactions between an electron and a specimen are described by the cross-sections for each kind of interaction. Nondamaging interactions are always elastic events, where the kinetic energy of the incident electron is conserved. Only a small fraction of elastic events are

Table 15.1 Dose of energy absorbed by protein samples, as a function of electron exposure and accelerating voltage

Exposure $(e^-/\text{Å}^2)$	100 kV (eVg^{-1}/MGy)	200 kV (eVg^{-1}/MGy)	300 kV (eVg^{-1}/MGy)	400 kV (eVg^{-1}/MGy)	1 MeV (eVg^{-1}/MGy)
1	$4.1 \times 10^{22}/6.6$	$2.8 \times 10^{22}/4.5$	$2.3 \times 10^{22}/3.7$	$2.1 \times 10^{22}/3.4$	$1.8 \times 10^{22}/2.9$
5	$2.0 \times 10^{23}/33$	$1.4 \times 10^{23}/22$	$1.2 \times 10^{23}/18$	$1.0 \times 10^{23}/17$	$9.0 \times 10^{22}/14$
10	$4.1 \times 10^{23}/66$	$2.8 \times 10^{23}/45$	$2.3 \times 10^{23}/37$	$2.1 \times 10^{23}/34$	$1.8 \times 10^{23}/29$
25	$1.0 \times 10^{24}/160$	$7.0 \times 10^{23}/110$	$5.8 \times 10^{23}/92$	$5.2 \times 10^{23}/84$	$4.5 \times 10^{23}/72$
100	$4.1 \times 10^{24}/660$	$2.8 \times 10^{24}/450$	$2.3 \times 10^{24}/370$	$2.1 \times 10^{24}/340$	$1.8 \times 10^{24}/290$

destructive, causing the so-called "knock-on damage" that dislocates atoms from their chemical bonds (Glaeser et al., 2007). The vast majority of damaging interactions are inelastic events, where some of the kinetic energy of the incident electron is ultimately transformed into heat. The cross-sections for both damaging inelastic and informative elastic interactions decrease with increasing accelerating voltage, maintaining a ratio of around 3:1 damaging events per useful event, with each inelastic interaction depositing an average of ~20 eV into the specimen, regardless of the accelerating voltage (Henderson, 1995; Langmore and Smith, 1992). Therefore, changing the accelerating voltage of an electron microscope will not alter the number of useful elastic events per unit radiation damage. However, a higher electron exposure will be required at higher accelerating voltage for the same number of scattering events to occur. This decrease in the cross-sections of interaction at higher voltages means that the probability of an electron being scattered more than once within the sample also decreases with increasing voltage. Multiple scattering events do not provide information about the structure of the specimen and avoiding them provides some advantage in EM. Therefore, for thick specimens where multiple scattering degrades image quality, high-voltage microscopy can produce better images. Also, due to the increased momentum of higher energy electrons, any distorting effect of specimen charging on the electron beam will be decreased at higher voltage. Other factors that may influence the preference for accelerating voltage include decreased curvature of the Ewald sphere at higher voltages (e.g., see DeRosier, 2000; Wolf et al., 2006), and the specific behavior of different electron detectors, which may be better or worse with electrons of different energies (e.g., see Faruqi and Henderson, 2007).

With these constraints in mind, the exposure to electrons at a given voltage should be adjusted to produce an allowable amount of radiation damage for any experiment. Inspection of the LET values given above suggests that equivalent exposures scaled to the exposure at 200 kV, are 68% at 100 kV, 122% at 300 kV, 133% at 400 kV, and 155% at 1 MV. Similar estimates of equivalent exposures may be obtained by comparison of the elastic cross-sections for electrons with carbon at different voltages (Yalcin et al., 2006). Guidelines on how to select an electron exposure at 200 kV to optimize images for cryo-EM of 2D crystals, single particles, and electron tomography are given in Sections 9 to 11 of this chapter.

In addition to increasing electron exposure at higher voltage, in order to obtain the same signal-to-noise ratio (SNR) one must also increase the defocus of the microscope so that the contrast transfer functions (CTFs) are similar at low resolution. To match the low-resolution CTFs, the product of microscope defocus and electron wavelength should be kept constant at different voltages (we thank Robert Glaeser for pointing out this behavior of the CTF). Electron wavelengths can be calculated with the expression

$$\lambda = 1.22639/\left(V_0 + 0.97845 \times 10^{-6} V_0^2\right)^{1/2}, \qquad (15.2)$$

where V_0 is the accelerating voltage in volts and λ is the wavelength in nanometers (Spence, 2008). At 100, 200, 300, 400 kV, and 1 MV electron wavelengths are 3.70, 2.51, 1.97, 1.64, and 0.87 pm, respectively. This requirement for increased defocus negates the apparent advantage of an improved spatial coherence envelope at high voltage for microscopes with the same source size. The improved temporal coherence envelope of a high-voltage microscope would only be beneficial if one could take images closer to focus than is practical in cryo-EM.

5. PRIMARY AND SECONDARY DAMAGE TO PROTEINS DURING IRRADIATION

At hundreds of keVs, the electron energies used for cryo-EM are significantly higher than the covalent bond energies in biological specimens, which are on the order of a few eVs. Most beam damage occurs due to electrons that lose between $\sim$5 and $\sim$100 eV during interaction with the specimen (on average $\sim$20 eV) (Langmore and Smith, 1992). The deposited energy predominantly excites or ionizes the valence electrons that make up chemical bonds, breaking the bond and producing free radicals and causing emission of secondary electrons. This process constitutes the primary damage event in cryo-EM. Ionization of K-shell electrons and knock-on collisions both occur with such low cross-sections that they are not relevant to a discussion of radiation damage of biological samples (Glaeser et al., 2007). The electrons liberated during primary damage go on to break more bonds, while the free radicals generate a cascade of chemical reactions in a process known as secondary damage.

In comparison to electron beam damage, irradiation of proteins with X-rays leads to photoelectric absorption of the X-ray photons and, to a lesser extent in the relevant experiments, Compton scattering. Both of these processes break covalent bonds, produce free radicals, and cause the emission of electrons within the sample (Henderson, 1995). Therefore, both electron and X-ray irradiation produce similar destructive effects with a similar mechanism. At present, because cryo-EM of proteins rarely produces atomic resolution maps, understanding of the specific chemistry of primary and secondary radiation damage has depended on X-ray crystallographic studies. Cysteine, aspartate, and glutamate residues are particularly susceptible to radiation damage. Some of the disulphide bonds present in a protein structure are already significantly damaged after a dose of $\sim$10^7 Gy (Weik et al., 2000). In EM, even at 1 MeV, this dose has already been exceeded with an exposure of 5 e$^-$/Å^2 (Table 15.1). During radiation damage of disulphide bonds, one sulfur atom is lost near the beginning of the damage process while the second sulfur atom disappears from the structure soon afterward (Weik et al., 2000). Notably, while some disulphide bonds are destroyed at these low doses, other disulphide bonds are apparently resistant to damage even at doses of $\sim 10^8$ Gy. For disulphide bonds with different dose tolerances, there is no noticeable correlation between the solvent accessibility and damage during irradiation, suggesting that if the damage is caused indirectly by species produced in surrounding water, these species must be mobile both through solvent and protein (Ravelli and McSweeney, 2000). Damage to carboxylic acid groups appears as disproportionately high B factors for aspartate and glutamate residues after irradiation. B factors, which reflect the attenuation of high-resolution scattering from atoms, could be elevated either due to increased mobility or decarboxylation of the residue (Weik et al., 2000). In enzymes, active site residues also appear to be particularly sensitive to radiation damage (Weik et al., 2000). This sensitivity could be because the residues involved in catalytic sites are often in strained geometries that destabilize the amino acid structure.

6. TERTIARY DAMAGE TO PROTEINS DURING IRRADIATION

In tertiary or global damage during irradiation, protein crystals and single particle EM or electron tomography samples become distorted, with bubbles becoming apparent in the extreme case. X-ray and EM analysis have shown that these distortions are coincident with the production of gas within the sample during irradiation. Bubbling of specimens in the electron beam (Dubochet et al., 1988) is due to the buildup of hydrogen gas in frozen-hydrated specimens (Leapman and Sun, 1995). Similarly, aqueous solutions and organic solvents can be irradiated with X-rays to produce gases. Analysis of the gas liberated from X-ray irradiation of various liquids shows that hydrogen is the major constituent (Meents et al., 2010) (Table 15.2). Organic solvents produce significantly more gas upon irradiation than pure water or even water containing 10% lysozyme (Meents et al., 2010). Despite the low volume of hydrogen liberated by irradiation of pure water, Leapman and Sun (1995) have suggested that radiolysis of water occurs with the reaction:

$$H_2O \rightarrow H^\bullet + OH^\bullet. \tag{15.3}$$

They propose that in bulk water the $H^\bullet$ and $OH^\bullet$ free radicals readily recombine back to H_2O but near a protein or organic molecule, the radicals produced by radiolysis of water could react with hydrogen atoms to cause reactions such as

Table 15.2 The production of gas by X-ray irradiation of liquids suggests that hydrogen gas is the primary constituent in bubbles in cryo-EM samples, and that pure water has much less capacity for gas production than organic solvents or organic material in water

Liquid	Volume of gas produced (relative to hexane)	Composition of gas
Water	0.5%	Undetectable
Hexane	100%	100% H_2
Acetone	43%	83.2% H_2, 6.2% CO, 10.6% CH_4
Methanol	42%	98.4% H_2, 1.6% CH_4
Ethanol	40%	96.1% H_2, 1.3% CO, 2.6% CH_4
Ethylene glycol	37%	99.6% H_2, 0.4% CO
0.1 M Sodium acetate	22%	100% H_2
10% (wt/vol) Yeast in water	19%	96.4% H_2, 0.5% CO, 3.1% CO_2

Data is from Meents *et al.* (2010).

$$OH^\bullet + R\text{-}H \rightarrow RO^\bullet + H_2. \qquad (15.4)$$

This reaction scheme is consistent with the observation that more gas bubbles develop at the interface of protein with ice than inside protein itself (Glaeser *et al.*, 2007). The extent of hydrogen production is such that pressures within bubbles in vitreous ice may increase to as much as 10^3 atm (Leapman and Sun, 1995). X-ray irradiation of samples at different cryogenic temperatures and subsequent warming of the sample to room temperature have shown that there is little dependence of gas production on the temperature at which the irradiation is performed (Meents *et al.*, 2010).

7. CRYOPROTECTION AND OPTIMAL TEMPERATURES

Breaking of bonds by electrons occurs at any temperature. Therefore, cooling the specimen does not affect primary beam damage. However, it has long been known that cooling a specimen reduces the rate of fading of diffraction spots from 2D crystals (Hayward and Glaeser, 1979; Taylor and Glaeser, 1976). It is now thought that the mechanism of cryoprotection in both cryo-EM and X-ray crystallography is the mechanical restraint of molecular fragments by the ice matrix, preventing their movement so that imaging or diffraction can continue for longer (Henderson, 1990). In other words, the same quantity of reactive free radicals is formed during irradiation, regardless of specimen temperature, but the mobility of different species changes with temperature.

One might presume that colder temperatures lead to better cryoprotection. Indeed, for electron tomography of *Caulobacter crescentus* at 4.2, 16, and 80 K, it was found that the colder temperatures offered a modest decrease in fading of calculated diffraction spots from the bacterium's S-layer (Comolli and Downing, 2005). Similarly, calculated diffraction patterns from images of thin catalase crystals showed a reduction in fading of spots between 20 and 60 Å at 25 and 42 K compared to 100 K, although no difference in fading was seen for higher resolution spots at the three temperatures (Bammes *et al.*, 2010). Unfortunately, the choice of temperature for cryo-EM is complicated by the observation that ice can behave differently at different temperatures. Experiments have shown that contrast fades faster during imaging of cells, liposomes, and protein particles at liquid helium temperature compared to liquid nitrogen temperature (Comolli and Downing, 2005; Iancu *et al.*, 2006; Wright *et al.*, 2006). Interestingly, the difference in contrast between samples irradiated at 12 and 82 K can be removed by warming the sample to 82 K for an extended time (Iancu *et al.*, 2006). Upon irradiation at temperatures below ~ 50 K, the diffraction pattern from amorphous ice changes (Heide, 1984; Wright *et al.*, 2006) and some have hypothesized that this change in diffraction pattern and loss of contrast are due to an increase of $\sim 20\%$ in the ice density (Heide, 1984; Wright *et al.*, 2006). However, despite commendable effort, direct evidence for increased density, such as a significant decrease in ice layer thickness, has not been observed (Wright *et al.*, 2006). Instead, a peculiar "doming" effect was detected, where ice that had been cooled to 12 K, irradiated, and then allowed to warm to 82 K forms a surface that bulges in one direction out of the plane of the specimen grid (Wright *et al.*, 2006). Due to these inconsistencies and unexpected effects, it seems that the mechanism of contrast loss below liquid nitrogen temperature warrants more investigation. It should be noted that an increase in density due to a decrease in the ice layer thickness alone would not result in contrast loss because the projected mass-distribution of the sample would not change. Instead, lateral movement of water in the ice layer would be required to alter density projections along the direction of the electron beam.

Amorphous ice also becomes more fluid as it is cooled below 70 K (Comolli and Downing, 2005; Heide and Zeitler, 1985; Wright *et al.*, 2006). In electron tomography near liquid helium temperature, this fluidity can lead to changes in the dimensions of bacterial cells after 50–100 $e^-/\text{Å}^2$ of irradiation (Comolli and Downing, 2005). Finally, the mobility of hydrogen within the sample changes with temperature, sometimes producing undesirable effects. While distortion of the ice by large bubbles occurs later at lower temperatures, small bubbles trapped beside membranes can lead to apparent contrast reversal during irradiation below liquid nitrogen temperature (Comolli and Downing, 2005; Iancu et al., 2006).

The optimal temperature for cryoprotection is one that presents the best compromise by reducing the mobility of reactive species, minimizing distortions in the specimen due to hydrogen bubbling, and avoiding additional sources of contrast loss. X-ray crystallographic studies have found that an optimal temperature for studying 3D crystals is around 50 K (Meents et al., 2010). As described above, there might be an advantage to performing cryo-EM experiments around 50 or 42 K (Bammes et al., 2010); however, it is not clear if the additional costs and complications of cooling beyond liquid nitrogen temperature (~ 77 K) are warranted.

8. QUANTIFICATION OF BEAM DAMAGE WITH INCREASING EXPOSURE

As has been recognized by many investigators, analysis of radiation damage in thin crystals offers a way of selecting optimal exposures for many types of samples by capturing the relevant effects of beam damage. Radiation damage of thin crystals results in a loss of diffraction intensity. This fading of the diffraction pattern does not automatically suggest destruction of the individual molecules that make up a crystal but instead could suggest that the crystallinity of the sample is destroyed. Indeed, X-ray crystallographers have long known that the mosaicity of a diffraction pattern increases during irradiation as the unit cell of the crystal expands. Recent studies have suggested that the evolution of hydrogen gas is the primary cause for the loss of crystallinity (Meents et al., 2010). Similar effects can be expected for the study of 2D crystals. These effects are also relevant for single particle EM and tomography because movement of molecules and deformation of thick specimens will reduce the quality of images obtained from both types of sample. Therefore, analysis of the loss of diffraction intensity, because it captures the effects of primary, secondary, and tertiary beam damage, is a self-contained method for studying the limitations imposed by radiation damage.

2D crystals and thin 3D crystals have been employed extensively in the study of radiation damage by electrons (Baker et al., 2010; Bammes et al., 2010; Chen et al., 2008; Glaeser, 1971, 1979; Hayward and Glaeser, 1979; Henderson, 1992; Howitt et al., 1976; Stark et al., 1996; Taylor and Glaeser, 1976; Unwin and Henderson, 1975). The most common biological samples used for quantifying radiation damage are thin crystals of catalase or 2D arrays of bacteriorhodopsin. A straightforward method to crystallize bovine liver catalase (Dorset and Parsons, 1975) involves resuspending the protein to 60–100 mg/mL in buffer (28 mM Na_2HPO_4 and 5 mM KH_2PO_4) at pH 7.4. The pH of the suspension is then lowered to 5.3 using saturated KH_2PO_4 and the sample is incubated for several weeks at 4 °C. This procedure produces thin crystals that may be withdrawn from the slurry, washed with water, and applied to an EM grid for diffraction or imaging studies. Crystals adsorbed to a continuous carbon film may experience less specimen charging than those suspended in ice in holes. However, comparison of the fading of the calculated diffraction pattern from catalase crystals prepared with both substrates (Baker et al., 2010; Bammes et al., 2010) suggests that charging is not a significant factor in measuring radiation damage with thin crystals.

Direct recording of electron diffraction gives rise to spots on a reciprocal space lattice. The resolution of the information conveyed by a spot depends on the position of the spot in the lattice and the intensity of the spot is equal to the squared amplitude of the Fourier component at that position. Fourier transforms of images of 2D crystals or thin 3D crystals can also be used to measure this information, with the additional detail that in Fourier transforms, amplitudes, not intensities, are measured and the Fourier components are scaled by the CTF of the microscope. Empirically, the fall-off of spot intensity has been shown to behave as an exponential function (Hayward and Glaeser, 1979; Unwin and Henderson, 1975) with

$$\left| f_{\vec{k}}(N) \right|^2 = \left| f_{\vec{k}}(0) \right|^2 e^{[-N]/[N_e(\vec{k})]}, \tag{15.5}$$

where N is the total accumulated electron exposure of the specimen, $\left| f_{\vec{k}}(N) \right|^2$ is the instantaneous intensity at exposure N for a component with Fourier coordinates $\vec{k}$, $\left| f_{\vec{k}}(0) \right|^2$ is the initial instantaneous intensity of the component, and the total accumulated electron exposure at which the intensity of the component decreases to e^{-1} (approximately 0.368) times its initial value is defined as the critical exposure, $N_e(\vec{k})$ (Unwin and Henderson, 1975).

Critical exposures can be measured experimentally by fitting diffraction intensities (e.g., Downing and Li, 2001; Hayward and Glaeser, 1979; Unwin and Henderson, 1975) or squared calculated Fourier amplitudes (e.g., Baker et al., 2010; Bammes et al., 2010) as a function of accumulated electron exposure to the model given by Eq. (15.5). Once the critical exposure for each Fourier component is known, a plot of critical exposure as a function of resolution can be produced, as shown in Fig. 15.1. Critical exposures provide a quantitative way to compare radiation sensitivity for different resolutions, temperatures, and imaging conditions.

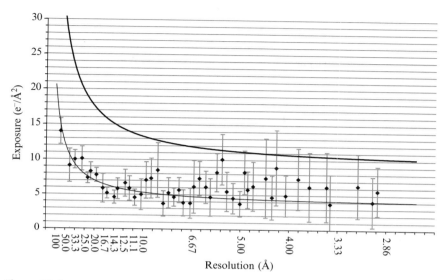

Figure 15.1 Critical and optimal exposures as a function of resolution in imaging experiments. For images acquired at 200 kV, experimental critical exposures (points) were fit as a function of resolution (thin solid line) (from Baker *et al.*, 2010) and the fit was extrapolated to optimal exposures (thick solid line) at 2.5 times the critical exposures, based on the signal-to-noise ratio (SNR) relationship derived by Hayward and Glaeser (1979).

 ## 9. OPTIMAL EXPOSURES FOR THIN CRYSTALS

For 2D crystals, the relationship between the exposure that maximizes the SNR at a resolution (i.e., the optimal exposure $N_{opt}(\vec{k})$) and the critical exposure, $N_e(\vec{k})$, depends on whether data is collected by imaging or diffraction. In diffraction experiments, SNRs depend on the height of the intensity peak above the background noise. Therefore, the optimal exposure varies with the intensity of the spot. Spots with large ratios of peak height to background intensity have optimal exposures well above $N_e(\vec{k})$, while those with peak-to-background ratios that are ≤ 1 have $N_{opt}(\vec{k}) \approx N_e(\vec{k})$ (Downing and Li, 2001). As a result, for diffraction experiments, an exposure should be used that is only slightly higher than the critical exposure of the highest resolution spot that one hopes to record.

During image acquisition, two competing processes are at work. First, the contribution of information from each subsequent electron decreases as the specimen is irradiated. However, even beyond the critical exposure, signal accumulates until the Fourier component has faded completely and information from high-resolution features stays recorded in the image even after the features themselves are destroyed in the specimen. Second, the statistics that describe the image improve as successive electrons are recorded. Hayward and Glaeser have modeled the improvement with cumulative electron exposure and combined this information with the exponential loss of information to show that, in imaging mode, the optimal exposure to maximize the SNR for a spatial frequency, $N_{opt}(\vec{k})$, is $\sim 2.5 N_e(\vec{k})$ in the absence of strong object and detector noise (Hayward and Glaeser, 1979). Therefore, for imaging of 2D crystals, an exposure should be selected that is ~ 2.5 times the critical exposure of the highest resolution feature that the experiment aims to observe (Fig. 15.1).

10. OPTIMAL EXPOSURES FOR SINGLE PARTICLE SAMPLES

For imaging of single particles, the SNR relationship derived by Hayward and Glaeser holds true. With these noncrystalline specimens, one must select the resolution at which the SNR is to be optimized in an image. While it may seem prudent to optimize the SNR at the highest resolution that the experiment aims to obtain, doing so may produce suboptimal SNRs at lower spatial frequencies needed to align particles so that images can be averaged coherently. Because critical exposures change rapidly at low resolution (Fig. 15.1), it is impossible to optimize data collection for all resolutions with a single exposure. Figure 15.2 shows a

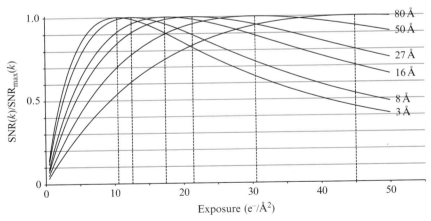

Figure 15.2   The effect of changing the electron exposure on the signal-to-noise ratios (SNRs) of individual spatial frequencies. Relative SNRs for selected resolutions are plotted as a function of exposure for imaging experiments at 200 kV (from Baker et al., 2010). SNRs were scaled between 0 and 1, with SNR(N_{opt}) = 1. The dashed vertical lines indicate the optimal exposure for each resolution. Note that the increase in SNR up to the optimal exposure occurs faster than the decrease in the SNR after the optimal exposure.

plot of SNR as a function of exposure for different resolution Fourier components. As seen in Fig. 15.2, the SNR increases quickly as exposures approach $N_{opt}(\vec{k})$ and then decreases slowly with exposures beyond $N_{opt}(\vec{k})$. Therefore, it is probably safer to slightly overexpose the specimen than to underexpose it. Depending on the size of the molecule and the expected resolution of the final reconstruction, at 200 kV an exposure between 10 and 30 e$^-$/Å^2 could be used to optimize the SNR at resolutions between 3 and 50 Å (Fig. 15.1). For large protein complexes, for which orientation parameters can be robustly determined and building a high-resolution map is feasible, a lower electron exposure is optimal (e.g., 10–15 e$^-$/Å^2 at 200 kV). For smaller particles where one still aims to obtain a high-resolution map, a slightly higher exposure (e.g., 15–20 e$^-$/Å^2 at 200 kV), while sacrificing some SNR at high resolution, may provide better SNRs at the spatial frequencies needed to align images. For particles where only a low-resolution map is feasible, a higher electron exposure (e.g., 20–30 e$^-$/Å^2 at 200 kV) may provide the best chance of successfully building a reliable 3D map. Figure 15.3, showing the fraction of the best possible SNR obtained at each spatial frequency, illustrates how different spatial frequencies are affected by the choice of exposure for an image.

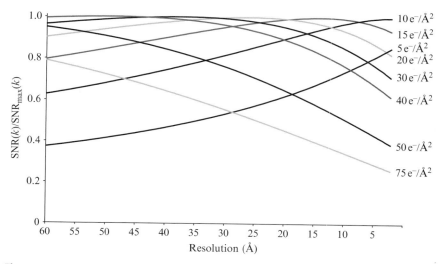

Figure 15.3 The effect of electron exposure choice on the signal-to-noise ratio (SNR) as a function of resolution. The ratio of SNR to SNR(N_{opt}) for selected electron exposures was plotted as a function of resolution for imaging experiments at 200 kV. Note the low SNR at most resolutions for low exposures.

11. OPTIMAL EXPOSURES FOR TOMOGRAPHIC SAMPLES

For electron tomography, target resolutions are usually 20–60 Å, which is generally lower than for single particle EM and electron crystallography, and therefore higher electron exposures can be tolerated. The total electron exposure required for a given level of statistical significance at a given resolution in a 3D tomogram is the same as the exposure required for the same significance at the same frequency in a 2D image (Hegerl and Hoppe, 1976; McEwen *et al.*, 1995). Therefore, the exposure that optimizes the SNR for a resolution in tomography can be estimated from the $N_{opt}(\vec{k})$ versus resolution curve in Fig. 15.1, the total exposure being divided over the images in the tilt series. Exposures between 20 and 100 $e^-/\text{Å}^2$ at 200 kV optimize SNRs at resolutions between 25 and 100 Å. As with the other experiments, the slow fade-off of the SNR at exposures greater than $N_{opt}(\vec{k})$ suggests that it is probably better to slightly overirradiate a specimen rather than underexpose it when collecting tomographic data. Empirically, some investigators have found that they produce their best tomograms using total exposures of around 200 $e^-/\text{Å}^2$ at 300 kV, which corresponds to ~ 180 $e^-/\text{Å}^2$ at 200 kV (Briegel *et al.*, 2008). Although bubbling in the specimen is most often not observed until exposures above 30–45 $e^-/\text{Å}^2$ for ice over carbon and above 100 $e^-/\text{Å}^2$ for ice in holes (Baker *et al.*, 2010), the presence of organic compounds in the vitrification buffer or in the tissues and cells themselves could cause bubbling at lower exposures. This bubbling could be a particular problem for tomography experiments, which usually involve relatively high cumulative exposures.

12. CONCLUDING COMMENTS

As described above, radiation damage imposes strict limitation on the electron exposure to which biological specimens can be subjected. Recent experiments have suggested that there is little advantage still to be gained from further optimization of specimen temperatures for the reduction of secondary damage. Radiation damage forces the experimentalist to work at conditions where every bit of extra signal in an image can improve a 3D model and every additional source of noise can reduce the chances of success in an experiment. Luckily, more than 40 years of study have given the cryo-EM community the tools to quantify and understand the effects of radiation damage, and the techniques needed to work around this inherent problem in biological EM.

ACKNOWLEDGMENTS

We thank Robert Glaeser and Richard Henderson for many informative discussions and a critical reading of this chapter. L. A. B. was supported by a Vanier Canada Graduate Scholarship from the Natural Sciences and Engineering Research Council of Canada and J. L. R. was supported by a New Investigator award from the Canadian Institutes of Health Research (CIHR). This work was funded by operating grant MOP 81294 from the CIHR.

REFERENCES

Baker, L. A., Smith, E. A., Bueler, S. A., and Rubinstein, J. L. (2010). The resolution dependence of optimal exposures in liquid nitrogen temperature electron cryomicroscopy of catalase crystals. *J. Struct. Biol.* **169,** 431–437.

Bammes, B. E., Jakana, J., Schmid, M. F., and Chiu, W. (2010). Radiation damage effects at four specimen temperatures from 4 to 100 K. *J. Struct. Biol.* **169,** 331–341.

Briegel, A., Ding, H. J., Li, Z., Werner, J., Gitai, Z., Dias, D. P., Jensen, R. B., and Jensen, G. J. (2008). Location and architecture of the *Caulobacter crescentus* chemoreceptor array. *Mol. Microbiol.* **69,** 30–41.

Chen, J. Z., Sachse, C., Xu, C., Mielke, T., Spahn, C. M., and Grigorieff, N. (2008). A dose-rate effect in single-particle electron microscopy. *J. Struct. Biol.* **161,** 92–100.

Comolli, L. R., and Downing, K. H. (2005). Dose tolerance at helium and nitrogen temperatures for whole cell electron tomography. *J. Struct. Biol.* **152,** 149–156.

Conway, J. F., Trus, B. L., Booy, F. P., Newcomb, W. W., Brown, J. C., and Steven, A. C. (1993). The effects of radiation damage on the structure of frozen hydrated HSV-1 capsids. *J. Struct. Biol.* **111,** 222–233.

DeRosier, D. J. (2000). Correction of high-resolution data for curvature of the Ewald sphere. *Ultramicroscopy* **81,** 83–98.

Dorset, D. L., and Parsons, D. F. (1975). Electron diffraction from single, fully-hydrated, ox-liver catalase microcrystals. *Acta Cryst.* **A31,** 210–215.

Downing, K. H., and Li, H. (2001). Accurate recording and measurement of electron diffraction data in structural and difference Fourier studies of proteins. *Microsc. Microanal.* **7,** 407–417.

Dubochet, J., Adrian, M., Chang, J. J., Homo, J. C., Lepault, J., McDowall, A. W., and Schultz, P. (1988). Cryo-electron microscopy of vitrified specimens. *Q. Rev. Biophys.* **21,** 129–228.

Faruqi, A. R., and Henderson, R. (2007). Electronic detectors for electron microscopy. *Curr. Opin. Struct. Biol.* **17,** 549–555.

Glaeser, R. M. (1971). Limitations to significant information in biological electron microscopy as a result of radiation damage. *J. Ultrastruct. Res.* **36,** 466–482.

Glaeser, R. M. (1979). Prospects for extending the resolution limit of the electron microscope. *J. Microsc.* **117,** 77–91.

Glaeser, R. M., Downing, K. H., DeRosier, D. J., Chiu, W., and Frank, J. (2007). Electron Crystallography of Biological Macromolecules. Oxford University Press, Oxford.

Hayward, S. B., and Glaeser, R. M. (1979). Radiation damage of purple membrane at low temperature. *Ultramicroscopy* **04,** 201–210.

Hegerl, R., and Hoppe, W. (1976). *Z. Naturforsch.* **31,** 1717.

Heide, H. G. (1984). Observations on ice layers. *Ultramicroscopy* **14,** 271–278.

Heide, H. G., and Zeitler, E. (1985). The physical behavior of solid water at low temperatures and the embedding of electron microscopical specimens. *Ultramicroscopy* **16,** 151–160.

Henderson, R. (1990). Cryo-protection of protein crystals against radiation damage. *Proc. Biol. Sci.* **241**, 6–8.

Henderson, R. (1992). Image contrast in high-resolution electron microscopy of biological macromolecules: TMV in ice. *Ultramicroscopy* **46**, 1–18.

Henderson, R. (1995). The potential and limitations of neutrons, electrons and X-rays for atomic resolution microscopy of unstained biological molecules. *Q. Rev. Biophys.* **28**, 171–193.

Howitt, D. G., Glaeser, R. M., and Thomas, G. (1976). The energy dependence of electron radiation damage in 1-valine. *J. Ultrastruct. Res.* **55**, 457–461.

Iancu, C. V., Wright, E. R., Heymann, J. B., and Jensen, G. J. (2006). A comparison of liquid nitrogen and liquid helium as cryogens for electron cryotomography. *J. Struct. Biol.* **153**, 231–240.

Langmore, J. P., and Smith, M. F. (1992). Quantitative energy-filtered electron microscopy of biological molecules in ice. *Ultramicroscopy* **46**, 349–373.

Leapman, R. D., and Sun, S. (1995). Cryo-electron energy loss spectroscopy: Observations on vitrified hydrated specimens and radiation damage. *Ultramicroscopy* **59**, 71–79.

McEwen, B. F., Downing, K. H., and Glaeser, R. M. (1995). The relevance of dose-fractionation in tomography of radiation-sensitive specimens. *Ultramicroscopy* **60**, 357–373.

Meents, S., Gutmann, S., Wagner, A., and Schulze-Briese, C. (2010). Origin and temperature dependence of radiation damage in biological samples at cryogenic temperatures. *Proc. Natl. Acad. Sci. USA* **107**, 1094–1099.

Ravelli, R. B., and McSweeney, S. M. (2000). The 'fingerprint' that X-rays can leave on structures. *Structure* **8**, 315–328.

Spence, J. C. H. (2008). High-Resolution Electron Microscopy. 3rd edn. Oxford University Press, Oxford.

Stark, H., Zemlin, F., and Boettcher, C. (1996). Electron radiation damage to protein crystals of bacteriorhodopsin at different temperatures. *Ultramicroscopy* **63**, 75–79.

Taylor, K. A., and Glaeser, R. M. (1976). Electron microscopy of frozen hydrated biological specimens. *J. Ultrastruct. Res.* **55**, 448–456.

Unwin, P. N., and Henderson, R. (1975). Molecular structure determination by electron microscopy of unstained crystalline specimens. *J. Mol. Biol.* **94**, 425–440.

Weik, M., Ravelli, R. B., Kryger, G., McSweeney, S., Raves, M. L., Harel, M., Gros, P., Silman, I., Kroon, J., and Sussman, J. L. (2000). Specific chemical and structural damage to proteins produced by synchrotron radiation. *Proc. Natl. Acad. Sci. USA* **97**, 623–628.

Wolf, M., DeRosier, D. J., and Grigorieff, N. (2006). Ewald sphere correction for single-particle electron microscopy. *Ultramicroscopy* **106**, 376–382.

Wright, E. R., Iancu, C. V., Tivol, W. F., and Jensen, G. J. (2006). Observations on the behavior of vitreous ice at approximately 82 and approximately 12 K. *J. Struct. Biol.* **153**, 241–252.

Yalcin, S., Gurler, O., Gultekin, A., and Gundogdu, O. (2006). An analytical expression for electron elastic scattering cross section from atoms and molecules in 1.0 keV to 1.0 MeV energy range. *Phys. Lett. A* **356**, 138–145.

Author Index

Subject Index

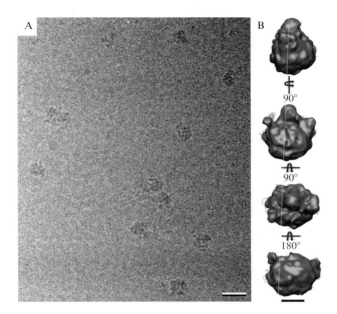

Deborah F. Kelly *et al.*, Figure 4.2 Ribosomal complexes prepared from *E. coli* cell extract using monolayer purification. (A) Image of vitrified ribosomal complexes prepared by monolayer purification by means of a His tag on the rpl3 subunit. Scale bar is 30 nm. (B) Different views of the 3D reconstruction of the 50S subunit (gold surface) with the fit crystal structure (red; Klaholz *et al.*, 2003). Scale bar is 10 nm.

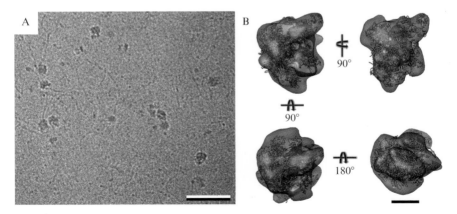

Deborah F. Kelly *et al.*, Figure 4.7 Isolation of mammalian RNAP II using an Affinity Grid in combination with His-tagged protein A and a specific antibody against the largest subunit. (A) Image of vitrified RNAP II particles attached to the Affinity Grid through His-tagged protein A and a specific IgG. Scale bar is 40 nm. (B) Different views of the 3D reconstruction (gray surface), which accommodates the crystal structures of the yeast RNAP II (green; Kettenberger *et al.*, 2004). Scale bar is 5 nm.

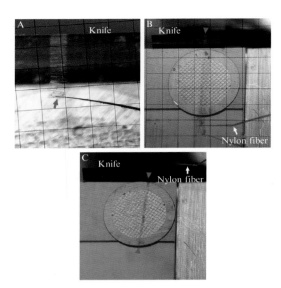

Mark S. Ladinsky, Figure 8.7 Steps in micromanipulator–assisted cryosectioning. (A) When sectioning begins, the leading end of the ribbon is picked up with the manipulator-controlled fiber. The first few sections should wrap around the fiber (arrow), allowing the ribbon to be lifted up and away from the knife. (B) As more sections are cut and the ribbon lengthens (arrowheads), the manipulator-controlled fiber is gently pulled back to hold the ribbon away from the knife, above the underlying EM grid, and under slight tension. Sectioning is stopped when the ribbon is slightly longer than the diameter of the grid (∼3.5 mm). (C) In a three-step process, the micromanipulator is used to lower the ribbon (arrowheads) onto the grid, where it is subsequently detached from the knife and from the manipulator-controlled fiber.

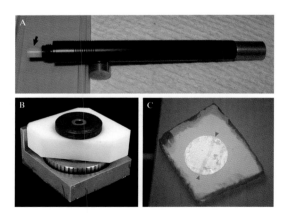

Mark S. Ladinsky, Figure 8.8 Devices for pressing cryosections onto the EM grid. (A) The most common pressing device, a pen-like instrument with Teflon (arrow) or aluminum tips at either end. The instrument is held by the microtomist to apply pressure onto the grid. (B) An earlier pressing device designed for the UltraCut-E/FC-E system. A grid with cryosections is placed on the slotted wheel and rotated under the spring-loaded block. The block is screwed down to apply pressure to the grid. (C) A simple, but very effective pressing device consists of a small chip of glass placed over the grid. A pair of precooled fine forceps is used to apply direct and even pressure along the ribbon of cryosections (arrowheads), firmly affixing it to the grid.

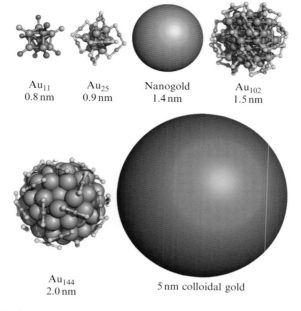

Christopher J. Ackerson et al., Figure 9.1 Representative inorganic gold clusters. The organic components of the ligand layer of each cluster are not shown. Orange, yellow, purple, and green represent gold, sulfur, phosphorous, and chlorine, respectively. The images of Au_{11}, Au_{25}, and Au_{102} are created from X-ray crystallographic coordinates. The image of Au_{144} is from a density functional theory model (Lopez-Acevedo et al., 2009). A 5-nm diameter colloidal gold particle is shown for comparison.

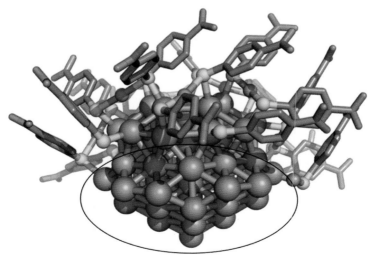

Ligand layer hidden to reveal crystalline core

Christopher J. Ackerson et al., Figure 9.3 An image generated from $Au_{102}(p$-mercaptobenzoic acid)44 ligand protected cluster X-ray crystal coordinates. Some of the ligand layer is hidden to reveal the crystalline core which is fcc (bulk) gold. Orange is gold, yellow is sulfur, gray is carbon, and red is oxygen.

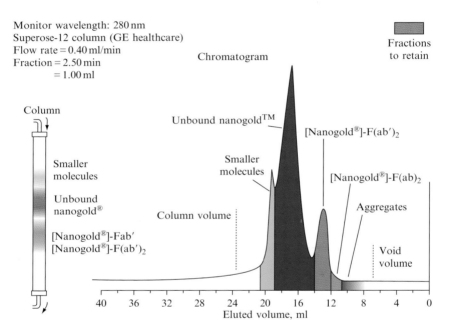

Christopher J. Ackerson *et al.*, Figure 9.5 Chromatogram showing the separation of Nanogold-labeled Fab′ from larger species, excess Nanogold, and smaller species by gel filtration over Superose-12 gel filtration media. Column volume = 16 mL (50 cm length × 0.67 cm diameter).

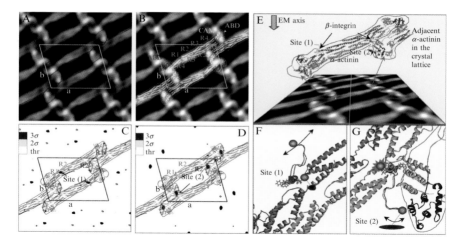

Christopher J. Ackerson *et al.*, Figure 9.6   Use of Nanogold-labeled β1-integrin to identify the binding site on α-actinin (Kelly and Taylor, 2005). (Left) (A) 2D projection map of the β1-integrin:α-actinin arrays. α-Actinin dimers are indicated by the red outline; the unit cell is indicated in yellow. (B) Docking of the α-actinin atomic model (Liu *et al.*, 2004) into the EM projection map. P2-symmetry-related dimers were also generated. (C) 2D difference map between unlabeled β1-integrin:α-actinin arrays, and arrays labeled with the Nanogold attached to C758, that is, in the middle of the integrin peptide sequence. The binding in this case occurs at site (1) on α-actinin. The most significant differences are at the level of 3σ (blue). (D) 2D difference map between unlabeled β1-integrin:α-actinin arrays and arrays with Nanogold labeling at C779. Labeling in this case occurs at site (2) on α-actinin. The density levels in both difference maps were truncated at a threshold corresponding to the values of 2σ (two times the standard deviation) for each map, respectively. 2σ peaks are colored sky-blue in (C) and (D). (Right) (E) Model of the unlabeled β1-integrin bound to both potential sites on α-actinin between the first and second spectrin repeats. The view direction chosen to show the atomic model is not perpendicular to the arrays, but rather at an angle that is approximately parallel with the local twofold axis of the R1–R4 domain. The effect is accentuated by the foreshortened appearance of the projection image shown below the model. This direction is maintained in (F) and (G), but because the local twofold symmetry of R1–R4 is only approximate, the integrin peptides are not seen in identical orientations. Arrows in (F) and (G) indicate the general degree of disorder in the gold label when attached to C779. (F) Integrin binding at site (1), while (G) refers to the integrin binding site (2) related by the local twofold of the R1–R4 domain. (F) The Nanogold label at C758 in the integrin peptide (red) fits into the α-actinin crystal lattice in an ordered fashion to give rise to a significant signal seen in the first difference map. The Nanogold label at C779 in the peptide (magenta) is too disordered to detect at a significant level in the second difference map. W755 is shown in stick rendering here and in (G). (G) Integrin cytoplasmic domain binding at site (2). When the Nanogold label is bound to C758 in the peptide (red), the fit is very tight because the inserted amino acid shifts the position of W755 toward the neighboring α-actinin molecule. No difference density is detected for this peptide when bound to site (2) suggesting that the peptide cannot bind under these conditions within the 2D lattice. However, the Nanogold probe positioned at C779 (green) of the integrin peptide fits with less restriction into the α-actinin lattice. The red line in (G) delineates an adjacent α-actinin molecule in the 2D lattice. Blue disks in (F) and (G) indicate the approximate location of the difference peak if the model were projected in a direction perpendicular to the plane of the 2D arrays (gold sphere = Nanogold).

Ariane Briegel *et al.*, Figure 13.1 *C. crescentus* cells with fluorescently labeled chemoreceptors (mCherry-tagged MCPA) are seen immobilized on a poly-L-lysine-coated holey-carbon support film (Quantifoil R2/2). The letters and numbers on the finder grid help locate targets in the light and electron microscopes (bottom left). FLM images (center right) are recorded of favorable regions of the grid using standard FLM instruments. Oil-immersion lenses with high-NAs can be used, since the sample is imaged at room temperature in the FLM. Here individual cells appear as dark, curved rods with bright red spots at their tips spanning the circular holes. After the sample is imaged in the FLM, the grid is retrieved from underneath the cover slip, plunge-frozen, and inserted into a cryo-EM. Tomograms of the same individual cells imaged previously by FLM are recorded, providing views of structures in nearly native states with macromolecular resolution (upper left corner). The positions of the fluorescent spots are seen to correlate perfectly with the locations of large protein arrays in the cryo-tomograms (red arrow), confirming that these structures are in fact the chemoreceptor arrays.

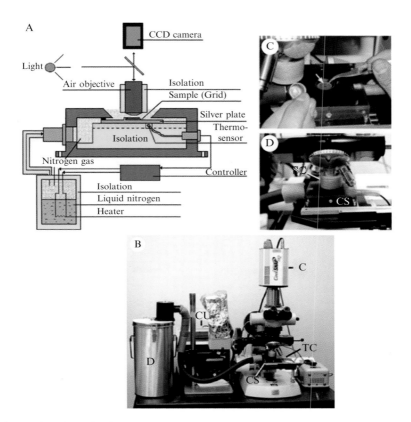

Ariane Briegel et al., Figure 13.7 The cryo-stage schematic and the LM set up from Schwartz et al. (2007). (A) Schematic diagram showing the three parts of the cryo-LM stage: The cryo-stage itself with the silver plate that holds the specimen grid, the Dewar that contains liquid nitrogen (*dark blue*) and a heater that blows cold nitrogen gas (*speckled blue*) over the grid, and the temperature controller (*red*) that reads the temperature from the thermosensor and controls the heater within the Dewar. (B)–(D) Pictures of the cryo-stage attached to a Zeiss Universal microscope. The Dewar (*D*) is on the left and is connected to the cryo-stage (CS) via thermally insulated tubing. The grid area is protected by a Styrofoam cylinder (SC) during imaging. The temperature control (TC) probe is attached to the right side of the stage and feeds back to the controller unit (CU). The digital camera (C) is used to collect the images. Reproduced from Journal of Microscopy with permission (Schwartz et al., 2007).

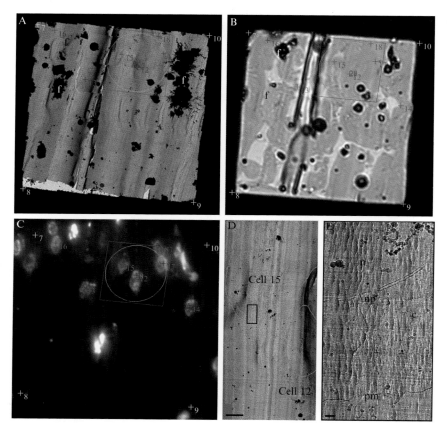

Ariane Briegel *et al.*, Figure 13.9 SerialEM can be used to correlate electron microscopy (EM) maps to light microscopy (LM) maps. (A) EM map of a grid square of
interest. (B) Brightfield LM map showing location of grid square for registration
purposes (yellow numbers at each corner). Frost (white f's where frost is present,
black f's where frost is absent) can move around during transfer from the LM to the
EM and consequently, frost present in the LM may be absent in the EM or vice versa.
(C) Fluorescence LM map showing locations of Hoescht 33342 positive nuclei (red
numbers). (D) High-resolution EM map of two cells found using the fluorescent map in
C (area marked by green circle and blue square). Cells are outlined with the plasma
membrane in green, nuclei in blue, and mitochondria in red. (E) Close-up of the black-
boxed area in D. The nucleus (n) and cytoplasm (c) are clearly evident with a nuclear
pore (np) found in the nuclear envelope. The plasma membrane (pm) is present. Scale
bars: D = 2 μm, E = 300 nm.